A PSYCHOLOGIST'S GUIDE TO EEG

Sara Miller McCune founded SAGE Publishing in 1965 to support the dissemination of usable knowledge and educate a global community. SAGE publishes more than 1000 journals and over 800 new books each year, spanning a wide range of subject areas. Our growing selection of library products includes archives, data, case studies and video. SAGE remains majority owned by our founder and after her lifetime will become owned by a charitable trust that secures the company's continued independence.

Los Angeles | London | New Delhi | Singapore | Washington DC | Melbourne

A PSYCHOLOGIST'S GUIDE TO EEG

THE ELECTRIC STUDY OF THE MIND

MICHIEL SPAPÉ

Los Angeles | London | New Delhi
Singapore | Washington DC | Melbourne

Los Angeles | London | New Delhi
Singapore | Washington DC | Melbourne

SAGE Publications Ltd
1 Oliver's Yard
55 City Road
London EC1Y 1SP

SAGE Publications Inc.
2455 Teller Road
Thousand Oaks, California 91320

SAGE Publications India Pvt Ltd
B 1/I 1 Mohan Cooperative Industrial Area
Mathura Road
New Delhi 110 044

SAGE Publications Asia-Pacific Pte Ltd
3 Church Street
#10-04 Samsung Hub
Singapore 049483

Editor: Donna Goddard
Editorial assistant: Esmé Carter
Assistant editor, digital: Sunita Patel
Production editor: Rachel Burrows
Copyeditor: Sarah Bury
Proofreader: Brian McDowell
Indexer: Gary Kirby
Marketing manager: Camille Richmond
Cover design: Wendy Scott
Typeset by: C&M Digitals (P) Ltd, Chennai, India

Library of Congress Control Number: 2021932631

British Library Cataloguing in Publication data

A catalogue record for this book is available from the British Library.

ISBN 978-1-5264-9076-6
ISBN 978-1-5264-9075-9 (pbk)

CONTENTS

YOU, ME, AND THE EEG: A PREFACE

Let me tell you a secret. When you were young, did you get *that question*: What do you want to be when you grow up? I wanted to be a wizard, said I, a reader of minds and conjuror of fireballs, at the very least.

That is not entirely true, but only because the question is a trick-question. There are two answers: the true one and the supposedly grown-up, 'real' one. What may come to mind first is not usually practicable. What you want to be, surely, is either fantastically ambitious, a famous rock-star, say, or comprehensively unproductive, like ice-cream taster or bed tester. These seem like completely reasonable desires to me, but the adult mind disapproves and keeps pushing you with quizzical looks and leading questions until you learn to *not want* what you want, but *to want what you should want*. So I wanted to be a storyteller, a philosopher, and finally a psychologist. A storyteller to some extent conjures fictional worlds, a philosopher is suitably arcane and comes with logical formulae and a majestic beard, and a psychologist can read minds.

So, in the late summer of 1999, when war, disease, and the irrational forces of history seemed squarely behind us and the internet was becoming the great, unifying, super information highway, in other words, when life was full of hopes and flowers and dreams, I entered the gates of the School of Psychology.

That moment marked the beginning of disillusionment. Indeed, few buildings are as devastatingly rational as the University of Leiden's Faculty of Social Sciences with its sound, 1960s architecture constructed around a massive, central staircase. Even the staircase itself was foreboding, flanked as it was by nets of strong Dutch colonial style ropes, allegedly there to protect people from being struck by the not infrequent cases of attempted suicide. This down-to-earth environment soon taught me that a Psychologist, unfortunately, does not get to learn how to read people's minds.

Sure, I am a silly sort of person for believing this, but I am not alone. Ask your peers what they expected and you'll find most have similarly exaggerated expectations, like thinking they will learn to know themselves (whatever that means), or to help themselves (as psychotherapists like to tell their clients), or to help others (nice idea, but psychology isn't psychiatry). And still, whenever I meet a random stranger from outside the field (a civilian, so to speak) and they ask what *do I do?*, I get two responses to *Psychology*: 'Oh, I know this weird person', and 'Oh no, you must have been reading my thoughts!' So, we all have the same idea what Psychology *should* be about.

But inevitably, education, experience, and accumulation of cynicism will slowly cause us to realise that we cannot read minds. We might predict what a person will do, given their personality and given this or that situation, which is pretty good as superpowers go, maybe like clairvoyance. But personality, as in a Big-5 model, does not predict individual decisions very well (Mischel, 1990). And we're now finding out that situational factors have limited effects: for example after reading some words associated with ageing (Bargh et al., 1996), one does not suddenly start acting like an elderly person. Well, one does if told by lab assistants that this should be expected (Doyen et al., 2012). Or what if we rephrase reading a mind as the power to observe non-verbal cues to tell us what they *really* think of us, given that 70% of all communication of liking is non-verbal (Mehrabian, 1971)? But now it turns out that this finding is based on an extreme over-interpretation of the data, and the number is largely meaningless (Spapé, Harjunen, et al., 2019). Even hypnosis seems hardly more than asking someone to do something and being amazed when they do so (Spanos, 1986), which is either scary Milgram conformity, or people being actually nice. So yes, perhaps we can do telepathy and telekinesis even, if that means listening to verbal instructions or lifting a spoon in the air with the power of the mind's control over the hand. Great abilities to be sure, but depressingly inconsequential.

I became a bit of a cynic, I suppose. But then I learned about electroencephalography (EEG), and things went different. The study of EEG is quite closely aligned to Psychology, the study of the functions of the mind, and yet in almost every way far removed from it. The language, of potentials, dipoles, oscillations, and so on, is *electrifying*, and feels more like engineering than psychobabble. Because many of us are, however, not engineers, we are constantly confused by our own equipment; a toolkit that is radically different from the psychologist's normal box of tricks. Even our way of dealing with subjects is completely opposed to the hands-off, safe-space, talking-cure approach characterising the rest of Psychology. Before you know it, you have your hands buried deep in the hair of strangers and desperately trying to restyle their make-up (removing it, that is), while distracting them by chattering away like a terrible hairdresser with medical ambitions.

And indeed, the electric study of the mind is the most ambitious one out there. Using our trusty yet ill-understood equipment, we don't ask people what they are seeing, but we measure the brain as it goes about its merry business. You show it a picture and you can trace the signal as it traverses our body and brain, coursing from retina, the basal ganglia, and visual cortex into the more dangerous and lesser-known tributaries of the brain that govern the uncharted territories of perception, memory and consciousness. It's almost like visualising an 'aura', but without the mystical mumbo-jumbo! We can trace EEG signals even if the person *owning* said brain isn't aware of seeing a picture to begin with. Finally, we can know something about the subject's subjectivity objectively.

There are of course obstacles, side-notes, limitations and challenges to interpreting mental states from electrical fields as measured from our scalp. Eye movements, blinking, muscle movements, just moving, blood pumping, random noise, boredom, human factors, bad design, dirt, and dreadlocks are all problems. Some of these problems can be resolved or circumvented by following sage advice from wise, old scholars. I, too, will try to help. But practical advice and tomes with medical or technical jargon can easily kill the beauty in the initial dream.

And the original dream, of what to be when you grow up? I still don't know myself, but I think becoming a wizard is a great idea, actually. And as much as I am still struggling with the fireballs, I do happen to know how you can read minds. This book will teach you how. Of course, if you continue reading, things will get a bit more complicated (as brains are) and hard (though brains aren't), but it will help if you repeat it as a mantra: mind-reading is worth it. Because by the end of the book, you will be very different from anyone you know: you will be able to read people's minds.

YOU, ME, AND THE EEG: AN OVERVIEW

This is a book about EEG and biosignal analysis aimed at Psychology undergraduates, postgraduates, and colleagues who want to start doing EEG but ~~cannot be bothered~~ don't have time to read the more serious and much thicker works of academic knowledge on the topic. It is designed with a very wide breadth of scope, covering pretty much all aspects of EEG from history, fundamental theories, and event-related potential components (Chapter 1), via experiment design (2), programming (3–4), lab setup and data acquisition (5–6), data preprocessing (7–8), and analysis (9–10), to writing up your research results (11–12). I use a simple trick to give an introduction to all these diverse topic in the space of a reasonably slim book: I discuss everything in the context of a single, simple study, which looks at what happens in the EEG when you recognise a person.

STUDENTS, ME, AND THE EEG

That is not to say you should study with only this very limited application in mind; I rarely follow my own rules, so I don't see why you should. However, as we explore this 'mind-reading', I hope you will get to understand my way of thinking a bit (tell me if you do – I wouldn't mind the insight!), such that you will gradually learn how to design your own research questions and analyse your own data. For example, there is no reason why my 'recognising persons' can't be exchanged for 'emotional' pictures in, say, a study on people with subclinical levels of symptoms associated with autistic spectrum disorder. Such topics are not really my cup of tea, but while I keep nattering on about recognising people, try to imagine how you would phrase the research questions for *your* topic of interest.

Most of my colleagues like to use a 'dipping' strategy when it comes to reading books such as this – fishing out the useful things, and skipping everything else. I make it easy for them to work like this, since I cover EEG research analogous to how it is conducted, from setup to analysis. And indeed, they might have a point: some content, like experimental design, may well be deemed below the level of postgraduates and lecturers in psychology. However, they would be wrong (I can safely say this as this audience would never read a preface anyway). Research design in EEG requires much more attention to detail than other, more woolly psychology fields, or even ostensibly 'hard-science' imaging tools like fMRI. But if you, reader, are an inveterate *dipper* fearing you may slip from the straight path, you may find it useful to know that all material needed to start on a chapter is available online with SAGE.

TEACHERS, ME, AND THE EEG

This book gives a start-to-finish introduction to EEG and was based on a 10-week course of two-hour seminars for third-year undergraduate students at Liverpool Hope University. Designing a curriculum with this book is easiest if you think of the book as a typical science paper:

Introduction: Chapters 1 and 2 cover the general background and the process of setting up a specific study to test a hypothesis.

Method: Designing experiments (Chapter 3 or 4) provide tutorials for developing a study in E-Prime (3) or OpenSesame (4). The former is a popular, commercial software package, which remains popular in the field, while the latter is similarly easy to work with but Open Source (it also runs 'on' PsychoPy).

Data acquisition (Chapter 5 and 6): This part of the book naturally runs parallel to having students record their own data, which raises the stakes and feels more personal than just using mine (online). I recommend presenting a big 'EEG training' session in the lab showing how to record EEG, but I found recall is limited. In addition, therefore, use small groups (e.g. three people), with one student being responsible for the equipment during a next meeting ('EEG data acquisition'). Usually this prompts more active involvement and preparation for the second meeting in which data are recorded, but I would still advise to have someone look after them anyway. The equipment is not cheap.

Results (Chapters 7–10): After data are recorded, students should be able to use the book to guide their data analysis, as long as MATLAB has been installed on the computers they use. At this stage, it is necessary to urge exploration rather than blindly following the instructions.

Conclusion and discussion (Chapters 11–12): Chapter 11 includes the inferential statistics part of the results section, outside of MATLAB, which a second- or third-year undergraduate should be familiar with. If the student has basic knowledge of statistical software, it should be easy to complete in a free-guided fashion, with additional time to be spent towards writing of reports. Chapter 12 is a conclusion of the book of sorts and, simultaneously, an introduction to writing conclusions.

On the topic of reports: I can recommend using a video diary (or 'vlog') method. Each week, have students record their work in progress using their phones. Students seem quite capable of that nowadays, and quickly start making recordings of themselves explaining things by talking over screen capture software, PowerPoints, or over videos of setting up the EEG. By the time of the discussion, have them edit all their recorded bits into one big 'documentary', which becomes essentially a portfolio. In other words, students will already at the start invest much more effort in the process. Vlogs are also more fun to mark (I watched them in class as part of a 'red carpet première'), and leave a tangible end product, which can be used as promotional material, for example https://youtu.be/HLXoB9_w9XA

Note: Throughout the book, I will reference additional material that can help in designing experiments and analysing data. These are all available from the supplementary information website hosted by SAGE: https://study.sagepub.com/Spape

ACKNOWLEDGEMENTS

Although this is probably the most personal book on EEG in print, I am in debt to quite a large number of people. For information on EEG systems, I would like to thank Mogens Christianson (cephalon.eu), Aaron Ellefson (MEQnordic), Robert Honsbeek (BioSemi), Lassi Juottonen (Clinilab), Katia Krane (MagstimEGI), Alexander Lechner (G.Tec), Fredrik Rådebjörk (JoR), and Stefanie Rudrich (Brain Products). For the use lab equipment, I appreciated the help of Jaakko Kauramäki and the Helsinki University CBRU lab, and Niklas Ravaja of the EIEH lab. On the software side of things, I am grateful to Sebastiaan Mathôt for making OpenSesame openly available to me and the community, and Psychology Software Tools, Inc.'s Tony Zuccolotto and Gretchen Brauch for decades of facilitating experimental psychology with their product, E-Prime. And thank you Arnaud Delorme and Scott Makeig, for allowing me and the authors of more than 14,000 cited studies the use of their open source EEGLAB software.

I also had access to amazing amounts of human resources that facilitated and supported this book in many ways. Thank you, friends and colleagues from Liverpool Hope University Department of Psychology for your faith in my bonkers, vlog-based seminar 'EEG acquisition & analysis', which ultimately formed the seed of this book. Thank you, Donna Goddard, Esmé Carter, Marc Barnard, and Katie Rabot, for your patience while that seedling certainly took its time. Thank you Imtiaj Ahmed and Ville Harjunen for being amazing PhD students and literally offering their brains to science. Finally, I cannot hope, however, to repay the debt I owe to her, who not only gave me her own brainwaves, but produced a personally customised and by now highly mobile brain; thank you for being an Ursula von Bülow to this Hans Berger, thank you Zania Sovijärvi-Spapé.

ONLINE RESOURCES

Visit https://study.sagepub.com/Spape to find a range of additional resources for both students and lecturers, to aid study and support teaching.

FOR STUDENTS AND LECTURERS

Experiment scripts in E-Prime and OpenSesame, including all stimulus material, facilitating problem-solving in experimental programming environments by presenting a clear standard.

Datasets provide students with additional opportunities to practise and help generate meaningful information to support mastery of EEG and biosignal analysis.

Analysis scripts enable deeper insight into MATLAB coding and EEGLAB dataprocessing while implementing a clear path from psychological question to cognitive neuroscience answer.

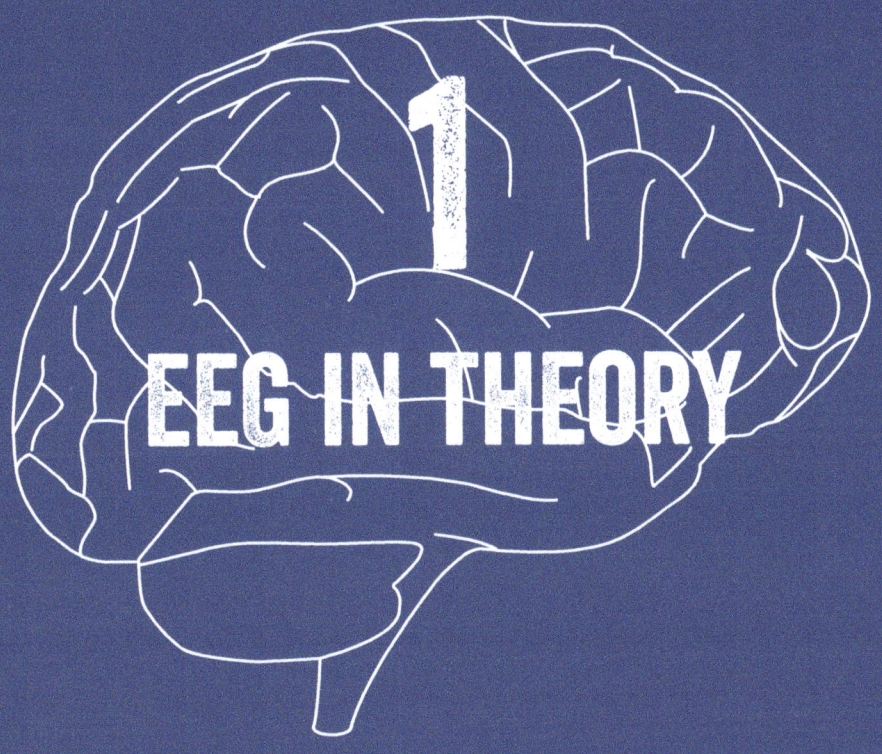

EEG IN THEORY

IN THIS CHAPTER, YOU WILL LEARN ABOUT:

- The history of EEG
- The neurobiology of EEG
- What EEG can tell us

As a student in Cognitive Psychology, I was often present at talks – officially colloquia, or fancy words such as symposiums and seminars, but those in the know simply call them 'talks' – about EEG. I was interested in cognition myself, particularly attention and executive control, and I had a hard time understanding why researchers in this field could get so excited about a bunch of squiggly lines. There's a bump, followed by a jump, and in *this* condition, the ski slope seems a bit steeper, so what? After all, the lines were only slightly more fashionable than my normal weapon of choice – the mighty reaction time – while the fancy, new tool of fMRI was all the rage, grabbing headlines with beautiful blobs showing exactly where in the brain the action was happening. True, EEG was also associated with the brain, but, in the words of one colleague, the spatial resolution was so poor that one might as well put a giant, single electrode on the head like a toilet plunger, and just go with that. It is fair to say that it took a while before I saw the magic of EEG research.

In this chapter, I will try to explain what is so great about EEG. Most of my students are able to explain by paraphrasing the usual textbook truism that 'while fMRI has good spatial resolution, EEG has excellent temporal resolution'. It is technically correct, but fails to capture the essence of what is so great about EEG. It's like saying that the computer is a great invention because you can move bits about really fast, unlike with the competing pen-and-paper technology. Imagine you had no idea what a computer was, would that explanation allow you to imagine iPads, VR goggles, social media, first-person shooters, or online pictures of photobombing animals?

Instead of thinking of EEG as 'squiggly lines with a good temporal resolution', then, think of it as *the fireworks in the mind*. In this chapter, I will explain why this is not nearly as ridiculous a simile as it appears at first.

THE HISTORY OF EEG

The Brain Mirror: Hans Berger waves

Any reasonable book on EEG begins with Hans Berger, the first inventor of alpha and beta waves, and the first one to observe EEG in healthy humans. That is to say, previous to Berger, EEG was already observed in dogs (Pravdich-Neminsky, 1912), and people with seizures (for a review, see Brazier, 1959), but very little in terms of mental operations had been related to the electrical activity, and only by decades of work refining the instruments did it become possible to really observe recordings of EEG.

Curiously, the discovery of EEG happened in the absence of reasonable insight to the biophysical properties of the brain, although to some extent this remains true to this day. Berger himself was, like many scholars of his age (such as William James; Blum, 2006), perhaps more fascinated by spiritualism than biology or electro-engineering. As a young man serving in the cavalry, Berger suffered a near-fatal accident. That same night he received a telegram from his father enquiring after his health, prompted by his sister who, despite being many miles away, had an overwhelmingly ominous feeling regarding her brother. The conclusion was straightforward: Hans's intense feeling of terror must have taken physical form and somehow reached his sister telepathically.

Despite this occurrence, which for many a credulous academic would have made an exciting start for a career in quackery and ghostbusting, Berger remained grounded in the natural sciences. Leading physicists of the 19th century were much enamoured by the principle of conservation of energy, applying it not merely to thermodynamics, but also psychological phenomena. Indeed, the leading light of thermodynamics, Hermann von Helmholtz, was also a founding father of psychology. Thus, Alfred Lehmann argued that 'the brain, like all organs, produces a store of chemical energy that it derives from various metabolic processes, which it converts into three major forms of energy: heat, electricity, and [...] 'P-energy,' – or psychic energy (Lehmann, 1912). Thus, Berger reasoned, one should be able to open a window into the interaction between mind and brain, if one only manages to measure brain metabolism, temperature, or electricity.

Early experiments showed the incredible possibilities. Measuring the volumetric changes in blood flow from a patient with an 8 cm circular hole in his skull caused by attempts to dislodge a bullet, Berger revealed that various mental tasks elicited miniscule changes in cerebral blood flow (Berger, 1904). Using precise thermometric measurements, Berger was further able to show that the brains of previously narcotic patients slightly heated up upon awakening (Berger, 1910). When he turned to the electric side of the energetic model, however, he ran into problems. First, due to the imprecise methods of the time on the one hand, and the great amount of noise and artefacts that remain with us even today (see Chapter 8), the results were varied. Indeed, his contemporaries in the faculty and elsewhere found the entire proposition preposterous, and his not unreasonable self-doubt – given a distinct lack of training in electrical engineering – contributed to a steadily more isolated life in the laboratory. I imagine him there as a kind of historical Frankenstein, madly aiming to measure the quality of the soul with his 'brain mirror', and tinkering away at new instruments to achieve the phenomenal sensitivity required to record brain activity from the scalp.

Yet despite the psychological, technological, and social problems – not to mention a world war – he continued perfecting the technique. After having managed to successfully record the EEG from patients, employees, himself, and his poor son Klaus, he convinced himself the recordings did not reflect something trivial, publishing 'Über das Elektrenkephalogramm des Menschen' (Berger, 1929). Quite wisely, he left out the spiritualist ontogenesis of his invention, publishing them without relaying any theory, merely reporting the concrete fact of the observation – which was probably for the best, as no one had the faintest idea of what caused the EEG. So there it was, the first squiggly lines, one simulated zig-zaggy line with a period of 100 ms and one recorded alpha wave, which looked like a similar 10 Hz wave (Figure 1.1).

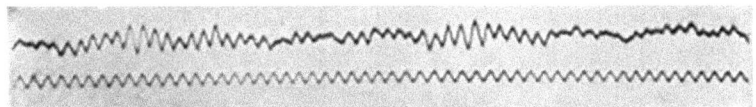

Figure 1.1   Recordings of Klaus Berger's EEG

Sadly, the ground-breaking publication was not immediately recognised as such in his own country, where localisation – arch nemesis of EEG (p. 8) – was popular, or abroad, where his colleagues did not always read German. International recognition came finally in 1934, when a Cambridge scholar replicated Berger's findings, naming the alpha waves 'Berger waves' in his honour.

The Elektrenkephalogramm remained somewhat of a niche interest, however, mainly for those who studied epilepsy. Beyond that, alpha waves seemed mainly indicative of a *lack of thought*, or closing one's eyes, rather than a relation to specific cognitions. Enthusiasm for the invention thus remained a fringe obsession, and perhaps this is why I have yet to hear someone actually use *Berger waves* in their conversation.

Advances in electroencephalography from gothic horror to cyberpunk

One man, however, certainly was impressed: William Grey Walter. Having read 'Über das Elektrenkephalogramm', he met up with Berger, and, even though he was still only a graduate student, became involved in the English efforts to use the new invention for clinical epileptology (Bladin, 2006). Like Berger, Grey Walter was very much taken with the possibilities of EEG, but unlike him, Grey Walter was less driven by the spiritual aspects than by the practical application of the methodology, which he matched with a knack for electrical engineering. Instead of using the clay electrodes and string galvanometers of Berger's day, Grey Walter almost completely redesigned the EEG measurement equipment from the ground up, undertaking all tasks from assembling and constructing to developing and testing. Critically, he implemented a system using silver-chloride electrodes (for higher conductivity), amplifiers (to boost the tiny potentials recorded from the scalp), cathode ray oscilloscopes (to visualise, and measure, the signal), and cameras (to record the signal) (Walter, 1937). With this highly flexible setup in place, Grey Walter felt confident to apply the study of electricity in the brain to anything he could think of.

To give an idea why Grey Walter is such a hero among EEG technicians, a non-exhaustive list of inventions and ideas is in order. After Berger's observation of Berger's alpha and beta waves, Grey Walter first named the **delta** (Walter, 1936, 1937) and **theta waves** (Walter & Dovey, 1944), and pointed to their relation to anaesthetics (ether or nitrous oxide causing delta) and their diagnostic value in localising cortical (delta) and sub-cortical (theta) tumours. Such **frequency-based analysis** was, over time, made much easier by his invention of ANNIE, the automatic frequency analyser (Walter, 1943), which removed much of the necessity of experts scrutinising recorded traces and using rulers to analyse signals. Visualisation was yet more improved by the invention of **scalp topography** or, in beautiful 1950s retro-futurist speak, the 'toposcope' (Walter & Shipton, 1951), which shows a snapshot of the electrical activity across the entire head, now a mainstay of EEG methodology. To me, however, the most interesting contribution he made was the discovery of the **contingent negative variation** (Walter et al., 1964), which not only was the first cognitive event-related potential to be discovered, but also provides the first clear explanation, rationale and demonstration of the **event-related potential technique** itself. I will discuss both in detail later in this chapter.

All that may be very heroic, but where does the cyberpunk come in? As I went through the history of EEG, I noticed that, like Berger, Grey Walter was more than a bit eccentric, putting himself at odds with the serious, academic communities of the time. Calling his automatic frequency analyser ANNIE could be an innocent quirk, but Walter went much further, populating his professional and domestic life with products of a mechanical sort of romance. He is particularly well known for his *machina speculatrix*, the robot tortoises Elmer (ELectroMEchanical Robot) and Elsie (Electromechnical robot, Light SensitIvE), which used only two sensors and mechanisms – moving towards light but away from touch – to create complex behavioural patterns. For example, by equipping Elmer and Elsie with lamps that were extinguished with too bright lights, the robots performed shy, courtship-like manoeuvres, circling around one another, but fleeing swiftly upon the accidental touch. Naturally, such gadgets worked wonders at parties (Bladin, 2006), but seemed vulgar to his serious, scientific contemporaries. This was not helped either by Grey Walter's application of EEG to anything from lie-detection (but see Farwell, 2012; Meijer et al., 2013) and marriage counselling to managing world peace (Walter, 1968). By making EEG the solution to almost anything in life, he envisioned a new world of cybernetic opportunity, becoming a kind of futurist prophet (though without the contemporary entrepreneurial-libertarian ethos of today). In the words of Hayward (2001: 616), he took on roles from 'robotics pioneer, home-guard explosive expert, wife swapper, TV-pundit, experimental drug user, and skin-diver to anarcho-syndicalist champion of leucotomy and electro-convulsive therapy'. If you ever need someone to show you that EEG is not merely a niche science hobby of staring at squiggly lines, but an exciting, almost psychedelic endeavour, Walter is your man.

THE NEUROBIOLOGY OF EEG

A common misconception held among laypersons, and early-career cognitive neuroscientists, is that a neuroscientist is just a fancy word for psychologists who want to sound less like tell-me-about-your-mother clinicians and more like rocket scientists. What they mean is cognitive neuroscientist – which is roughly anyone who use neuroscience methods to study cognition – but not neuroscientist – those who use all sorts of scientific methods to study neurons. While this may sound vague, this difference is sometimes apparent in faculty (e.g. psychology versus physics and biology), object of interest (emotion versus endocrine level), type of research (experiment versus clinical trial), not to mention type of participant (student versus zebra fish). Why am I telling you this? Well, aside from the slight annoyance of hearing people call themselves neuroscientists when they mean psychologists, it is also an elaborate way of apologising for the simple fact that I am not the best person to ask if you want to know about the precise, causal biophysics underlying EEG.

Strange as it may seem, this situation is not as unique as one might hope – leading lights Berger and Walter had very little reasonable idea either, and that didn't stop them. Indeed, I once heard a famous cognitive neuroscientist explain that we only know two things for sure when it comes to EEG: we know *that* it is caused by the brain's activity, and we know *when* it does that. And that may seem pretty poor, but it actually beats fMRI in terms of *that* (the 'activity', being 'blood-oxygen-level dependent', which is an indirect measure of activity) and *when* (the BOLD only rising after thousands of milliseconds). So there.

Neuronal activity and action potentials

Now let me stop with the petulance and explain at least what we can agree on in terms of what causes EEG. As you will remember from your neuropsychology, biopsychology, or indeed high school biology class, the brain is composed of neurons, and neurons themselves comprise dendrites (these 'branches' receiving input from neurons), soma (the cell body), and an axon (the long tail that activates other cells through its terminals) (Figure 1.2 A). Neurons commonly maintain a negatively charged electric polarity within the interior of their membrane against their positively charged exterior during their **resting state (Figure 1.2 B)**. As the cell receives input through its synapses, it gradually **depolarises**, causing the membrane potential to rise. If multiple inputs are received – whether simultaneously from multiple other neurons or in quick succession from the same one – this depolarisation will continue, until a point of no return is reached, which is called its **threshold potential**. Once reached, the membrane potential fires its **action potential**, an extremely rapid depolarisation to a positive electric charge immediately followed by a quick **repolarisation** into negativity, commonly to a value beyond its resting state. During this hyper-negative **refractory period**, the neuron will not be as responsive to incoming stimulation. The cycle of the action potential is extremely quick (c. 1 ms), and many neurons show such activity in an almost continuous manner at a rate of 10–100 potentials per second.

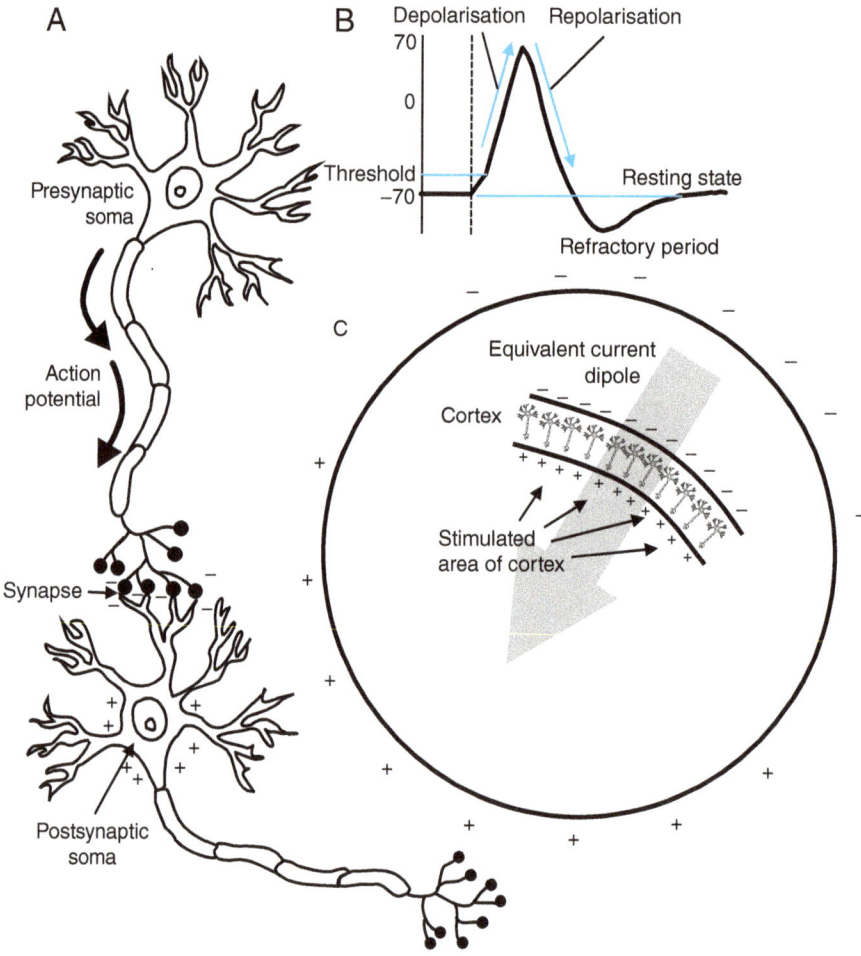

Figure 1.2 Neurons, activity, and dipoles. A: Two neurons connected. B: Stages of a neuron's action potential. C: Similarly oriented cluster of neurons fires synchronously and creates an equivalent current dipole.

Source: adapted from Luck, *An Introduction to the Event-related Potential Technique*, MIT Press (2005), p. 30

However, unless you manage to stick electrodes inside someone's brain, it is unlikely you will pick up action potentials using EEG (scalp-based). As someone who is habitually holding electrodes, weird, gooey gels, and needles in front of nervous subjects, I sometimes need to comfort them that *of course we do not stick electrodes straight into people*. Well, we don't, although there are those who do *in vivo* recordings, which happens with ECoG, or **electrocorticography**, which generally entails plugging a grid of electrodes through an opening of someone's skull onto the brain. Do note, of course, that this is only possible because such an opening already exists, most likely due to brain surgery. While ECoG will measure both action potentials and other potentials, by sticking micro-electrodes straight into individual neurons, so-called **single- and multi-unit recordings**, one can isolate the action potentials and analyse the activity of neurons directly. As a psychologist who is only dabbling in cognitive neuroscience, rather than a real neuroscientist, this goes beyond my expertise, but occasionally we cross paths with such people, for example when it comes to the famous mirror neurons (Di Pellegrino et al., 1992). Consequently, I know these people are excellent at imitating the 'sound of a neuron' as it spikes into rapid activity – a certain *tic------tic-----ticTICcerticTIC*! Neurons, as far as we know, do not make any sound, but it's a better approximation of what neurons 'do' than EEG waves.

Postsynaptic potentials, dipoles, and EEG

In any case, with the action potential occurring, the electric charge travels through the neuron's axon to its terminals, where neurotransmitters are released. These travel across the **synapse** (from Greek, 'joining together'), the structure between the source and target neurons, binding onto the target's receptors. Personally, I try to think of it as a kind of ferry system between two cities that are separated by a river, or a strait like the English Channel. Dropped onto the parking lot, people (neurotransmitters) enter the ferries (neurotransmitter transporters) that move towards the harbour (receptor) on the other side, which causes activity to build up, until they are thankfully moved by trains towards their more central destination. This graded change in the activity after arrival is called the **postsynaptic potential**. As this change is happening gradually, and in the dendrites and cell body, rather than extremely rapidly, and moving across the axon as action potentials do, these allow the postsynaptic potentials to summate, rather than cancel out one another (Luck, 2014). What that means is that if the conditions are just right, then many incredibly tiny electric currents from many bits of skull-space can beat all the odds and be measured across relatively great distance (the brain, skull, skin, and so on), such that they are recorded in the EEG. Perhaps this is a bit like news organisations in the UK generally not picking up the news in Helsinki, unless something that draws huge crowds is happening, such as the 1952 Olympics.

While many cognitive neuroscientists, in particular psychologists such as myself, normally ignore the above story, the important part is exactly in these conditions. Our best guess is that certain neurons, such as the cortical pyramidal cells, are aligned in such a way that their dendrites are conveniently oriented[1] so that they all point in the same direction like the hair of a big crowd at a rock concert. Thus, when multiple postsynaptic potentials of a single neuron occur at the same time, the negativity involved is oriented towards the same location (i.e. up), contrasting with the relative positivity of the cell body. This creates a miniature **dipole**, a pair of charges of opposite

---

[1]It seems unlikely that we are evolved to make the life of EEG researchers easier, or that God has a special place in His heart for us, but it's a fine bit of luck.

polarity. Of course, if two neurons have opposite orientations, the average of the charges would cancel out, so only if thousands, if not millions, of neurons are aligned in similar way (hair up, feet down, so to speak), do we register anything at all. If we combine the concert of activity of similarly oriented, similarly activated postsynaptic potentials, a large cluster of activity with a negativity on one end and a positivity on the other is formed, which is called the **equivalent current dipole** (Figure 1.2 C). This is the source of what we eventually measure with EEG.

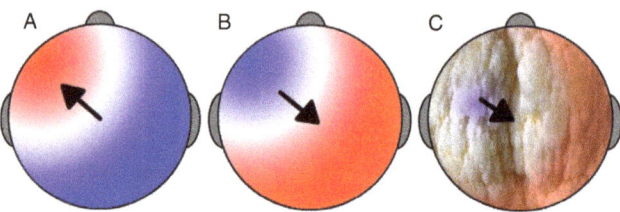

Figure 1.3 Equivalent current dipoles. Imagine a neural source roughly in the middle of each arrow. Despite very similar source location, the electric potential measured in B is the opposite of A. Of course, the brain is not like a sphere, having a strongly folded structure within its much simpler encasing of the skull. C illustrates this as a cauliflower inside a head model.

In other words, an EEG squiggle does not correspond to 'activity' from somewhere near the electrode, but EEG always reflects a (equivalent current) dipole with an orientation. This means that (1) the positivity and negativity we measure are always relative to some other point and (2) the location of the dipole does not correspond to where the activity is measured. Thus, for example, in Figure 1.3, the activity in A and B originates from roughly the same location, while their scalp recordings are completely different. Making matters worse is that the brain is not like a muddy balloon with grey matter (the cell bodies and dendrites) neatly on the outside, but has a folded structure like a walnut or a cauliflower (see C).

This essentially illustrates why source localisation, in EEG, is such a difficult problem, known in mathematics as an example of an **inverse problem**. Imagine, for example, that you manage to wiggle a tiny electric torch inside a balloon. Now, because you know where the torch is and which way you are pointing it, it is relatively easy to calculate how the balloon will light up. This is called the **forward problem**: if you are aware of the causes, you can easily calculate the consequences (the forward problem is, confusingly, not much of a problem). However, if you only see the outside of the balloon and someone else is holding the torch, you will be unable to decide whether they are holding the torch far away, or are just using a weak light, or perhaps they are not even holding one torch but many torches at the same time! In other words, because many different causes can result in the same consequence, we cannot calculate which of the possible causes elicited the present result.[2]

A final source of frustration for those who are interested in EEG – as opposed to ECoG – is the fact that we normally place electrodes on the scalp, not in the brain. What lies between? The brain itself is more or less floating around inside a triple bag of membranes, the meninges, with liquid (cerebrospinal fluid) between them, which together give some protection from concussions.

---

[2]Unless by oversimplifying, disregarding many possible causes or taking some assumptions for granted, such as how many dipoles there are. This kind of modelling is necessary for source localisation techniques (see LORETA, p. 241).

Onwards, the skull is about 80 times less conductive than its contents, which diminishes the signal (Manola et al., 2005). Then we get to the skin: This is a pretty conductive part, but that has its own problems, because this causes the EEG to be smeared – blurred across the scalp. Finally, between the skin and the electrode, various other sources of frustration still appear: dead skin flakes, grease, and more of the stuff people in the know commonly call *muck*. However, as will be explained in Chapters 7 and 8, we have ways of dealing with these problems.

WHAT EEG CAN TELL US

It may feel a bit disheartening, reading about all the caveats of EEG. You cannot locate activity, we are not measuring neurons firing but abstract things like dipoles, the skull is not cooperating (some people literally have *thick skulls*!), the skin blurs your EEG, and so on. But look at it this way: is it not beyond fantastic that despite all these problems, we can still say so much about the mind using EEG? I feel a little bit like Carl Sagan here, who in *Pale Blue Dot* wrote inspiringly about the vastness of the universe against the almost infinite tininess of our lives on this Earth: all the forces of physics combining in such an incredibly unlikely way to come up with the very presence of what it is to be you. Surely then, the enormous, vast complexity of the brain and the incredible improbability of potentials aligning does not diminish the enterprise, but rather adds to the excitement of this, the electric study of the mind?

I know I am getting a little ahead of myself. Let us consider what EEG has already taught us.

The event-related potential

If we want to know what is happening in the brain, we normally mean one of three types of brain activity, each typically characterised by a different time-course and different methodology used to investigate it. If we want to know what makes the brain activity of one group of people different from others – say people with schizophrenia from controls – *regardless of a specific situation*, we normally ask about **spontaneous activity**. Such studies may, for example, require people to lie in an fMRI scanner, close their eyes, and have their resting activity scanned, which typically shows ongoing activity in areas now called the default mode network (Greicius et al., 2003). Most studies, however, prefer a stronger amount of control and investigate our response to stimuli, scenarios, or other externally observable events. If we ask whether, say, viewing fearful faces causes activity in the amygdala, then fMRI can again be informative, as viewing fearful faces over a few seconds normally **induces activity** in the amygdala (Morris et al., 1996). The third type of question, whether viewing a fearful face **evokes activity** is almost the same, but asks whether the fearful face *prompts* a certain neuronal response.

In day-to-day language, the words 'causing' and 'prompting' are used interchangeably, yet the type of effect we are looking for, on a psychological level, is rather different. Watching a horror film may *induce* (cause) a sort of low-level dread that grows, even as various scenes may offer breathing (hunt-for-snacks) spaces, which relax. However, as it nears its climax, it will likely offer cinematic tricks that are designed to *evoke* (prompt) immediate frights with jump-scares. Here, the appearance of a demon suddenly appearing in a mirror is a clearly defined stimulus, which the director is well aware is evocative. You could say that while fMRI is is best used for its ability to measure induced responses (breaking news! *Brain boffins find watching horror causes your insula to light up*), EEG captures evoked responses much better (*Neuroscience: Jump-scares leave killers cold!*).

In that regard, it is true that the main advantage of EEG is in its excellent temporal resolution, but while easy to recall during an exam, this does not mean much to anyone except EEG researchers. Practically, it allows three benefits over fMRI. One, it allows you to measure an immediate response, even if the response does not continue over a lengthy period of time (such as tiny, unconscious jump-scares). Two, it allows much faster measurements, which means you can include more repetitions in an experiment. Showing a single stimulus takes up to 8 seconds to give a good BOLD response in fMRI, while most evoked responses in EEG take less than a single second. This means we can repeat the same physical stimulus many times in EEG, improving the reliability (more trials, less noise) and validity (fewer effects due to novelty of stimuli) of our findings – although it should be said that participants rarely enjoy such excellent statistical properties. Third, it allows us to say something about *when* something happens in the brain. For example, EEG research has shown that we identify whether a stimulus is a face or not at about 170 ms (Bentin et al., 1996) but, amazingly, that an angry face is differentiated from a neutral face at around the same time (Blau et al., 2007), if not earlier (Zhu & Luo, 2012). In other words, I know you are angry before I even know that you are there.

I think that is mind-blowing. But maybe you are not into emotions or faces, so let us have a look at other known potentials, what kind of paradigms they show up in, and what kind of exciting insights they have provided. I will start with an overview of event-related potentials that are classically referred to as related to stimulus-processing, as they tend to have strong, temporal definitions defined by the stimulus onset. Some stimuli, however, are less strictly defined, and may be delayed or amplified as a function of complex mechanisms. I call them *cognitive potentials*, perpetuating a category error common in psychology (there is nothing cognitive about psychology – it is the psychology of cognition!). The third type of event-related potentials is more clearly related, particularly in time, to reactions, rather than the stimuli evoking them, and are therefore called response-related.

Terminology of sense and time

Event-related potentials (ERPs) are defined by their relation to externally observable events, be they visual, auditory, tactile stimuli (visual evoked potentials, auditory evoked potentials, somatosensory evoked potentials: VEPs, AEPs, SEPs), motoric responses (motor-evoked potentials, response-related potentials), or more situationally defined (error-related potentials). Conventionally, they are named by their polarity (negative, N; or positive, P) and onset. Somewhat confusingly, the onset can either be in milliseconds (P200 on the right being the positive potential at 200 ms after the stimulus onset), or in order of appearing (P3 being the third positive potential, generally after P1, N1, P2 and N2). Thus, some authors refer to the P3 as P300 if it happens to occur at 300 ms. All this works in theory, but the practice tends to be harder as we normally use multiple different electrodes. Furthermore, even though earlier potentials tend to appear as distinct peaks, later ones are more commonly defined as a difference between conditions; for example, the N2 in Figure 1.4 isn't even strictly speaking *negative*, but it might show negativity against a more positive control condition. We will get to such problems in Chapters 9 and 10.

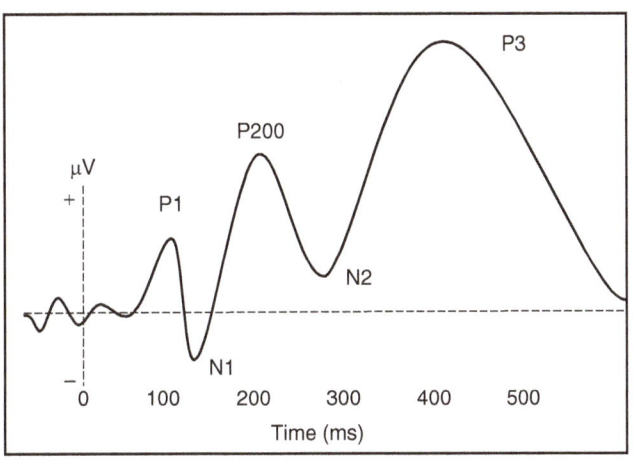

Figure 1.4

Stimulus-related potentials

Extremely early potentials result from stimulating any sense, although olfactory and gustatory evoked potentials are unpopular (frequently involving disgusting stimuli) and technically challenging (as evoked potentials require a rapid onset of the stimulus). With audition, the first observable potentials within 8 ms after the onset of a click sound can be attributed to activity within the inner ear itself, and the brainstem (Picton et al., 1974), whereas mid-early latencies (up to c. 40 ms) may involve the thalamus and, increasingly, the primary auditory cortex. With somatosensory potentials (related to touch, pain, or electricity), for example, evoked potentials can be observed already between 15 and 30 ms (Larson et al., 1966) after a 1 ms electric current is applied to the wrist. Such stimulation can be picked up in EEG around the neck after c. 20 ms (presumably indicating thalamic activity), and only at around 50 ms near the somatosensory cortex (Ravaja et al., 2017). Processing of visual stimuli is slightly slower, and the first clear potential arrives from 65–80 ms, most likely generated from the primary visual cortex (Jeffreys & Axford, 1972).

The extremely early onset of these potentials makes the task of explaining which brain areas are involved easier, as it restricts the number of possible neural structures that can anatomically be involved in processing. However, that simultaneously limits their interest to psychologists and cognitive neuroscientists. Take a normal phone: it can do very clever things with visual input using its camera – for example, it can apply all sorts of filters to a selfie. Yet, these filters are not part of the camera module itself, and certainly not of the lens. Whether you take a selfie or take a picture of a blue sky is not going to change the way the lens deals with the input, although it might adjust for the difference in distance. Likewise, the lens is not bothered by the degree to which you *want* to look good in the selfie, that is a task for whatever apps you have running. In other words, very early **bottom-up** processes do not usually show higher-level differences between stimuli and **top-down** effects (like expectations) have few if any discernible effects at this stage. Psychologically more interesting processes rapidly come into play in all other potentials.

P1 and N1

Have you ever noticed that even though you are not listening to a conversation, you instantly become aware if someone speaks your name? Somehow, this factoid seems to be known as the Cocktail Party Phenomenon, *which it isn't*. The Phenomenon refers merely to our ability to suppress (or 'gate') information, even though irrelevant sensory information is physically as present as relevant information is. In fact, people *rarely hear their own name*, and if they do, it may merely suggest they have low working memory capacity (Conway et al., 2001). The P1 and N1 are the first ERPs that show selective attention to visual (Van Voorhis & Hillyard, 1977) and auditory (Näätänen & Picton, 1987) input. Thus, for example, if a participant is told to attend to the left visual field without moving their eyes, the N1 is enhanced to stimuli that occur left, rather than right. Interestingly, pleasant and unpleasant pictures have been found to increase the N1 even in unattended situations (Schupp et al., 2003), suggesting that evolutionarily salient stimuli break through the attentional suppression – in the way that your own name doesn't.

N170/VPP

Most humans are excellent at remembering thousands of different faces (Jenkins, Dowsett, & Burton, 2018). We get to enjoy the fact that our visual system is great at this kind of pattern recognition when we see faces in the clouds, although less so when we mistake a shadow in the night for an intruder Jeffreys (1989) compared VEPs to abstracted faces to other types of stimuli and noticed a distinct positive potential over the top of the head, which he termed the vertex positive potential (VPP). Bentin and colleagues (Bentin et al., 1996) showed a similar difference, now between faces, scrambled faces, and cars, but as a negative potential at 170 ms (N170), over the right and left occipito-temporal cortex. The two potentials likely reflect two sides of the same coin (Joyce & Rossion, 2005), with an underlying generator proposed in the fusiform face area (Ghuman et al., 2014). Consistent with the idea that our visual system is over-eager in perceiving faces are findings suggesting that N170s occur whether we pay attention or not, and even before we are aware that there is a face (Rossion, 2014).

Terminology of location

When we are talking about the location of ERPs in terms of where we measured them, we mean the scalp distribution, or ERPs topology, which follows neuroanatomical conventions. Thus, ERPs can be frontal (towards the face), central (on top of the head), occipital (on the far back), parietal (between occipital and central), or temporal (to the side). We also use terms like medial and lateral (respectively in the middle or the sides of a horizontal plane), superior (above something else), and inferior (below). English has a long tradition of preferring Latin terms in science, which makes us sound wise and unintelligible. I could recommend practising localisation terminology whenever you are asked 'have you seen my keys?', but have evidence this is not great for relational harmony.

Cognitive potentials

Probably the most famous paradigm in EEG research is called the **oddball** paradigm. A fast sequence of stimuli of whatever modality (in vision, the term **rapid serial visual presentation** is sometimes used), featuring many *normal* types, and infrequent *different* ones (the oddballs). What 'different' means depends on the type of cognitive function one is interested in, but the noticing of something being *weird*, and how much such a weirdness matters, is inherent to the study of many cognitive potentials.

MMN and the N2s

Näätänen and colleagues (Näätänen et al., 1978) first noticed a negative potential (from c. 150–200 ms) in response to auditory tones being different in frequency from the preceding ones, whether the tones were attended to or not. Although different theories exist, the most regular interpretation is that this 'mismatch negativity' (MMN) occurs as a result of the present auditory stimulus being compared to the memory trace of the preceding ones, and found, well, mismatching. Other negativities in this temporal range that is generally referred to as the N2 family are similar to the MMN in that they tend to reflect that something is different from something else. However, while with N2s the 'unlikeness' is relative to the previous stimuli, N2s may be different in other ways: different from surrounding stimuli or behaviourally weird in the sense that they promote incompatible responses. For example, a central, task-relevant arrow pointing to the right, flanked by two irrelevant arrows pointing to the left (e.g. <><<) commonly evokes an N2. What that means, however, is a matter of ongoing debate. It could be that this N2 reflects a mental comparison process similar to the MMN. Alternatively, N2s such as this are generated in response to a need for control, or response inhibition due to the conflicting stimuli (Folstein & Van Petten, 2008). Whoever is right, it remains true that N2s, in general, are elicited by events that aren't quite right, and in that sense, they show remarkable conceptual similarity with the ERN and N400 (see below), which has given some cause to argue that all later negativities are related, as are all positivities (an interesting framework was proposed by Kotchoubey, 2006).

Late positivity and the P3s

The P3 remains one of the most studied ERPs in the dictionary of potentials, so it is surprising we remain quite unsure as to its relation to cognitive functioning. Typically, one reads about P3s in the context of oddball paradigms – even more so than N2s – but in the 1960s, these were not as standardised as they now are. Thus, in 1964, Chapman and Bragdon observed that while participants were doing simple numerical problem-solving, the appearance of numbers, as opposed to flashes of light or non-numerical symbols, evoked late positive potentials over the central-parietal area that were hardly affected by stimulus luminance. Sutton and colleagues (Sutton et al., 1965) are more usually seen as providing the first 'true P3'. In their design, cues were shown to indicate a sound, a light, or either one would follow. With the latter, uncertain case, a strong, positive potential at 300 ms was observed in response to the sound or light. Thus, one of the first proposed theories of the P3 suggested that it is evoked whenever a stimulus reduces uncertainty, or, in other words, contributes information (Sutton et al., 1967).

However, the design used in these early experiments made it difficult to dissociate whether the P3 was due to stimuli being improbable, informative, attended, memorable, meaningful, novel, or relevant. If you think of any vivid memory you can easily bring to mind, it will likely concern an event that was different from normal occurrences, and novel to you then, not to mention unlikely, important, and meaningful: clearly these adjectives correlate in day-to-day life. The UCSD group (Courchesne et al., 1975; Squires et al., 1975) was therefore quite ingenuous when they managed to disentangle novelty and relevance in what has become the gold-standard oddball design in P3 research. Oddball designs now distinguish between targets (commonly stimuli that require extra attention, e.g. they need to be mentally counted, or an explicit response) and novels (stimuli that stand out in some other way). The former typically incur a more parietal P3b, while the latter prompt the 'novelty' P3, or P3a.

So what neural structure, or cognitive function, exactly causes P3s to happen? Its neural origin remains somewhat unclear, but likely reflects synchronised activity from multiple different structures (Linden, 2005). Similarly, the significance of the P3 to cognitive functions is under debate, with competing theories suggesting a functional relationship with stimulus-processing, response-selection, or memory. For example, the earliest integrated theory on the P3 held that the P3 indicates the working of a 'context updating' mechanism (Donchin, 1981) – a kind of re-evaluation of our environment that means a stimulus, which initially provokes a P3 as it is unexpected, becomes less and less surprising as it is repeated. Alternative conceptions point out that the onset of the P3 predicts the reaction time, and therefore may be more related to response-selection (Verleger, 1997; Verleger et al., 2005). Another popular theory is that P3 is evoked when stimulus processing *results in* memory operations (Polich, 2007).

As the oddball paradigm is used as the central experiment of this book, I will return to these theories when they become less abstract. For now, it will be enough if you could try to remember the main distinction between N2s and P3s in the following, over-simplistic way:

*While we could think of N2s as 'No'-responses, no*gativities*, P3s are about the 'So?' question, so*sitivities*.* If you ask people, 'are you OK?', most will answer yes, and those saying *no* will not match expectations (evoke an N2). If that changes things for you, you need to remember and do something about it ('no, I'm not OK, I actually need an ambulance'), in other words, *if it matters, the no becomes a so* (and evoke a P3).

Language-related potentials: N400, P600

As the oddball paradigm took the world of psychophysiology by storm (a world that fits in a pretty tiny tea-cup), many *different* sorts *of differences* were rapidly investigated. Thus, when Marta Kutas (Kutas & Hillyard, 1980) first showed words, one at a time, instead of flashes or sounds, she fully expected improbable words to incur P3s. However, showing a sentence like 'He spread the warm bread with socks', a late negativity at 400 ms, rather than a P3, was observed to the word 'sock'. The oddness in meaning has since then been replicated hundreds of times and the consensus is now that it is related to semantic processing: when words are difficult to understand, or when they are hard to integrate with the rest of the sentence (Kutas & Federmeier, 2011).

Following the discovery of the N400, another type of language-related potential was found in words that were odd in the sense that 'The spoilt child throw the ball aside' does not parse. Instead of the previously observed N400s, such syntactically anomalous words (Hagoort et al., 1993) prompt a positivity at about 600 ms, commonly called P600s or SPSs (syntactically positive shifts) – despite the terrible reputation of the latter abbreviation among students. To me, it seems counter-intuitive that, somehow, our understanding of what a word means occurs before we understand grammar, as we know that grammar often enough changes meaning. However, a syntactically ill-fitting word might evoke some sort of desperate attempt to re-evaluate the preceding sentence: 'The plural verb suggest an agreeing subject, so might *The Spoilt Child* be a group noun? Indeed, they sound like a fabulous, glam-rock sort of band!' This theory would explain why the P600 shows such remarkably similarity with what we know about P3s (Kaan & Swaab, 2003).

Response-related potentials and other non-stimulus evoked potentials

Not all ERPs are stimulus-evoked potentials, however. While, for example, the P3 already shows a less strict relationship with the stimulus (as it occurs between 250 and 700 ms) than the earlier potentials, another class of stimuli have a much stronger relationship to performance and action than the nature or complexity of the stimulus. The border between ERPs that are stimulus-related, cognitive, and response-related is to some extent writ in water and I would recommend paying less attention to this division than to the ways they can inform us about psychology.

CNV

The contingent negative variation (CNV) is a prime example of an evoked potential that is critically dependent on the response, though not evoked by it. Grey Walter, the swashbuckling EEG pioneer of the beginning of the chapter, was interested in why repeating flashes or other simple stimuli showed stable early effects (regardless of the number) while late ones (possibly the then unnamed P3) reduced dramatically. To investigate how expectations are formed, he repeatedly presented subjects with a warning click followed a second later by a series of flashes that produced their typical VEPs. However, if the series of flashes could be ended by a button press, then the interval between the warning and the button press suddenly showed a very strong negativity (Walter et al., 1964). If the series of flashes were removed, however, then gradually the negativity would disappear, whereas if participants were lied to, and told the series of flashes would reappear, then suddenly the CNV returned. Thus, the CNV is a useful indicator of conditioning – or expectation, if you are not a behaviourist. As behaviourism gave way to cognitivism, note how later ERP studies made stronger claims on how components related to stage of cognitive processing, whether they be stimulus-related, central cognition, or response-related.

RP/LRP

In any listicle of best brain potentials, somewhere near the top should appear the readiness potential (RP), or grander-sounding *Bereitschaftspotential* (try buh-<u>right</u>-shafts-poe-ten-tziahl for maximal academic appeal). The (unsurprisingly German) discoverers were the first who came up with the simple idea that, rather than analysing stimulus-evoked potentials, they could also look at evoked potentials to the response onset (Kornhuber & Deecke, 1965). The idea is to have a participant not do anything much at all, until they feel like pressing a button, with the EEG before and after the response being the focus of the analysis. Interestingly, they found pre-response negativity over the left fronto-central side of the head if the response was right-sided, and right if left-sided. This negativity generally builds up over a second (if not seconds) until the actual response and allows us to detect whether an action is being prepared.

A fundamental issue with the RP is that it is not easy to disentangle which part of it concerns the more 'cognitive' decision *to move*, and which part concerns the pure 'motoric' preparation of *where to move*. As it occurs on both sides, it arguably reflects both. However, as you will probably remember from biology, the left motor cortex largely controls the right hand while the right motor cortex controls the left. To deal with this problem, Michael Coles (Coles, 1989) thought of an elegant way to separate the lateralised part of the RP from the ERP that is a more general kind of preparation. Given a task that requires both left- and right-handed responses, one would expect contralateral activity (on the opposite side of the response) more than ipsilateral activity (on the same side of the response). Thus, for example, a 'press left' type of stimulus should evoke more right than left activity. As shown in Figure 1.5, we can use this information, subtracting contralateral from ipsilateral activity to precisely define when activity is pointing to the correct as opposed to the incorrect response direction, or, in other words, when the readiness potential becomes lateralised.

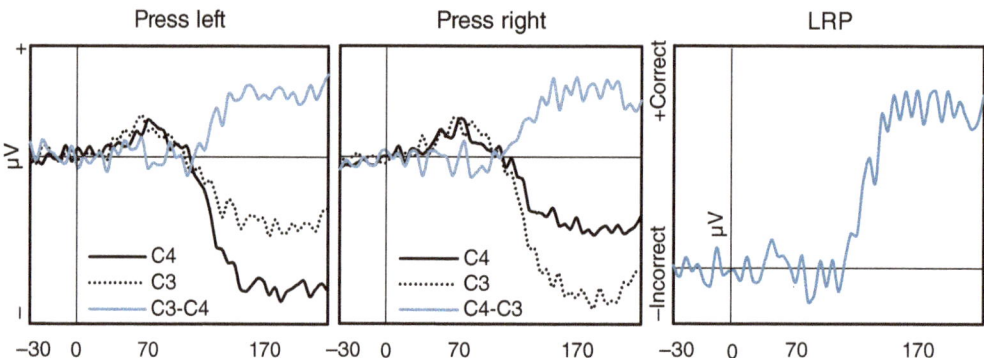

Figure 1.5 Calculating LRPs. Participants either see 'press left' or 'press right'. With the former type, more negativity will be observed over right (C4) than left (C3) electrodes, while the latter stimulus shows similarly contralateral activity. Subtracting contralateral from ipsilateral activity gives a similar outcome for both stimuli (grey lines), and summing (or averaging) these presents a figure in which we can determine when motor activity starts to build up for the correct response direction, the lateralised readiness potential.

Why go through all this trouble? The great thing about the LRP is that it can be interpreted a bit like a reaction time, but before the actual reaction happens. This shows, for example, that when we *almost, accidentally, but not quite* respond wrongly, such as with the previously mentioned flanker stimulus (<<><<), this *almost wrong* activity shows up in the LRP. But the best example of the use of the readiness potential in Psychology is no doubt Benjamin Libet's famous studies of free will. Libet and colleagues (Libet et al., 1983) asked people to look at the hands of a clock counting seconds, and, whenever the urge struck them, to *voluntarily press* a button. Having free will intuitively suggests that we make a decision, then tell the brain (so to speak) to set the hands to move, and then finally the hands press the button. Indeed, Libet reported that an urge to press the button occurred about 100 ms before the button was actually pressed.

Strangely, however, the readiness potential could already be measured 300 ms earlier. So even if we feel like we think before acting, it seems as if the brain is already preparing to move before we know it. Maybe consciousness is therefore not in control at all, but merely a happy story we tell ourselves?

ERN, Ne, Pe, FRN, MFN

The final main type of response-related potential is not so much about the presence or absence of a reaction, but about the quality of the action. Like the RP and LRP, the error related negativity (ERN) shows up best in relation to a response, but unlike the other potentials, it is specifically evoked by *incorrect responses*. Thus, Michael Falkenstein and colleagues (Falkenstein et al., 1991), had participants respond by pressing the F or J key as soon as they saw these letters. Normally such a task is pretty easy even on a Monday morning, so in order to cause people to commit errors, the researchers kept on pushing participants to be faster. This no doubt caused enough frustration, but also produced a solid ERP effect: a kind of frontal negativity can be observed roughly 80 ms after bad responses. This error negativity (Ne) was found delayed if attention is not focused upon one modality, and occurs even if the participant is unaware of their error (Nieuwenhuis et al., 2001), although a slightly later occurring *positivity* after errors (the Pe) was found to occur only if a participant was aware (Shalgi et al., 2009). In other words, sometimes the brain notices something is awry but fails to inform 'us'. Such a story attracts cognitive neuroscientists like flies towards honey: after all those years of mocking Freudians for their fuzzy science, we can finally contribute something useful about the unconsciousness!

Error negativity, or error related negativity, is a deceptively simple potential that stands at the root of a complex family. The problems already start with the word 'related': whether you use Ne, Pe as terminology, or ERN, seems to indicate whether you follow the Falkenstein et al. (1991) tradition or that of Coles et al. (Gehring et al., 1993; Holroyd & Coles, 2002). Furthermore, the ERN may be related to the feedback-related negativity (FRN), which is evoked by stimuli that classify your performance as bad (Hajcak et al., 2006). Finally, the medial frontal negativity (MFN) is a generalised version of the FRN that occurs for any event that is worse-than-expected (Cohen et al., 2007), such as when somebody treats you unfairly (Boksem & De Cremer, 2010). Notice that in all these cases, the ERN, Ne, FRN and MFN are not all that different from N2s, N400s, all the way to the MMN: in each of these cases, something is amiss. However, although it is tempting to argue that every negativity is a kind of mismatch, this ignores the large differences in time, topology, and experimental design. Instead, we should perhaps regard them as a family with several branches. Thus, for example, Gehring and Willoughby (2004) argue that mismatch Ns and higher level 'control' related N2s are dissociable both functionally and topologically.

Beyond ERPs: Frequency analysis and other biosignals

While to some, the study of EEG has become almost synonymous with the study of ERPs, the investigation of the various waves, such as the previously mentioned alpha (Berger), delta and theta waves, has steadily progressed as well. To understand the distinction, try to see EEG as music inside the brain. As with EEG, music happens both in time (e.g. rhythms) and pitch (the bass), and in both time and frequency (melodies). Any continuous signal, whether it be music from *The Top of the Pops* or brainwaves from the top of your head, can be analysed in both time and frequency, by **decomposing** the signal, a process known as the **Fourier transform** (after the 19th-century French maths wonder Joseph Fourier).

I already discussed alpha waves as a signal that can even be spotted by the naked eye and looks a bit like a steady wave going up and down about 10 times per second, a bit like the sinus in Figure 1.6 A. By adding an additional sinewave with a 3 × shorter period to the first, the two signals together will look like B. Using the Fourier transform on this signal should therefore decompose it as having high amplitudes at 10 and 30 Hz, in other words to have a **spectrum** as visualised in C. To show that any continuous signal can be decomposed into its constituent frequencies, I walked up to my loud cat and recorded some of her vocal ambitions, the result of which is shown in D as a time-series, and E as a time/frequency decomposition.

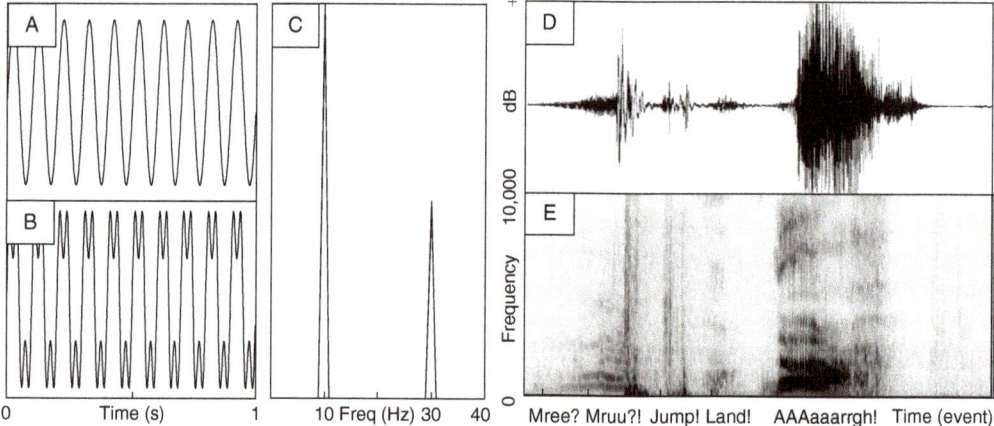

Figure 1.6 Time and frequency analysis. A: 10 Hz sine wave as a time series. B: A 30 Hz sine wave added to the 10 Hz. C: Spectrum Analysis of B. D: Time series of audio recording of a cat. E: A time-frequency analysis based on D. Note that here the Y axis is frequency and the X axis is time, the darkness of each spot indicating the amplitude at the specific time. Further note my annotation below E, describing how the cat made some meowing while looking at a hanging plant. As I recorded, I made a convenient perch, so she jumped on my shoulders to get at the plant. Very sharp claws evoked a loud human scream with both high and low frequency components.

Oscillations and frequency analysis

Let's say that you record my brainwaves, do a simple Fourier analysis, and find that I produce a lot of activity in the 40–100 Hz range. What does this mean? You do an internet search and find that this 'high frequency, like a flute' (brainworksneurotherapy.com) wave is related to 'fight/ flight-mode', 'multi-tasking' (emotive.zendesk.com), 'being "in the Zone"' (mindvalley.com), and, of course, 'spiritual emergence' (brainworksneurotherapy). Apart from the first statement, which is just nonsense (it is more similar to a bass drum), these kinds of ridiculous explanations turn a reasonably exact science into some sort of wellness-inspired marketing prattle that makes me want to tear my hair out.

One of the problems is that even if carefully controlled experiments may show that visual consciousness is associated with increased gamma activity (Crick & Koch, 1990), it does not follow that if I find increased gamma activity, there must be more consciousness. For example, merely frowning will cause muscle activity that contaminates EEG recordings in the gamma spectrum, so perhaps increased activity means less that someone is in touch with the universe, and more that they're simply puzzled. Another line of research throwing mud over the gamma-consciousness relation was a set of findings suggesting that miniature saccades, such as may occur if people are concentrating strongly, affect gamma readings (Yuval-Greenberg et al., 2008), which means that gamma itself could be a mere by-product. This illustrates the dangerous fallacy known as **reverse inference**: just because visual consciousness may be associated with increased gamma, does not mean that observing increased gamma should indicate visual consciousness. Thus, for example, training gamma to improve consciousness (as some of the websites promote) may be as useful as modifying a thermometer to stop global warming.

Now we have this out of the way, what kind of information *might* frequency analysis provide if we were able to control for all other explanations? You will remember that alpha waves (between c. 8 and 12 Hz) show up when people close their eyes, are inactive, and bored – you should be able to summon alpha waves in your subject as well (Chapter 6). While the *more alpha power means less activity* is too simple an inference, it holds reasonably well in terms of concentration: your subjects will no doubt generate more alpha during the end of the experiment than the beginning. A common suggestion is that this inference holds at a neural level, i.e. measuring less alpha around a certain area means more activity. This is a common assumption of the idea that certain emotional states, such as approach motivation, may be measured as the difference between the alpha power over the left and right frontal cortex (**frontal asymmetry**; Coan & Allen, 2004).

The other frequency bands have likewise been associated with various cognitive functions. Delta waves (0.5–4 Hz) can indicate deep, or *slow-wave*, sleep, and are thus commonly studied in sleep research. Theta (4–7 Hz) waves are associated with hippocampal activity (Lega et al., 2012), and account, to a large extent, for event-related potentials (filtering out the theta band removes the P3, as I unfortunately found out: see p. 105). At the upper end of alpha are also mu-waves (10–13 Hz), that are known to be involved in voluntary movement (Pfurtscheller & Neuper, 1994), and more controversially in mirror neuron function (Ramachandran & Oberman, 2006). The beta frequency band (13–29 Hz) has been implicated in motor control studies, and the more spectacular phase of sleep research, the rapid eye movement stage, in which dreams are best. Finally, there is of course gamma (30+ Hz), which means *being in the Zone*.

I have noticed that once you try frequency analysis, things quickly start to escalate in terms of complexity. As illustrated in Figure 1.6, while a time series form of data can always be Fourier-transformed in its entirety to the frequency spectrum, it is also possible to run the analysis over a very short (e.g. 1 second) window and then slide this window in what is called a **time/frequency transformation**. As I will explain in Chapter 10, this allows one to do **event-related spectral perturbation analysis**, enabling us to investigate not only when a potential occurs, but also what frequencies it involves. While doing this, you may notice oscillations in EEG often occur at the same time, and are correlated at certain frequencies (**coherence**), which may involve meaningful networks of neural communication. But as you start correlating each channel with every other one, complexity escalates. While things were simple enough when the investigation concerned 'there is alpha', imagine having 100 times, 100 frequencies, 32 channels, and each of the channels may show coherence with any of the other channels: that's *4,960,000 dependent variables*! Maybe if you're high on gamma (*in the Zone*) you can cope, but otherwise don't do it unless you have a really good idea what you are looking for.

Other biosignals

EEG equipment is expensive (see Chapter 5), but once you have an EEG kit, you can usually add other signals at little extra cost or effort. Such 'auxiliary' signals can be most other kinds of time-varying signals. You should even be able to plug in an electric thermometer and literally measure your work climate, even if this is perhaps not the best way to go about it. The point is that once you have mastered EEG, the same kind of analysis you do on EEG can, with some minor tweaks, be applied to many other kinds of physiological and non-physiological sensors.

For example, applying two electrodes on the skin above muscles, with about 1 cm of space in between them, is a common strategy to measure **electromyography** (EMG, muscle activity). By doing the same, but on the muscles that control frowning, smiling, and nose wrinkling, you get facial EMG, or fEMG (see the standard guide by Fridlund & Cacioppo, 1986), which is a common measure in emotion research. Another popular use of EMG is by combining it with transcranial magnetic stimulation (TMS), *zapping* the motor cortex and measuring the degree to which involuntary muscle contractions occur, which can inform you of how excitable the cortico-spinal system is (Bestmann & Krakauer, 2015). However, some people are worried that EEG has a sort of magical switch that 'reverses the flow of the current', instead of measuring volts from the brain, producing volts in the brain. It is a big relief for them that I can say that I do not normally *zap* brains.

Much safer (TMS is really also very safe) are the truly old-school methods of measuring excitement, alarm, or arousal – autonomic activity – by measuring heartbeats (**electrocardiography**) and sweaty hands (**electrodermal** activity). Various suppliers of EEG tools will also happily sell you disposable or reusable sets of two or three electrodes, which can take a pretty good ECG along with your EEG. After calculating the heartbeats per minute (or interbeat interval, IBI), which provides a general kind of excitement level, one can additionally find out how people emotionally respond to sudden stimuli by measuring the immediate change in IBI *evoked* by stimuli. Similar to this **orienting response** in ECG (Graham & Clifton, 1966) is a sudden increase in sweatiness levels (the galvanic skin response, GSR), called the **phasic** part of **electrodermal** activity (EDA) in order to separate it from very slow, **tonic**, changes in skin conductivity. The fun part of using EDA is that it gives pretty visible evoked results, so if you measure it, you can see an immediate effect of clapping your hands, or getting your subject to lie, as the method is commonly used for the polygraph. So, if you are one of those who chose the study of Psychology or Cognitive Neuroscience because you wanted to solve crime using brain science, then you are getting pretty close.

To be honest, though, I won't record EEG or other biosignals from participants who do not wish to cooperate; that would be unethical. It is also exasperating to work with people who are there to prove a point, whether it be their innocence in some sort of Cluedo-EEG or something about their personality. For example, some people may want to show how tough they are by proving their brain is wholly unaffected by emotional pictures. I doubt such ability to begin with, but as an experimenter you ultimately have little control over fanatics who will just close their eyes or hit keys at random. Such data will invariably be Consistently Related to Absolute Piffle (an acronym). So let us stick to reading minds from people who actually want to be mind-read, starting with the next chapter.

ESSENTIAL EXERCISES

Essential exercise: Interpreting EEG

Consider the four EEG figures presented in A, B, C, D below. Each shows a very different type of effect:

- One of the figures shows an EEG effect related to a difference in intensity (of a soft versus hard vibration).

- One of the figures shows induced alpha negativity followed by the beta negativity that precedes an expected stimulus.

- One of the figures shows an ERP effect related to semantically relevant and irrelevant words.

- One of the figures shows an ERP effect related to seeing something unexpectedly *unfair*.

Based on the timing of the effect(s), can you determine which figure fits which description?

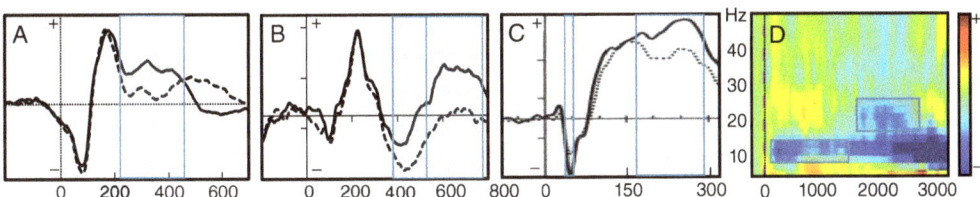

Figure 1.7

EXPLORE

There is a lot more to read when it comes to the history and science of EEG aside from the references appearing in the text. I would like to mention a few sources in particular, though.

Millett, D. (2001). Hans Berger: From psychic energy to the EEG. I am much indebted to Millett for digging up a ton of facts about Hans Berger in this entertaining and insightful article.

Walter, W. G. (1961), *The Living Brain*. If you want to discover why the *alpha* rhythm is the rhythm of love, and why it is not quite as necessary to have similar alpha rhythms as suggested in the first edition of the book (apparently, Walter's new wife's rhythm was not quite as compatible as the first), then this second edition is your book. If you are more interested in what the author was on about, I recommend Rhodri Hayward (2001) The tortoise and the love-machine: Grey Walter and the politics of electroencephalography.

Luck, S. J. (2014), *An Introduction to the Event-related Potential Technique*. This book provides a fantastic introduction to ERPs that is mentioned almost every time an ERP study is being reviewed. Its comprehensive status as *gold standard* is, I believe, due to the degree that Luck's deep insights of the technique are displayed in an insanely readable and entertaining way, which makes it the second-most captivating book on the market. So much so, indeed, that I failed to return the edition my EEG tutor kindly lent me (I'm sorry Guido, I'm still embarrassed about this!). For this chapter, I borrowed heavily from both of you.

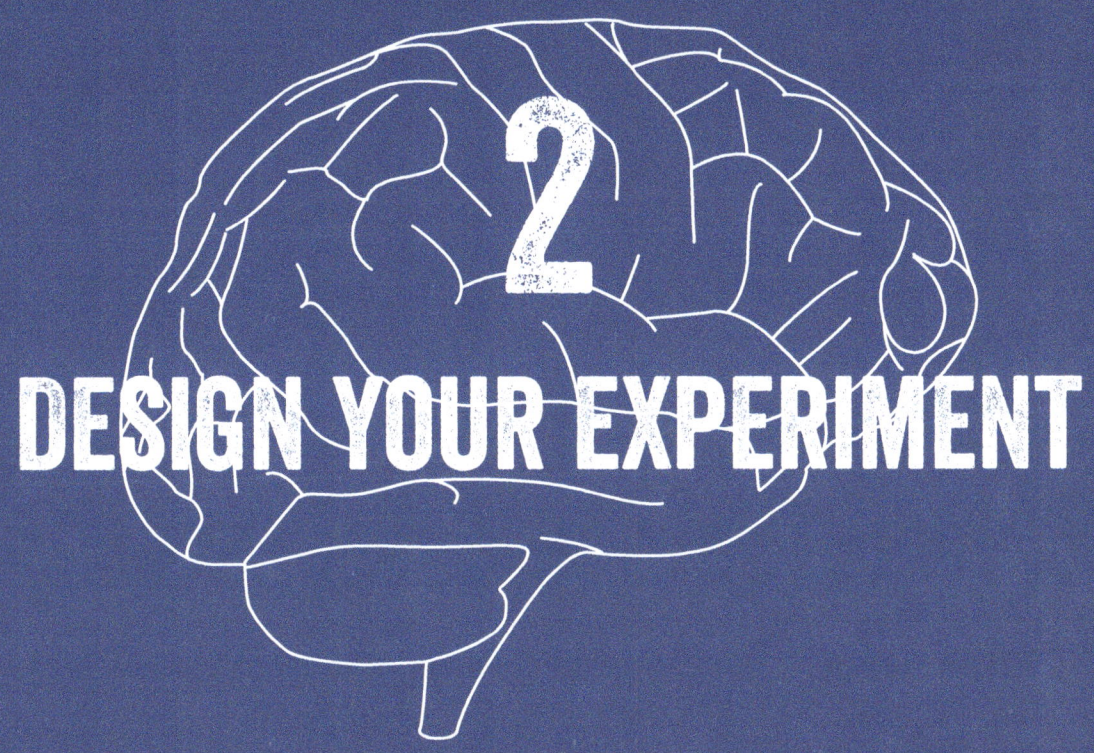

2
DESIGN YOUR EXPERIMENT

IN THIS CHAPTER, YOU WILL LEARN ABOUT:

- Measuring mind-reading
- Predicting effects on brain potentials
- Designing the study on paper
- Generating your own questions

In Chapter 1, I set out to convince you that EEG is not just squiggly lines, and is more like exploring the fireworks of the mind. However, despite the peculiar origin story of Berger's Brain Mirror and the revolutionary advances brought about by Walter Grey, things admittedly got pretty dry. Making post-synaptic potentials as exciting as they are excitable is not an easy job, though, so thanks for sticking around. Fundamentally, you should now know what event-related potentials are, know of an N or 2 and a P or 3, understand the oddball paradigm, and get what the general idea is behind frequency analysis.

With these basics out of the way, let me assure you that doing EEG is not quite as theoretical as Chapter 1 made it sound. We will be progressing from theory and reading to doing real work as fast as possible. Learning what a P3 might indicate is a very fine thing, but I believe it is useful to start with something more practical. For example, reading minds is an exceedingly practical skill, so let's set that as our goal.

In the present chapter, I will define this a bit better: what does reading minds mean? It could mean reading the unconscious, but I know enough about Sigmund to tell sane people to avoid him. Or it could mean figuring out some attitude someone might have towards you – this sounds very useful, but again, the results may not be flattering. So let's start with something simple: *the goal is to find out who someone is thinking about*. And, because I and plenty of others are pretty bad at remembering faces, let's concentrate on faces, and maybe learn a thing or two. Which specifies the goal:

The goal is to figure out who someone is thinking of by measuring their face-related EEG.

In other words, if you ever worried that your friend, let's call him Bob, seems suspiciously interested in someone we'll call Alice, rather than Bob's partner Carol, you might be able to help out! No, not seriously, of course, that would be quite unethical, and besides, what kind of cover story would you need to get the suspect to become a subject in an EEG experiment? It is a tempting proposition, though, so if you did not appreciate my lengthy previous chapter about EEG being the fireworks of the mind, then maybe this kind of practical application works for you. Don't blame me if you accidentally create some sort of dystopian nightmare by unleashing the *Braincrime Test™* onto the world.

MEASURING MIND-READING

How do we find out who someone is thinking of? To make it easier for ourselves, we need a limited number of potential candidates, rather than consider the entire world's population. It is much easier to guess the answer if there are a limited number of alternatives, as you know from any multiple-choice exam. In terms of the previously mentioned practical example, we have our subject Bob, whose interest in Alice we want to measure. Let's say we somehow manage to obtain a measurement of interest using EEG, and it says 6.04 Level of Interest. Is that a suspicious amount of interest? Of course, if we have standardised norms, a bit like IQ, of what amount of interest counts as above the level of an uncaring psychopath but below that of unseemly fascination, we would be able to say something; but alas, we don't.

Comparing brains and EEG norms

Why don't we have such norms? To some extent, the problem has to do with variance: one brain is simply not exactly the same as another. This may sound very woolly (we are all individuals!), but I simply refer here to the fact that brain structures are smaller and bigger between individuals. Indeed, a clinical case (Feuillet et al., 2007) showed that civil servants can even deal with a massive amount of brain volume completely missing. But then it gets worse: the neurons might be in the 'right spot', but slightly turned, thus presenting different dipoles (see Figure 1.3, p. 8). Likewise, people may just have thicker skulls, dampening the brain waves we can pick up (Chapter 1, p. 9), and brain waves in general are tens, if not hundreds, of times smaller than other things we pick up in the EEG. Finally, tiny differences in what we show to the subject can have extreme effects on our measurements, even if those differences have little to do with our measure of interest.

Even though *such norms* do not, or should not, exist, *some* norms do. The field of Quantitative EEG (QEEG) aims to use EEG measures in order to detect mental states and neurological conditions. It tends to use frequency analysis principles rather than the more stimulus-specific event-related potentials we will mainly be concerned with and, as such, tends to be used more for clinical applications like detecting brain injuries.

Our lack of norms means that inevitably we need to see how the measured 6.04 compares with others. This means we will have to show the target, Alice, among a limited number of other pictures: if we know Bob's interest in Alice beats his interest in coffee, hamburgers, and cats, all things he's terribly fond of, we will know that his interest crosses the line of what is reasonable.

Or would we? After all, for one, maybe Bob isn't nearly as interested in coffee, hamburgers, and cats as he says he is, or maybe he is a genuine people person, interested above all in (pictures of) people. More important, however, is the possibility that there's something in pictures of people that is generally not in other pictures, but which could conceivably give rise to a big evoked potential that somehow ends up in our Level of Interest measure. After all, EEG measures all sorts of things (see Chapter 1), mainly things that don't even have anything to do with brains (see Chapter 8), so our Level of Interest is never going to be a 'pure measure'. Let's say, for example, that Alice's particular flesh-tone is very bright, certainly brighter than the pictures of coffee, hamburgers, and the tabby cat we used. Alice will therefore evoke bigger early potentials, and likely altered later potentials also, just because she happens to be ginger. You simply cannot accuse somebody for being suspiciously interesting just because they happen to be ginger.

CONFOUNDS

This touches upon the most important topic in experimental design: confounding variables, or just ***confounds***. It's a word that generally elicits pained expressions in empirical researchers across fields. The idea is that your research design suffers from an additional variable, which sits on top of your independent variables like a parasite, sucking the explanatory power out of your study.

To use more classic terminology, a confound, in experimental design, is any inadvertent difference between your experimental and control conditions which can create an alternative explanation for your results. It is the one thing your teacher, your peers, or your reviewer will focus on most to figure out if your results, as you finally present them, can really be interpreted as you interpret them, or indeed, if you actually thought things through properly. I know at least one colleague who is famous for always, *always* raising multi-thronged questions at conferences, usually including the dreaded 'why did you not think of this *obvious* confound?'.

Try not to be him – the world, or at least the conference circuit, can only support one – but do ask yourself whether there might be any confounds at two select research stages. During the experiment design stage, it makes sense to brainstorm what confounding variables might possibly occur, so we can try to avoid them. During the interpretation stage, while writing the *limitations* section of your paper, you will have to ask yourself again: how certain of my results am I? Were there any confounds? Usually there will be one, and even though it will be too late by then to change your data (without fabricating them), it pays to have an answer ready: 'Of course, an alternative explanation could be, and with such and such experiment we could rule out this potential confound.' Let's focus for now, though, on what we can do to avoid confounds (see Chapter 12 about limitations).

STANDARDISING THE STIMULI

One of the best ways to avoid confounds at this stage is what Steven Luck calls the *Hillyard Principle*: 'To avoid sensory confounds, you must compare ERPs elicited by exactly the same physical stimuli, varying only the psychological conditions.' (Luck, 2014: 134). This is somewhat related to Occam's razor, or the principle that the more assumptions are required for an explanation, the unlikelier it is. Or, if a complex and a simple explanation account equally well for the same phenomenon, we should prefer the simple, more *parsimonious* one. As psychologists, we are naturally tempted to understand the world in psychologically salient ways, but as scientists, we should treat our own biases with scepticism.

As an example, consider a study which compares perception of a survival-related scenario – a picture of a snake – with a neutral condition – an abstract artwork (Figure 2.1).

 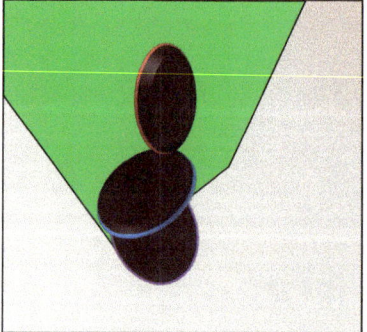

Figure 2.1   Snake (Diego Delso, delso.photo, License CC-BY-SA) and abstract art (designed by me after IAPS source). They are similar to images used in the International Affective Picture System (Lang et al., 1999).

The researchers find that the early posterior negativity (EPN) is enhanced, which they go on to say is attentional capture and explains that survival-related scenarios boost attention to promote the survival effect on memory (slightly based on Van Strien et al., 2014, although they have better control stimuli). Look closely at the figures for a bit and think of alternative explanations that you, as devil's advocate, could offer. Write them down, if possible.

Now let's compare our explanations. Did you think of: snakes are exciting (causing arousal), which may cause ERPs even in the absence of survival-related cues? Did you think artworks are beautiful in aesthetic ways (see neuroaesthetics as a field)? Maybe you noticed that pictures of snakes are more common than this particular artwork? That snakes lead to easy association? Great, you're a natural psychologist! Such psychological explanations are also the first to come to my mind, but if I try to be sceptical, the first difference we should recognise is that the *snake just happens to be a bigger picture*. Instead of thinking about the *meaning of the snake*, try to think physical differences first: the snake also has less saturated colours, more grainy details, a wider composition, lower contrast, and so on. If you try to think of the myriad ways in which the two pictures above are different, you will quickly notice that very few have anything to do with how we *feel* about the pictures, or what they *mean* to us.

In other words, we can never be sure of our psychological explanation unless we can rule out all the possible explanations that suggest a physical difference underlies our EEG effect. To help us in that, we should try to eliminate the physical differences by standardising the stimuli, making them maximally similar in every way, except for our psychological factor of interest. This seems not nearly as obvious to psychologists as one would hope – I assume because of the aforementioned preoccupation with psychological explanations – as it appears that many validated picture databases that I know of are riddled with inconsistencies. The two pictures in Figure 2.1, for example, are inspired by *the* most cited database in the fields of affective neuroscience and emotional cognition: the International Affective Picture System (Lang et al., 1999). So, what about that study you read that said this or that area lights up in the brain when you see a piece of art? Maybe it's the peculiar area in the brain that responds to smaller pictures.

In our previous example, then, we would want to compare Alice to a number of other stimuli who are as much similar as possible to Alice, without being Alice. Think of it as a police line-up: by comparing Bob's Alice-response to a number of other women of a similar age, skin and hair colour, we can get a somewhat better idea of just how suspicious his interest could be. But, because we are working with digital pictures (which is much easier than shoving all sorts of people in and out of Bob's view!), try to work out *at least* the following:

1. Are all the pictures the same size? If they are all shot with the same camera (ideal), they should be all the same number of megapixels (you can also check the properties of the image in your operating system or image manipulation software). Of course, the pictures should all be taken either in landscape or portrait mode (the latter being more obvious for selfies but less for widescreen displays).

2. Are the pictures similar in layout? For example, use a scheme with the eyes in the exact centre, keep 100 pixels space between the chin and the bottom of the image, 100 pixels between the top of the hairline and the top of the picture, and 500 between each side of the face and the side of the picture.

3. Are the pictures showing similar composition? Ideally, every participant should look straight ahead, towards the camera. My students tend to show themselves in more *Instagrammable* ways (diagonal, camera high up, and certain facial expressions). I think this is fine, but then all stimuli should look like this!

4. Do the pictures have similar lighting? Light-curves in software can sometimes help to equalise the pictures a bit, but it is always better to start with more similar material than to fix things later. Again, with the Instagram-crowd, try to go for a 'natural' effect, or apply sincere amounts of make-up to your entire group.

RANDOMISATION

Now the savvy reader should note that no matter our degree of standardisation, Alice may always be different in such a way to cause Bob to show an EEG effect which we take for Level of Interest, while being caused by something entirely else. After all, the Alice picture will always be a different picture from the Police-line of similar-but-not-quite ones, and maybe it's just this tiny difference that causes the *completely unrelated* EEG effect to confound with the Level of Interest. I will, for this reason, have to end the story of Alice, Bob, and evil cognitive neuroscientist Carol, because (1) Bob's excuses are starting to sound thinner by the minute; and (2) as long as Alice is a physically different stimulus, we will never be able to decide between the psychological (interest) and physical (a small foveal contrast). And Bob, saying your interest is just physical won't help.

Now, in order to rule out the psychological differences between the physically different stimuli, what we normally do in Psychology is to randomly distribute the one across the other. This would be a bit difficult to do for our previous scenario (Bob's suspected interest necessarily being confounded with a physically different stimulus), but for our general-purpose mind-reading instrument, this will work just fine.

What I mean with 'the one across the other' is this: instead of having a stable target, consider the situation in which our target – the face to be kept in mind – randomly changes from trial to trial. For example, we have 6 stimuli: A, B, C, D, E, and F. In trial 1, the participant is asked to keep A in mind. The participant sees random sequences of stimuli and the second one happens to be A. Imagine we see a big mind-reading response, what was formerly the Level of Interest. But, to find out if this effect is not due to a physical effect, we should ask the participant to keep someone else in mind, such as the old chap in trial 2 opposite in Figure 2.2. *If* the mind-reading response is due to the identity of the face kept in mind, we should expect in trial 2 the mind-reading response will be observed on stimulus position 5, and not on 3 or 2. That is, if the effect is still found on position 3 and not 5, there is likely to be a physical difference (the striking hair?), while if the effect is found on position 2, it may just be due to the ordering. This way we can disentangle physical, statistical, and psychological effects, fulfilling the *Hillyard Principle*.

Or do we? Instead of talking about single stimuli being different, we will now need to concern ourselves with how the group of targets is different from the others. Now it starts to become less about trying to look for the easily disregarded (like the difference in size)

Figure 2.2 Two oddball trials. In each trial, participants are requested to keep one face in mind, following which they see many faces in quick succession (five are pictured).

and more about statistics. For example, how do the targets differ from the non-targets (or normals)?

- Are targets shown as often as normals? Each target above is shown equally often to each type of normal, but the group of normal is more common. This has repercussions for what our data will look like, as the higher the number of instances (epochs) that go into an averaged ERP, the flatter it tends to become. We will discuss this in Chapter 10.

- Do targets follow one another or do we apply some form of pseudo-randomisation? If, for example, we want to avoid repetitions of stimuli (classic designs commonly do), it will mean that a target is *invariably after a normal*, while this is not necessarily true for a normal. This can cause the two stimuli types to have different baselines (see Chapter 9).

BEYOND THE BAREBONES ODDBALL DESIGN

Figure 2.2 already starts to look a lot like the oddball design. However, I like to add some elements to make things a little bit more interesting. These tweaks are suggested to make the experiment of more theoretical interest, and will enable you to do more challenging analyses later on. They also provide extra opportunities for learning, because they can be further implemented (Chapters 3–4), analysed (7–10), and interpreted (11–12) at each part of the book.

Cats are weird: Oddballs

What is an oddball design without anything really odd? I have four cats, and like most cats, they are indeed odd. So, like some sort of cognitive neuroscience Tyler Durden (Palahniuk, 1996), I like to splice in pictures of cats into my experiments. At the very least, it makes things more interesting for everybody involved. It also allows for interesting questions (see Hypotheses) to be investigated later by checking the differences between oddballs (cats) and normals. *But doesn't that violate Hillyard's Principle*, you may say – well, he isn't watching, is he?

Knowing me, knowing you

If you use your own pictures – which you definitely should – you will sometimes have to see your own picture. Will it cause some sort of automatic recognition, 'Hey, that's me!', or, if you hate the particular selfie, 'Argh! That's me'? If this is not a target, then to some extent it becomes a bit of an oddball, which may or may not evoke a 'had in mind' response. You may also wonder whether the *having in mind* interacts with the *me-vs-not me* effect: Is the effect of keeping yourself in mind bigger than the effect of keeping others in mind?

Sleeping subjects

It is not uncommon for participants to lose interest in their participation. Is it the excitement of the enterprise sapping their fortitude, or the relaxation induced by the hairdressing procedure (Chapter 6, p. 119) that induces this effect? Either way, it is not uncommon that halfway through an EEG experiment, participants start producing more alpha (see Figure 1.1, p. 3) than other brain activity, which makes it pointless to have them proceed. Ideally, you could sit next to them and cheer at every trial as they make it through, but even my most faithful lab assistants would kindly tell me to – well, let's not go there. The typical solution, therefore, is to pop in an extra task. Some people ask the subject to press a button whenever they see the target. This dramatically increases the 'mind-reading' response, but it makes the mind-reading slightly pointless. Another option is to have the subject silently count and, when they reach the end of the trial, to indicate the number of targets they observed. This also helps a lot, but it does make it more an experiment about mind-count-reading than mind-reading. My solution, then, is to quiz subjects at the end of each trial to indicate which person they had to keep in mind. It's not a super difficult task, but it does force them to at least keep a person in mind, even if they do not necessarily attend to the pictures on the screen. It might also allow you to study whether people who failed to keep someone in mind *still had those people subconsciously* in mind.

TRIAL PROCEDURE

With all the above, we have the experiment conceptually pretty much designed, but before we start coding (Chapters 3 and 4), we need to make many small decisions on various details. That is, you cannot ask computers to show you stimuli at 'a reasonable pace', as computers normally do not have great knowledge as to what you or I find reasonable. I hope I have just been giving all sorts of new information, and with this wealth of knowledge, it may be difficult to clearly see what details we are still missing in the design of our experiment. In order to find out, it may help to start writing a *Procedure* section of a journal paper relating experimental findings. A Procedure section needs to define at *each point in time* what a participant is undergoing. Here's a typical Procedure for an oddball study:

After reading instructions, and signing informed consent[1], the experiment began. Every trial started with a screen showing the target face, along with the instruction to keep this stimulus in mind during the trial. After pressing space, a 30 pt '+' character in Courier typeface[2] was shown for

600 ms to serve as a fixation screen before the rapid serial visual presentation (RSVP) was initiated. RSVPs comprised 62 sequentially displayed stimuli, each with a duration of 600 ms[3], and an inter stimulus interval (ISI) of 100 ms[4], showing a black square. Stimuli were 512×512 images[5] of either faces (targets, non-targets) or pets (oddballs), presented against a grey (RGB 50%, 50%, 50%) background[6]. RSVPs used a randomly selected sample of 5 non-targets from the total pool of faces, which was presented in random order along with the non-target, and repeated 10 times, along with 2 oddballs[7]. Following the RSVP, participants were shown the complete pool of faces and asked to indicate which of them was the target. Failing to accurately indicate the target would repeat the trial[8]. The complete experiment had 48 trials[9] and took about one and a half hour.

---

[1]Informed consents are necessary unless you work for the CIA.

[2]Fixation screens (or crosshairs, fixation crosses) are supposedly there to remind the participant where to look on the screen, and when to focus their attention, although this seems naïve: surely participants won't expect stimuli to appear in a random corner of the screen? Does it even work? Are we using an eye-tracker? The point seems to mainly tell the reader we are familiar with traditional experimental design principles, hyper-specifying a few pointless details to underline the point.

[3]Why 600 ms? Well, the longer the stimulus takes, the fewer stimuli we can show a subject before they need to go. And I mean that quite literally (see 8). Longer than 1000 ms will feel excruciatingly slow, but making stimuli very short risks that ERPs will start to overlap. That is, let's say 2 stimuli are 200 ms in duration (no ISI, see below) and we pick up a big positive ERP 200 ms after the second stimulus. Is it a P200 related to stimulus 2 or a P400 related to stimulus 1? There are ways to get around this problem (look up *deconvolution*), but we might as well avoid the problem. Given that the expected P300 is until about 400 ms, we want the stimuli to be at least 500 ms. I added 100 ms to be on the safe side.

[4]An inter-stimulus interval (ISI) is timing jargon. The stimulus onset asynchrony (SOA) is the time from the onset of one stimulus to the onset of the next. Duration is the time between onset and offset (end) of the stimulus, while ISI is the time between the offset of the first stimulus and the beginning of the next (SOA – duration = ISI). This time is filled here by showing a black square, which we will use as a neutral baseline that is the same across conditions (see Chapter 7).

[5]The size depends a bit on your display device, the screen resolution, and so on, but 512×512 pixels fits on pretty much any screen. It's usually pretty small, but that is no problem, as we don't want subjects to move their eyes across the picture anyway – ideally, the process we want to pick up should be immediate and not depend on eye movement.

[6]This 50%, 50%, 50% background refers to the amount of (computer-programmed) light in the red, green, blue hues. For example, 0%, 0%, 100% should show 'pure' blue (but see O'Regan & Noë, 2001), and equal values, as here, will provide white, greys, and black. Grey is nicely neutral, but it also makes the stimuli pop out a bit by being slightly darker, without giving too much contrast.

[7]The two oddballs are, of course, to fulfil the requirements of *cats are weird*, above. They do not need to be cats (some people like dogs), but students taking my classes know it is not safe to admit this. Another relevant question is why do I use 5 non-targets and 1 target? The reason is that each specific target occurs equally often as each non-target within the same trial, so that the difference between target and non-target is not driven by the rarity of the target image (which is 1 in 6, just like every other stimulus). Smaller probabilities give a stronger P3 response (Squires et al., 1977), but the fewer targets we present, the less reliable our result. One in 6 is therefore a pretty good trade-off.

[8]We repeat trials to avoid participants completely falling asleep. Of course, they *could* still remember the face they were to keep in mind without actually watching the RSVP, but most people are not *that* evil. But just to be sure, we will have to keep a watchful eye (see Chapter 6).

[9]With the specification as above, note that this means we should end up with $48 \times 50 = 2400$ non-targets, 480 targets, and 96 oddballs. How many trials you need per design cell depends on the experimental paradigm, but generally *many, many more than you first think*. A design cell is a combination of experimental variables (factors, see p. 191). Let's say in our experiment, we want to investigate whether the effect of seeing oneself is affected by the target vs non-target effect. That means there are 4 design cells: target/self, non-target/self, target/other, and non-target/other. Given that 1 out of 6 (roughly) stimuli are targets, and let's say that you appear as target only in 2 out of 48 trials, we know that the target/self design cell will appear $2 \times 10 = 20$ times, which is not a lot. Of these 20, you will need to expect a large number of trials to be contaminated with artefacts (see Chapter 8), and it is not improbable you will end up with about 15. Is that enough? It depends.

Can you make the experiment shorter?

Most students find my experiments boring beyond belief, and because they understandably don't want to torture their peers too much, they tend to try to convince me to make experiments shorter. They will tell me this is to avoid 'boredom' being a confound, which means that they didn't understand what a confound means. A confound between boredom/tiredness and our 'target detection' effect would have been if targets occur near the beginning and non-targets near the end of the experiment. Now we would be unable to decide whether anything observed would be due to tiredness or to target detection. Randomly mixing targets with non-targets means we will have as large a proportion of targets at the beginning (awake participants) as at the end (asleep participants). Boredom may still set in and reduce effects (seeing anything being hard while asleep), but a finding cannot be due to boredom.

A better argument is the plea for sympathy. Of course, experiments get repetitive sooner rather than later and you would prefer your peers not to hate you. But try to see it from this point of view: sure, an experiment taking an hour is shorter than one that is an hour and a half. What if, however, you find you cannot easily report your results as everything is too noisy? That could mean that all their time was pointless, as was the time of the lab assistant, as was your time in analysing the data. Was that worth saving half an hour? Better safe than sorry: make the experiment a bit longer and give a cup of coffee to the participant. Blame me if it makes you feel better.

HYPOTHESES

Now that the experiment design has been committed to paper, it becomes high time to decide upon the hypotheses: what exactly do we predict? That is to say, the theories were already there (Chapter 1), and since we are largely replicating others, the predictions are not difficult to imagine. But now that all the jargon has been discussed, it becomes much easier to precisely define what we will observe. Writing this down will help you later on, as you inevitably get bogged down in the quagmire of fine-tuning your analysis (Chapters 9–10), and could possibly help you against sceptical peers who say you are just making up things as you go along (see HARKing in Chapter 11, p. 268). So, looking back on Chapter 1, what we should at least observe is:

I.   Relevance – the target or 'to be kept in mind' vs non-target 'other faces' dimension – should affect the P3 component of the ERP, as has been found since the 1960s. The P3 starts normally around 300 ms, and commonly has a topography with a maximum over the centro-parietal electrodes (CPz, Pz).

II. Relevance should *not* affect early potentials or anything before c. 150 ms. In other words, the within-subject difference between target and non-target averaged ERPs should not be significantly different from 0. This is a kind of safety-check: if we find something here, we're probably doing something wrong statistically speaking. As you will find out, there are many steps that can go wrong, so it's good to keep in such checks. Alternatively, you might be able to falsify 60 years of research, which would also be interesting!

Then there are some findings we may expect, but the literature on them is not nearly as clear:

A. The same P3 might be observed while seeing one's own face (Ninomiya et al., 1998), which might be related to a known theory in cognitive neuroscience that suggests stimuli that are related to oneself are prioritised in multiple domains.

B. The P3 of oddball stimuli is very often preceded by enhanced N2s, resulting in some lines of research speaking of a N2-P3 complex instead of one or the other (mainly now with stop-signal tasks, e.g. Ramautar et al., 2006). If the N2 or mismatch negativity effect (Chapter 1) occurs in response to the relevance category, we might expect it here with targets. That is, targets normally follow non-targets, while non-targets follow other non-targets, so targets mismatch in that sense. On the other hand, if the N2 is more due to a perceptual difference, such as between human faces and animals, we would expect it to purely happen in out pet (oddball) category. The N2 is normally anterior (Cz, FCz, Fz, or Fpz) and at c. 200 ms.

C. Finally, the N170 is very often observed in response to faces (see Chapter 1), so you might expect an N170 here. However, it is also known that a strong repetition effect causes neural adaptation of the N170 (Eimer et al., 2010): because seeing one face engages certain brain areas, repeating a similar processing does not trigger as much renewed processing. This argument is common in fMRI research, but we are only recently seeing it in EEG research. Perhaps it might explain why I never get N170s in oddball face stimuli experiments? Two minor predictions, then: (1) if the effect is merely due to stimulus repetition, you should observe an N170 on the first stimulus of the RSVP, and (2) we should still likely be able to observe a difference between non-human and human faces, as they are perceptually not the same (depending on the hairiness of your human and non-human stimuli!). The N170 is normally tempero-lateral and parietal (T5/T6, P7/P8, PO9/PO10) and at c. 170 ms.

That wraps up all the main hypotheses. From the next chapter onwards, it will be time to stop with all the theory and start doing things. Here are some warm-up exercises to get accustomed to that.

How many trials is enough?

How many trials you need per design cell depends on who you ask and what your design is, but the general consensus is *more*. Consider one rule of thumb as around 80 for preference, but it depends on the potential and purpose. An anonymous reviewer once suggested to me *at least 25 for feedback related*

(Continued)

negativity experiments, and we know from BCI research that six gives a reasonable classification (Farwell & Donchin, 1988). On the other hand, it is impossible to get more trials after you finish recording, whereas it's never a problem to have too many. My own rule is that we cannot reasonably expect experiments to take longer than three hours before participants need to go (I know at least one PhD student who took the EEG equipment to the loo!). Remove from that about an hour involved in setting up, breaks, instructions, and such extraneous activities. Test your experiment: how long does a single trial take? Now, divide two hours by the time it takes to run a trial (slightly under a minute for the present study), and you have got a fair estimate of the maximum number of trials you might be able to run. This is what you keep in mind as you start negotiating with your peers, and likely you will end up with a number roughly between far too few and this theoretical session maximum.

ESSENTIAL EXERCISES

Essential exercise 1: Designing stimulus material

To start doing the experiment, we will need stimulus material. Make a picture of yourself, or a selfie, but do not overuse filters. Copy it, and edit the copy to have a portrait style composition and fulfil the requirements as outlined in the *Procedure* section above. My sample material (available online) is based on stimulus material I had lying around, so tweak the suggested layout and other aspects to work for you. The precise values are not particularly important. The only thing that matters is that the images should be as similar as possible in size, style, composition, and so on. If you do this as part of a group, make sure you compare your stimulus with those of others and together adjust the images until they are congruous and fit precisely defined criteria. If you are not part of a group, you can compare your stimulus with either a bunch of images you source from social media networks or alternatively use the images from the online material that is part of this book.

After the stimulus material is sufficiently standardised, add an image of your favourite animal as the oddball stimulus. Try to standardise this stimulus to conform to the previous specifications, but this may prove harder than expected.

Note: I find that people tend to underestimate just how difficult designing good stimulus material is. If you take less than an hour for the above, take a good look again at all stimuli in quick succession; sometimes a sneaky difference is easier to spot that way.

Essential exercise 2: Drawing procedure schematics

This is the best time to create a proper '*Schematic display of the trial procedure*'. All good experiment reports give a clear overview of the task and what the participant is seeing and doing. This is true for most published work in experimental psychology and cognitive neuroscience, but I know enough colleagues who appreciate it in terms of student grading as well. It may also be useful as part of the participant's instruction sheets (Chapter 6).

For example, Figure 2.3 is a procedure overview from a simple priming study (Spapé et al., 2017).

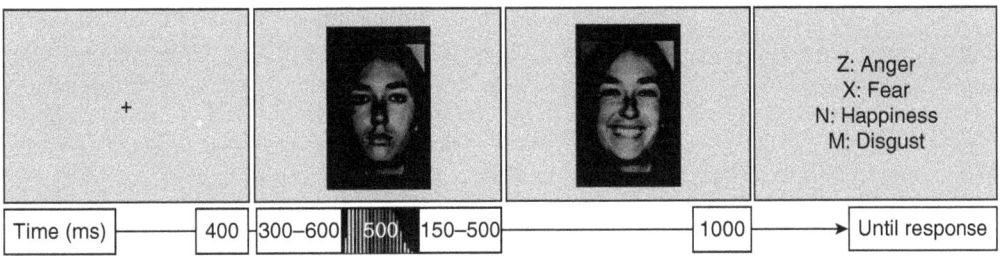

Figure 2.3 Trial procedure in horizontal layout with multimodal information. Note: the black box underneath the first face shows the tactile stimulus.

Source: Reprinted from *Biological Psychology*, 124, M.M. Spapé, Ville Harjunen, N. Ravaja, Effects of touch on emotional face processing: A study of event-related potentials, facial EMG and cardiac activity, p.10, Copyright (2017) with permission from Elsevier

Using very few words, it makes clear that there is a fixation screen, followed 400 ms later by a face with a neutral expression, 300–600 ms later by a tactile stimulus with a duration of 500 ms, and a further 150–500 ms later the face is replaced by an emotional expression. After seeing it for

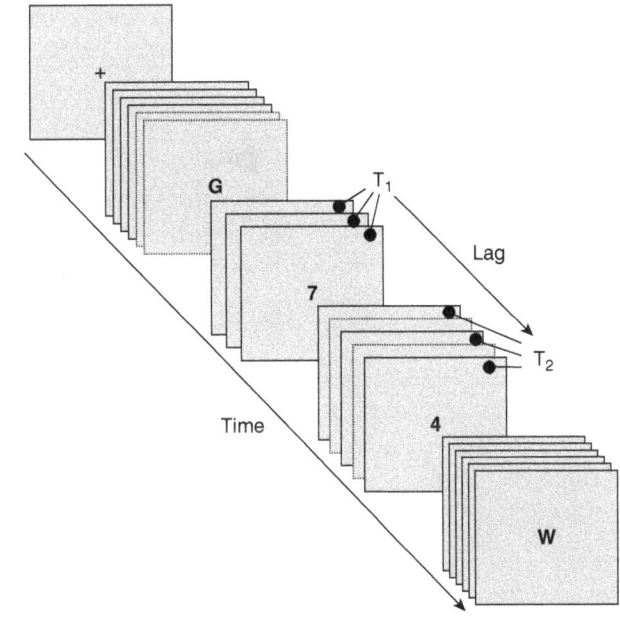

Figure 2.4 Trial procedure in diagonal layout with explanatory elements

Source: Reprinted from *Neurophychologia*, 46(13), Lorenza S. Colzato, Heleen A. Slagter, Michiel M. A. Spapé, Bernhard Hommel, Blinks of the eye predict blinks of the mind, p.5, Copyright (2008) with permission from Elsevier

1000 ms, participants classify the emotion (here: happiness) with a button (N). This has most of the details required to replicate the experiment. But oddball studies tend to use many images, so perhaps the trial procedure figure of this RSVP study (Figure 2.4) gives some inspiration.

Figure 2.4 is from a study (Colzato et al., 2008) in which people were to detect target numbers among non-target letters. The important variable of interest is the difference in time between the first (T1) and second (T2) target, which is explained well using the figure. However, some details could be better integrated (the 80 ms SOA), some details are quite trivial (T1 was either the 7th, 8th or 9th stimulus), and others are missing (where is the response screen?).

So, with these suggestions as inspiration, make a figure of the trial procedure for the present study. I would suggest using vector-based software (such as Adobe Illustrator or PowerPoint) as opposed to pixel-based software (such as Adobe PhotoShop or MS Paint), as quite often you will want to make small changes in such an image. Furthermore, it allows you to create high-resolution images that look good in print.

Wait, what, PowerPoint? Yes, the above two images were made with hopelessly uncool PowerPoint. Pro PowerPoint tips: (1) Go to Design > Slide Size > Custom Slide and use A4, portrait. (2) After drawing a square (the 'display') and tweaking it until fully satisfied, copy paste it and keep a uniform style. (3) After selecting a bunch of squares, use the Format > Align tools as much as possible, both distributing (even distribution looks better) and aligning. (4) Export by printing to PDF, then importing the PDF using Adobe PhotoShop or such (e.g. free software The GIMP) at high (600 or 1200 DPI) resolution. Finally, export to .EPSF, for a maximally high-resolution image.

Additional exercise: Your own research design

Make an overview of the specifics of both hypotheses discussed in this chapter (I, II), but try to be as specific as possible. Write things out in terms of a null-hypothesis (H0) and alternative hypothesis (Ha), if that's the way you roll. Furthermore, draw an idealised ERP, like Figure 1.4 (p. 11), but with multiple lines, indicating how the predictions *should* pan out.

Next, do the same for A, B, and C, but here it requires some additional reading from the references. As you are reading, think about what other predictions you might test. If, for example, seeing yourself is really automatic, it might show a reduced *target* effect (after all, irrelevant targets will also be *slightly* relevant). How about seeing your own pet? Dog-people and cat-people? Anything you can think of that you would find interesting to test should be written down as clear and precise predictions.

After you have done this, find out exactly how many stimuli would be available for each *design cell* of your hypothesis (see p. 31, note 9).

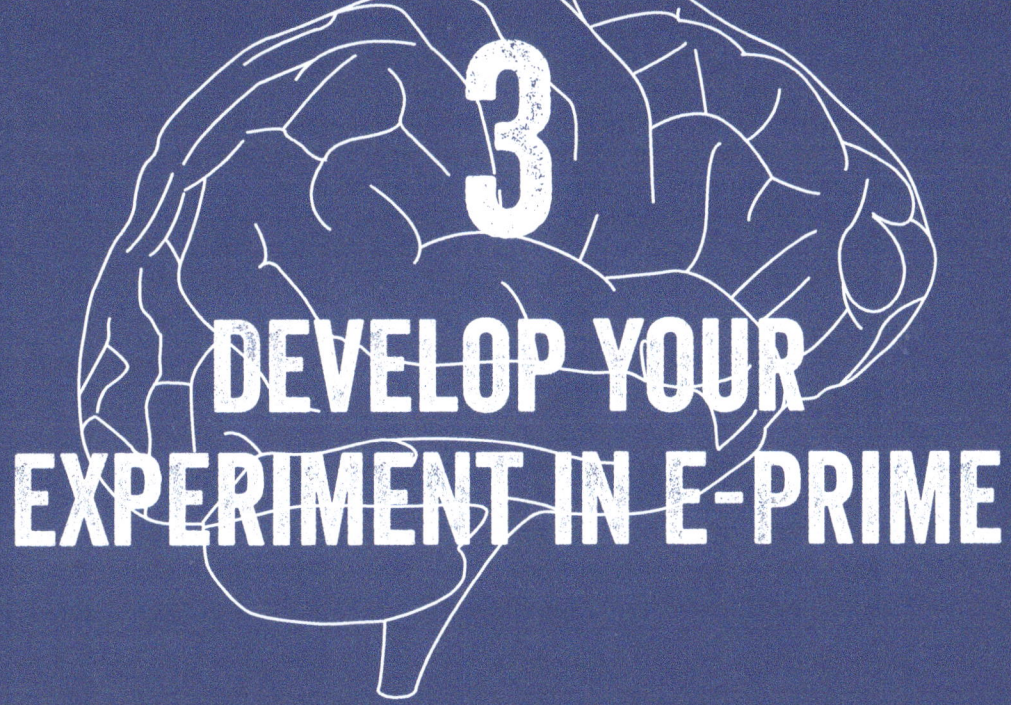

3

DEVELOP YOUR EXPERIMENT IN E-PRIME

IN THIS CHAPTER, YOU WILL LEARN ABOUT:

- Developing experiments in E-Prime
- Creating an RSVP study
- Synchronisation with external equipment

When I was about nine years old, I got a Commodore 64 for Christmas.[1] Imagine a laptop, or a phone, the size and weight of an old suitcase but without screen or storage space. Needless to say, I was ecstatic, as not only could you run computer games, loading them from cassette tapes or even recording them from the radio, but once you were bored with those, you could develop your own games in BASIC, which should be a simple coding language. It proved much harder than expected, so as I got stuck, I did what any other sane person would do: ask the local internet. That is, get some books from the library. These were invariably filled with lines and lines and lines of code, like:

```
READY
10 PRINT 'HELLO WORLD'
20 GOTO 10
```

…But then a lot more complicated. Imagine writing a thousand lines full of numbers, symbols, and formulae without having any clue what they mean, only to find out afterwards that either you or the original author made an error somewhere. Instead of having a fancy new and free game (if you forget about the hundred hours spent typing), you end up with a huge amount of frustration. So that is how I learnt to code. Or at least, to code as little as needed to develop tiny adventure-like games that worked a bit like Choose Your Own Adventure books.[2]

There are two points to my story. The first is that while instructions are useful, do not mindlessly follow them. In this chapter, I will provide many explicit instructions, although I will do my best to explain them as we go along. I would suggest not skipping the instructions, but if you are the type of reader who will do that anyway, then I suppose you won't have read this very warning. More importantly, I would suggest going over and beyond the instructions to find out what happens if you make small alterations. If it suddenly breaks the experiment, you go back and find out the exact breaking point – this is what debugging is about. Due to the book being more practically focused than going in depth about every subject, I cannot give as much explanation about the ins and outs of E-Prime as I would like. If you feel you want to learn more, I would recommend you take a look at *The E-Primer* (Spapé, Verdonschot et al., 2019), which gives a pretty thorough introduction to developing experiments in E-Prime. But then, I would say that.

The second point is that, at the end of the chapter, I hope you feel like you have created a kind of adventure, or at least have the experience that you have made something *real*. Programming can be frustrating, requiring more logic than our lazy minds provide. But then, step by step, you circumnavigate all the annoyances until a *product* finally comes into existence. This engineering type of learning trajectory is very different from, say, writing an essay, which I hope you will appreciate.

Before you begin: in this chapter, and the next, I will explain how to develop the experiment to run in your lab, either using E-Prime (this chapter) or OpenSesame (Chapter 4). I would recommend you choose one based on availability: E-Prime is proprietary software, and not cheap, but requires relatively little computer skill.

---

[1] I can safely relate this anecdote now that Silicon Valley geeks rule the world, but back then being into computer games meant you walked around with a sign around your neck saying 'beat me up, please'!

[2] Critical study: watch *Black Mirror: Bandersnatch* on Netflix.

GETTING STARTED IN E-PRIME

E-Prime is probably the most popular commercial software available for running experimental psychology studies. It is commercial and pretty expensive, but on the other hand you can depend on the developers – Psychology Software Tools Inc – to make a pretty intuitive bit of software. Or, if practice fails to live up to expectations, they will usually help you out. With intuitive, I mean that if you want, you should be able to design experiments by using its graphical interface and run them smoothly with solid timing, which is essential for EEG experiments. Finally, because it is such a standard piece of software for many psychology labs, it also tends to be well supported by third-party devices. That is, EEG hardware manufacturers will know what E-Prime is, and they will likely have a system in place which makes it easy to communicate between software (E-Prime) and hardware (EEG amplifier).

With software, I mean specifically a Windows PC, not a Mac. E-Prime thus far only operates on Windows 7/8/10. In Chapter 4, I will discuss OpenSesame, which does run on Macintosh OSX, as well as Android. For this chapter, I will assume you are running E-Prime 3, the latest version. It does not matter an awful lot if you are running an older version (E-Prime 2), but some of the figures in this book will look a little different for you.

Setting up

Make sure you have the following set up and ready to go:

- *E-Prime has been installed on your computer.* Evaluation versions can be downloaded for free from *pstnet.com*. These have, however, severe limitations (saving only a few trials, much less accurate timing) so they should not be used for the actual running of your experiment. They are very helpful if you want to do some extra training in E-Prime at home or on your own laptop.

- *You have a clear idea where, on your file-system, the experiment is located.* This can be your regular USB stick, the E-Prime default directory (Documents > My Experiments), the desktop, and so on. Create a new directory at this location, and call it something reasonable. I will refer to this directory as ExperimentDirectory. Do not forget to judiciously back up everything in ExperimentDirectory.

- *You have the stimulus material at your disposal.* If you went through the exercises in Chapter 2, you should have images of yourself and others, each 512 × 512 pixels. Make sure they are all renamed to single numbers. That is, image 1 will be called 1.jpg, image 2, 2.jpg, and so on. If they are not jpg files, do not rename them to jpg, but instead open them in your image manipulation software (e.g. mspaint), and save them to jpg format. Finally, do the same for the images of pets, but add 100, so we have 101.jpg, 102.jpg, and so on. Create a new directory within ExperimentDirectory and call it 'img'. Move all images to this directory.

- *If you have some stimulus material, but fewer than 12 persons or 24 pets, proceed to the preceding bullet point, then follow with the next.* Do not overwrite your own images with mine.

- *If you have no stimulus material, download my example images from the web* (see (https://study.sagepub.com/Spape). Open the compressed (.zip) file for this chapter, and extract the contents onto ExperimentDirectory. Check, afterwards, if ExperimentDirectory now contains a single folder ('img'), with all the images inside. If not, create the folder, and manually move all images to this folder.

So now you should have everything ready to go. The images are in their own 'img' folder, and they are clearly named and numbered (Figure 3.1):

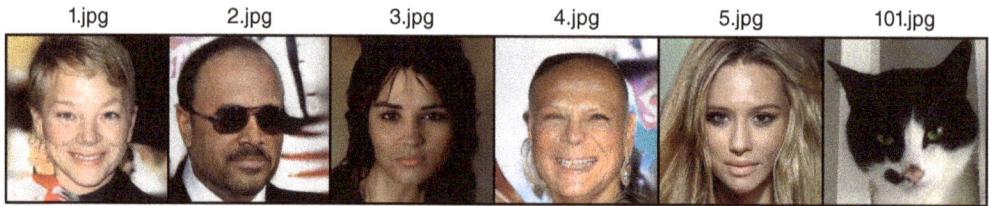

1.jpg   2.jpg   3.jpg   4.jpg   5.jpg   101.jpg

Figure 3.1 Named and numbered images from my own database. Persons (1–5 shown of 100) are actually not people, being generated by a fancy neural network, hence the occasional weird looks. Pets (101 shown of 101–30) are local celebrities who gave permission in return for treats.

Welcome to E-Prime

The part of E-Prime we use to design experiments is called E-Studio. Let's get a few basics working.

Step 1

⊞ = start

Start E-Studio (easiest by pressing start and typing 'E-Studio'), close the window suggesting various templates so we start from blank. It should look a bit like Figure 3.2.

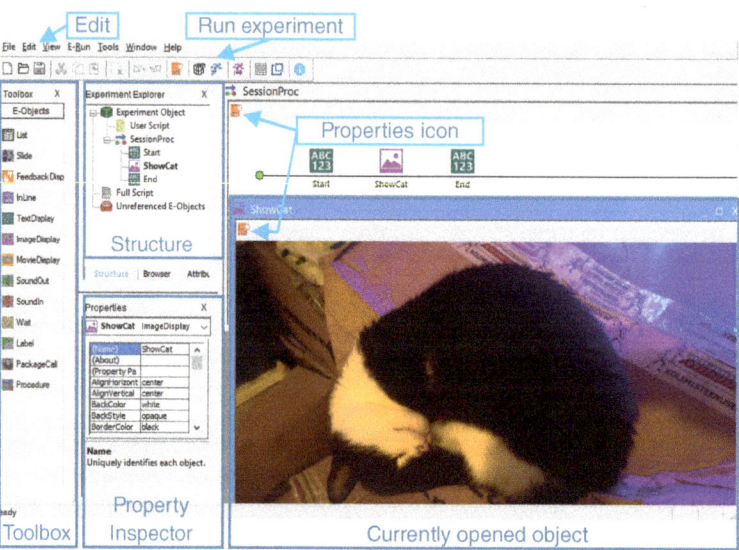

Figure 3.2 E-Prime screenshot with a few regions of interest pointed out

Step 2

Go to Edit > Experiment > Devices (tab). Unclick the mark next to Sound, as we're not using any sound in this experiment (it also tends to cause problems if you are running the experiment while the audio device is deactivated). Then select Display (do not uncheck), and click on Edit. Set 'Match desktop resolution at runtime' to 'Yes', and change the 'Default background color' to 'gray'. Apply and close the Edit Experiment windows.

By default, E-Prime runs experiments at a resolution that is slightly lower (1024 × 768 pixels) than most PC screens, or indeed even than most phones, which means that on a bigger screen things will look pixelated. Matching desktop resolution means E-Prime will conform to whatever Windows is currently running at (depending on your screen). It will make your experiment look a lot better, but the downside is that it may also result in unexpected changes when you move your experiment over from your home computer to the lab. Changing the default background colour is generally a good idea. This is particularly true in the age of LCD screens, in which the 'liquid crystals' have slower response times (the time to change from black to white to black again), which could lead to slight blurring in the EEG response. Although this problem used to be much worse, it is safe to say that our *eyes* have not changed much over the last couple of decades. Pictures will evoke a response (in EEG, but also pupils!) partially due to changes in luminance, so it makes a kind of sense to give a neutral baseline: not a sudden change from dark or from bright. For this reason, most of my experiments have a grey background. It is also easier on the eyes than white backgrounds, and keeps people more awake than black backgrounds (I hope).

Step 3

If you don't see something that looks like ⬤———————⬤ already, then double click on 'SessionProc' (see 3a right). Now drag a single TextDisplay from the 'Toolbox' to the Session-Proc, as shown with the red arrow 3b. Rename the TextDisplay (e.g. click on its name, press F2) to 'Welcome' (Figure 3.3).

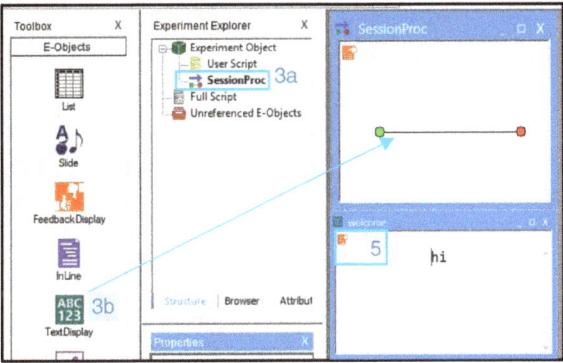

Figure 3.3

The SessionProc is our first Procedure. Procedures are symbolically represented as a kind of time-line. E-Prime runs everything from left to right. At this point, it will just run 'Welcome', although it is currently lacking in being welcoming. Let's change that.

Step 4

Double click on Welcome. You should see a cursor above a window saying <insert text here>, and you can immediately start typing. Create a friendly greeting. Run the experiment, by going to E-Run > Run (or pressing F7, or clicking on the running person) to see how it looks.

Chances are it doesn't look good enough: it will be very short (1 s), so you might not even have read the welcome, and the background is white. This simply won't do.

Step 5

Again, double click on Welcome if it isn't already open. Then have a look at its properties by clicking on the icon to the top left. Indeed, any of these orange hand icons will show the proper-

ties of the currently opened object. In the properties dialogue box, change the following: under the General tab, change the BackColor to 'silver'; in the Frame tab, change both the width and the height to '600' and the BorderWidth to '4'. The Duration/Input contains everything about timing and interaction. Go to this tab, change the Duration to '(Infinite)', then click on Add, select the keyboard, and press OK. Finally, change the Allowable to '{SPACE}' (note the accolades and caps!). Click on OK again to close the screen. Under the instruction you wrote previously (step 4), add something to the tune of 'Press SPACE to continue'. Run the experiment again to see what it looks like now.

Step 6

Drag two additional TextDisplays to the SessionProc, after Welcome, renaming the first 'Instructions', and the last 'Goodbye'. Edit the text of Instructions to make clear to the subject what is expected: they are to keep images in mind while looking at series of images. It is also a good idea to remind them of what is generally expected during EEG experiments (Chapter 6). Similarly, edit the text of Goodbye to make clear the experiment is now complete and participants may soon be free again. Work on the properties of Instructions and Goodbye to give a satisfactory look and feel. Run the experiment and keep changing the properties until you're happy with the end result.

Basic structure in E-Prime

Now you have the beginning and ending of the experiment, but no actual experiment. E-Prime works best with a hierarchical kind of structure in mind: one experiment is a single **session**, with a sequence as specified in the SessionProc. One session contains multiple *blocks*, which can mean a training part and a testing part (or, say, a questionnaire part and an experiment part). Each **block** has a sequence with a beginning and ending (e.g. 'training begins', 'training ends'), with many different **trials**. In most experiments, the trial is the smallest, most frequently occurring sequence that is described.

Or think of it the other way around. For example, an emotion processing experiment testing whether we can localise emotional stimuli (such as snakes) faster than neutral stimuli (like chairs) could show a fixation, followed by a stimulus (say, a snake), that requires a single response (depending on where this picture was presented, press LEFT or RIGHT). Given two pictures and two locations, we would have four possible combinations to create what is called a fully **orthogonal design**:

Picture	Location	Emotionality
Snake	Left	Emotional
Snake	Right	Emotional
Chair	Left	Neutral
Chair	Right	Neutral

(Note: this would be a poor experiment design as the *picture* is completely confounded with the *condition*, see Chapter 2)

I call variables like Picture and Location **manifest variables** while Emotionality is here the **latent variable**. That is to say, we wish to measure the effect of emotionality, but sadly E-Prime doesn't know emotions, so we have to explicitly tell it what we mean: the different pictures. In E-Prime we commonly define conditions like these as part of a list of trials, conventionally named TrialList, in which the same sequence is repeated in a kind of loop, except with different pictures and locations. Let's see how this works in our experiment.

Step 1

Drag a List from the E-Objects to the space just between the previous Instructions and Good-bye TextDisplay (see Figure 3.4, 1a). Rename it to 'BlockList'. Double-click on BlockList and underneath the Procedure column *type* (do *not* use the dropdown menu) 'BlockProc'. E-Prime will ask you whether you want to create the new Procedure (select yes), and whether you want it to default to this Procedure (again yes).

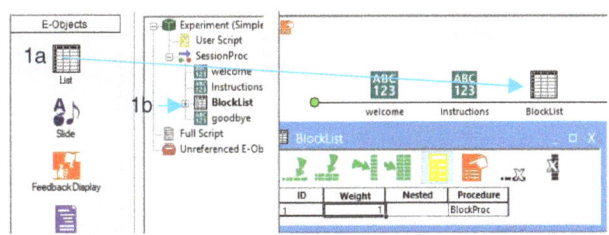

Figure 3.4

You will notice that within BlockList a new Procedure has appeared. If you do not, notice that next to BlockList in the Experiment Explorer (1b), a small plus appears, which allows you to expand the currently 'folded' BlockList hierarchy. If you see nothing next to BlockList, right-click on BlockList and click on 'Allow Collapsing'.

Step 2

Make an additional level in the structure such that it will start to resemble the screenshot in Figure 3.5 by double-clicking on BlockProc, adding a new list to it, calling it TrialList. Within TrialList, create the Procedure 'TrialProc' (like you created BlockProc in step 1). While still in TrialList, add 5 rows (2a) and 1 column (2b) called 'relpicture'. Change the values to go from 1 to 6 (as in the screenshot). Finally, open the TrialProc and move two new Slide objects from the Toolbox (just below from where you got new Lists) to the TrialProc. Rename the first one as 'KeepInMindSlide', and the second as 'KeepInMindRecog'.

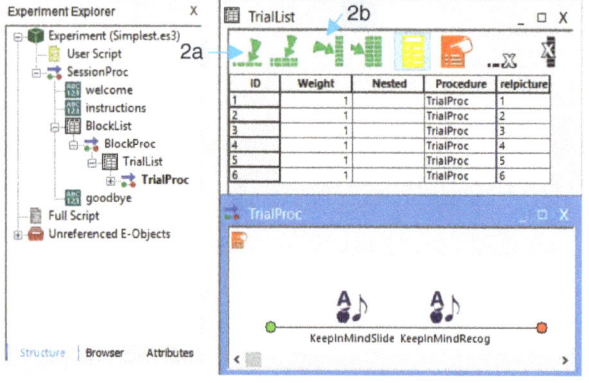

Figure 3.5

Slides are like TextObjects, but can *contain* text, images, and sound, as well as (in E-Prime 3) movies and interactive elements. Text and images take an amount of predefined space, the **frame**, which is then placed on a single page, the slide's **state**, which can be dynamically presented. This can be a bit confusing, but it may help to compare presenting a Slide to showing a book to someone. A book has multiple pages (the slide's *states*), each with different text and images. The state you present to the subject is the slide's **activestate**, which would be the actual page you show to someone (maybe the book's cover). For now, this is not super important, but it explains why the Slide has *two* property buttons: the **Subobject property pages** are about what you just selected (for the book's cover, perhaps its title header), whereas the **Normal property pages** concern that which is part of the entire slide (for a book, that would be the author, for example, but for E-Prime this is mainly the Input/Duration tab).

Step 3

Double-click on the KeepInMindSlide, go to its normal properties (see Figure 3.6, 3a). and set, similar to the Instruction screen you made before, the Duration to infinite, add a keyboard input, and set its allowable to 'c' (just c, no accolades or caps). OK the dialogue. Now click on the TextDisplay subobject once, then click anywhere on the slide (3b). This creates a new text frame. Edit its properties by right-clicking on it, and selecting Subobject property pages (or clicking on icon 3c). Change these such that: it says 'Keep in mind'; it is 25% in width and 10% in height; it is presented at X: 50%, Y: 10%. Do something similar for the 'Press C to continue' shown right, but have it shown at the bottom of the screen. To change the background of the slide, right-click anywhere outside the TextDisplay frames (e.g. on 3d), go to Subobject property pages, and change the BackColor to 'silver'. Now you will notice that the two TextDisplays look alarmingly white. Change both their Subobject property pages such that their BackStyle is 'transparent'.

Note: Percentages in E-Prime's width, height, X, and Y properties are relative to the frames they are placed on, or, if they are not placed on a frame, the display's height. So if the display's width is 1024, a slide with a width of 80% will be 819 pixels, but if this slide contains a further text display subobject with width at 80%, it will have a width of 655. If you leave out percentages, the positions and sizes will be absolute, in numbers of pixels. Both absolute and relative positions use the top-left of the screen as their origin: X = 100 being 100 pixels right of the left edge, Y = 100 being 100 pixels down from the top.

Step 4

To show the picture, click on the ImageDisplay subobject (4), and again anywhere in the slide. Then edit its properties such that it is 512 × 512 pixels, is neatly presented at the centre of the screen, and has a little border around it. Finally, change the all-important filename property, in the General tab, by clicking the browsing icon and finding any suitable picture you have (e.g. a funny cat you found on the internet). Don't think too much about the picture – we just want to get the basic structure working for now!

Step 5

Repeat this procedure, but apply it to KeepInMindRecog slide. It should work more or less the same, but ask the participant 'Were you able to keep the picture below in mind?', and 'Press Y or N'. Obviously, *two* response choices are now available, so instead of the 'c' you had before, change the Allowable global property of the KeepInMindRecog slide to 'yn'. If the subject did well, they would of course respond 'yes', so change the Correct response to 'y'. E-Prime suggests logging responses: this is always a good idea, so allow it.

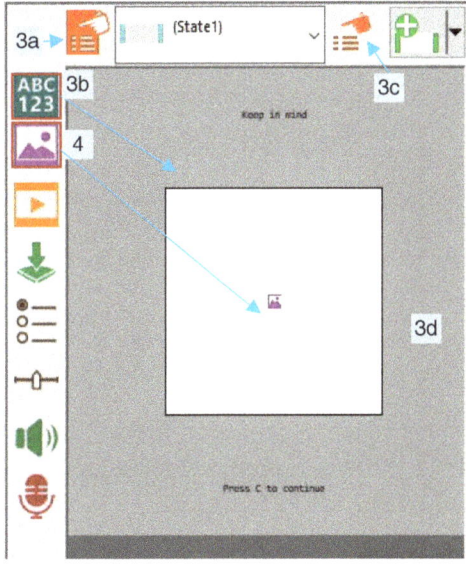

Figure 3.6

OK, now run the experiment to see if it works, that is, if it does not crash. If it does, it would be a good idea to go back and see what went wrong before continuing.

Implementing variables

Once you get this running, you might find the results are a bit underwhelming. You are just asked to keep the same picture in mind over and over and over again, what's the point of that? Then, no rapid serial visual presentation sequence (i.e. the oddball stream of images) is shown at all yet. Finally, the recall part is really naïve: what prevents a subject from lying? The problem is basically that everything is just far too static. We need more variation, more *variables*.

In E-Prime, the easiest ways to implement variables is through list *attributes*. Remember you added, in the TrialList, the column 'relpicture'? This is an attribute. Attributes can be used, or referenced, in almost every place you can think of in E-Prime by adding its name in between brackets, like [relpicture]. Let's see how that works:

Step 1

Change the Keep in mind TextDisplay Subobject on the KeepInMindSlide to say 'Keep in mind: [relpicture]'. Run the experiment and check that it now gives a counter, the first trial saying 'Keep in mind: 1', the next 'Keep in mind: 2'.

It does not do this magically by performing a trial counting operation. Think of it like this: While running the first trial of the list, E-prime does everything with a certain **context** in mind, as defined by the list. That means that at the first trial the procedure is 'TrialProc', and the relpic-ture is '1'. As it runs TrialProc, which includes KeepInMindSlide, E-Prime notices the brackets [relpicture], and will insert what is true in this **context**, namely 1.

Step 2

This is a bit abstract, so you might find it helpful to change a few numbers in the TrialList under relpicture: change the 3rd, for example, to everyone's favourite psychologist, 'William James'. Run the experiment and observe the result.

Now, this is, of course, a bit pointless: we need no trial counter, and though we will always need William ($\heartsuit$), we want to see the pictures that are to be kept in mind.

Step 3

On the KeepInMindSlide, go to the central Image Subobject properties, and change its filename to be img/[relpicture].jpg. Exactly like so:

Filename: | img/[relpicture].jpg

This may look cryptic, but it is in no way different from what you did before. Run it to see how long it works. If it crashes on the *third trial*: great! Remember, on the first trial, the image will be the first level of [relpicture], therefore, img/[relpicture].jpg becomes 'img/1.jpg', which points E-Prime directly to a picture in the 'img' sub-directory you created. On the second, it will be 'img/2.jpg'. On the third, it will be 'img/William James.jpg', which I doubt you have placed in that directory. So, go back to the TrialList, and fix the third trial to say 3. Also remove the reference to [relpicture] from the *text* above the KeepInMindSlide's SlideImage.

Step 4

Now try to extrapolate this thinking towards KeepInMindRecog. Add an attribute to the TrialList, call it 'recogpicture'. Enter, on the 1st and 2nd row '1', on the 3rd and 4th row '3', and on the 5th and 6th row '5'. Then, inspect the SubObject properties of the picture on the KeepInMindRecog, and change the filename such that it will retrieve pictures (see step 3).

Step 5

But, how do we know our participant isn't just pressing buttons randomly, and how do we keep a record of their performance? To do this, add yet another column to the TrialList, calling it 'correct'. Now change it such that when the picture to be kept in mind is the same as the picture that is to be recognised, the correct response would be 'y', and otherwise 'n'. In the Duration/Input tab of the global properties of the KeepInMindRecog slide, change the Correct property to '[correct]'.

If you run it again now, it should work, but it should feel extremely predictable. What we need to do is to, of course, randomise things, which is what a List in E-Prime does, as I will demonstrate with the next step.

Step 6

Go to the Properties of the TrialList, Selection tab, and change Order from 'sequential' to 'random'. The Reset/Exit tab is also of interest as it determines how many times new **samples** (here, trials) are drawn from the list. Change it such that you run 2 cycles of 6 samples (12 trials).

The RSVP

At the moment, it is of course rather missing the main part of the experiment: We were going to look at a sequence of images (the Rapid Serial Visual Presentation, or RSVP) while keeping an image in mind. Designing this repeats many of the previous steps, so I will keep the explanations short. If you run into problems, you might want to read back. We start with a new list.

Step 1

Create a new List, RSVPList, just after the Fixation and before KeepInMindRecog with a new Procedure (type 'SingleStim'), and let it have 6 rows. Add an extra attribute, 'Picture' and give it values of 1, 2, 3, 4, 5, and 6 for this attribute. Let it randomise, and make it have 10 cycles of 6 samples.

Step 2

Drag a TextDisplay (call it ISI) to the new SingleStim procedure. Let it be 512×512 pixels, have a black background and a 2-point black border. Give it a duration of 95 ms.

As you will remember from the procedure (Chapter 2, p. 30), the flash in between pictures was supposed to be 100 ms, so why do I write 95? Displays update at a constant rate – usually 60 Hz, or every 16.67 ms – which means that presenting an image in between two refreshes (say, halfway between 16.67 and 33.33 ms) could mean that for a brief moment, half a picture shows and half does not. To avoid this, E-Prime (and OpenSesame as well as many videogames) synchronise presentation such that the display receives its information *just before* the new refresh. In E-Prime, see under the ISI TextDisplay's Sync properties the 'vertical blank' (on by default). This means that even if I tell E-Prime that my textdisplay *should* have a duration of 95 ms, it will wait a little bit extra before the next stimulus is shown anyway. That means that if a timing issue occurs (and they do), and my ISI is 4 ms late, the next stimulus will still be at 100 ms. If, on the other hand, I entered a (hoped for) duration of 100, and a 2 ms timing occurs, then the next stimulus will suddenly occur one refresh later, or at 116.67 ms. So it's good practice for timing critical stimuli to have a duration that is slightly shorter than their prescribed duration, keeping in mind the refresh rates. For example, a refresh rate of 70 (refresh every 1000/70 = 14.2857 ms) could have an ISI of 86 (14.2857 × 6 = 85.71), or 100 (14.2857 × 7 = 100), but not 90. Entering a duration of 95 will cause E-Prime to have the next image shown at 100 anyway.

Step 3

Drag an ImageDisplay (call it 'Stim') just after the ISI. Let it be 512 × 512 pixels, a 2-point black border, a duration of 595 ms, and let it *reference* the attribute you made in step 1, using the same strategy you used in Implementing Variables step 3.

If you run the experiment, you can see you're nearly there. However, the information of 'Keep in Mind' keeps showing through the RSVP, which is distracting.

Step 4

Just after the KeepInMindSlide, insert a TextDisplay, call it 'Fixation', make it 695 ms in duration, 100% × 100% in size, and with a grey background. The text should just show a 30-point '+' character in Courier typeface. Note: since Fixation is 100% in width and height, while the subsequent ISI and Stim are just 512 x 512 pixels, everything *around* the stimuli will continue to show Fixation's grey colour, as described in the trial procedure (pp. 30–31).

Run the experiment and see for yourself how you can improve it. As such, it should basically work, but it is really the minimum of what I would expect from a student. What can we do better?

- There are only 6 images in use, always the same ones.

- It is still extremely easy to guess the answer, and there's no penalty for doing so.

- There are no cats.

IMPROVING THE EXPERIMENT

The last issue is obviously the most pertinent one, so let's start there.

Two cats

Remember that the pets had a similar naming scheme to humans, except the pictures were named with numbers above 100. In the procedure, I describe 62 images being shown, 10 of each person, and 2 oddballs. That gets difficult with our list which randomises 6 images 10 times. The easiest way to sort this is by doing the following:

Go to the RSVPList, and change the number of cycles back to 1. Then, change the weight of each stimulus to 10. Add two rows, with each Weight 1, and change the value of their Picture attribute to 101 and 102.

Why? Previously, we mixed a stew of 6 ingredients (images) 10 times, meaning that after eating a dish in some random order (e.g. 132465), the ingredients were remixed. What we now have is a pot of 62 ingredients, that is mixed just the once, with some ingredients (1–6) 10 times as much in the pot as two others (101–102). This means that sequences may feel less random than they were: A sequence like 1112144 used to be impossible, but is no longer. This may be problematic: after noticing something relevant once, you might not have the same evoked potential upon repetitions. Some experimental designs therefore try to avoid repetitions.

One way to do that would be to create 4 RSVP lists with 4 cycles, and just the people (1–6), *then* include the list with cat 1, then another 4 RSVP lists with 4 cycles, and the single list with cat 2.

A further possibility is to look ahead inside the list, and kind of randomly replace a human with a pet. Each possibility has its own issues, though, so I will keep it like it is here, as what we have now is simpler to implement and less restrictive for the randomisation.

But there are still just 2 cats being shown (I have 4 cats myself), and 6 humans (I do not have 6 humans) throughout the experiment. That is insufficient.

Advanced stimulus randomisation

The problem is that we have a large number of images, but within a single trial, we are only showing 6 different images (and 2 cats). This is part of the design: if we would always show *all* images within a trial, it would be much easier. However, this would severely impact the degree to which we control for differences in stimuli, and how well we can generalise towards other stimuli (see Chapter 2). Also, if, say, we would have, 25 different images, and 1 of them would be relevant, then only 1 in 25 stimuli would show the image to be kept in mind. In other words, participants would mainly be spending time on watching images that were *not* kept in mind. Of course, we could increase the weight of the relevant image to 4, such that the stimulus to be kept in mind would occur in 4/25 stimuli, which is still about once every 6 stimuli. This is not ideal either, however, as the stimulus that is kept in mind would then be showing over and over again, while every other picture would show very infrequently. In other words, it would now confound *frequency* with *relevance*, making it impossible to say afterwards which of these would cause the observed effect.

Ideally, we therefore pick 1 random picture that is the relevant one, and 5 irrelevant ones each trial, then continue using only these throughout a single RSVP. There are many ways to accomplish this in E-Prime, but my current favourite is the following:

Step 1

Change the TrialList to include five extra attributes, named 'irr1' to 'irr5' (see Figure 3.7, left List). Whenever the relpicture has a number, fill the irr1 to irr5 attributes with the pictures that could be irrelevant. Continue the list so that each picture can be relevant: I went until 6 below (because I'm lazy), but if you have 20 people in your group, then you should continue until 20. No pictures should be irrelevant if they are not also at some point relevant, so if the relpicture is 16 (and there are 20 people in your group), for example, the irrelevant pictures should be 17, 18, 19, 20, 1, and 2. This type of design is sometimes referred to as a **Latin square** design, and popular in textbooks on experiment design.

ID	Weight	Nested	Procedure	relpicture	recogpicture	Correct	irr1	irr2	irr3	irr4	irr5
1	1		TrialProc	1	1	y	2	3	4	5	6
2	1		TrialProc	2	1	n	3	4	5	6	7
3	1		TrialProc	3	3	y	4	5	6	7	8
4	1		TrialProc	4	3	n	5	6	7	8	9
5	1		TrialProc	5	5	y	6	7	8	9	10
6	1		TrialProc	6	5	n	7	8	9	10	11

ID	Weight	Nested	Procedure	Picture
1	10		SingleStim	[relpicture]
2	10		SingleStim	[irr1]
3	10		SingleStim	[irr2]
4	10		SingleStim	[irr3]
5	10		SingleStim	[irr4]
6	10		SingleStim	[irr5]
7	1		SingleStim	101
8	1		SingleStim	102

Figure 3.7

Step 2

Change the RSVPList so as to match the List to the right in Figure 3.7. Notice that instead of absolute numbers, the values below picture now are themselves attribute references. Thus, in trial 3 of the TrialList, for example (given that we wouldn't be using randomisation), the Pictures should retrieve the values 3, for the first row of the RSVPList, 4, 5, 6, 7, and 8, for the irrelevant pictures, and 101, 102 for the cats. Pets 103–112 are now MIA, which is unfair, but we'll get to that in due time.

Feedback and avoiding guesses

One niggling problem remains that it is extremely easy to guess right. This is an issue, because it might lead people to just sit there, press a button, then dig out their phone and start doing something else completely, because even without looking, they will get the right answer half of the time anyway, and even if not: so what? So let's make it a bit harder to guess the answer and let's be a bit cruel about their lack of concentration and punish wrong answers in the harshest way I can think of: make them do the trial again. First, however, we need to tell the participant off so they can avoid future mistakes.

Step 1

Move from the Toolbox a FeedbackDisplay (call it 'Fbk') to the end of the TrialProc. Then move an InLine from the Toolbox just after the FeedbackDisplay. Finally, move a Label to the beginning of the TrialProc and call it TrialStart. Your structure should now look like Figure 3.8.

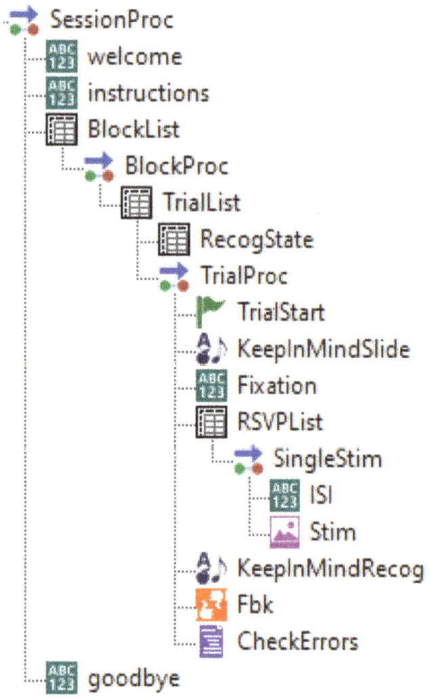

The FeedbackDisplay is very easy to work with. It is a bit like a slide, but with four states that are shown as tabs to the bottom of the FeedbackDisplay. Notice them, they should say Correct, Incorrect, NoResponse, Pending. The Correct state is the one showing if a response is accurate, the Incorrect if inaccurate, while NoResponse and Pending are unused in this experiment (and indeed, usually). If you now try to run the experiment, however, you will notice E-Prime will display a 'Fatal error message', which sounds rather ominous! It isn't. E-Prime simply doesn't know what Fbk is supposed to show feedback *about*. You want to tell E-Prime: give feedback about the subject's response. 'OK,' says E-Prime, 'But the subject's response to *what exactly?*'. This is what is called the Input object of the FeedbackDisplay: it's the stimulus to which a participant responds to which E-Prime gives feedback.

Step 2

That is, participants respond to the recognition picture, which is called KeepInMindRecog (above). Point this out to E-Prime by going to the global properties of Fbk, General

```
SessionProc
    welcome
    instructions
    BlockList
        BlockProc
            TrialList
                RecogState
                TrialProc
                    TrialStart
                    KeepInMindSlide
                    Fixation
                    RSVPList
                        SingleStim
                            ISI
                            Stim
                    KeepInMindRecog
                    Fbk
                    CheckErrors
    goodbye
```

Figure 3.8

tab, and typing 'KeepInMindRecog' in the Input Object Name field. Run the experiment again and see if it works. It should, but notice how tremendously ugly the feedback is! It's full of pointless information, a white background, and so on. Edit this to make it look better. I tend to show just a centrally displayed 40-point smiley against a black background depending on the feedback. It looks minimal and classy – unlike myself.

However, a sad smiley is not nearly as effective in punishing behaviour as the threat of having to stay in an EEG lab until the cows come home. Currently, the participant may just keep on pressing random buttons, leaving us with useless data. It would be better to get more data out of bad behaviour while also getting our subjects to behave!

Step 3

Edit the InLine script CheckErrors by writing the following.

```
If KeepInMindRecog.ACC = 0 Then GoTo TrialStart
```

In regular English, this instructs E-Prime that if the accuracy of KeepInMindRecog is equal to 0 (i.e. incorrect), to go to a point on the current (TrialProc) timeline that says TrialStart – the label we placed earlier. Try and see if it runs. Does it work? If so, you have successfully written actual programming code (in this case E-Basic).[3]

Even more avoiding guesses

Still, a 50% chance that a guess will end up with a correct answer is really a bit high. It would be much nicer if the participant would see all 6 faces and recognise the one they were to keep in mind, lowering the chance of a successful guess to only 1 in 6. Since we know which picture is relevant and which are irrelevant, we can now also implement this in the KeepInMindRecog slide. Let's make the experiment a bit prettier and use a button.

Note: if you have an older version of E-Prime, you will not have access to the new button interaction element. Instead of using 6 buttons, follow the instructions 1–8 as much as possible but use SlideImages instead of buttons and ask the participant to indicate the person they were to keep in mind by responding with a key, 1–6. This requires you to also make little labels above each image, so that they know how the keys correspond to the images. The trick with

(Continued)

---

[3]Apologies if I sound condescending, maybe you are an experienced coder? It is just that (1) most people sadly do not, and (2) all this dragging/dropping/clicking does not sound like Serious Coding to most people, although they are quite wrong about that (try LabVIEW!).

the 6 different States (step 5) does not easily extrapolate, so you may want to skip steps 5–8. Alternatively, have a look at the online material to see how a similar design can be achieved in E-Prime 2.

Step 1

Double-click on the KeepInMindRecog and remove the central picture. Instead, move a SlideButton subobject where the picture used to be. Go to the subobject properties of the SlideButton. In the General Tab, change the name to 'PictureInMind' and notice that E-Prime tells you 'Responses reported by this SlideButton shall report '{PictureInMind}'' for the RESP property – this will be important later on. Now go to Frame tab, change the Width and Height to '256', the X and Y to '30%'. Then, in the Theme tab, you'll notice sub-tabs (a crime against the laws of user interface design). Here, in the sub-tab Text, change the Text field to be blank (instead of Button). Then in the Image sub-tab, change the File to 'img/[relpicture].jpg', and the Stretch to 'Yes'. Use Color Key should be set to 'No'. In the General sub-tab, the BorderWidth should be set to '1'.

These are all general aspects of the button, but we can also change specific aspects to how we interact with the button. For example, just above where you just changed Image, change the Circumstance to Hover. Below it, change the BorderWidth to 2. For Circumstance 'Down' (which refers to clicking and holding the mouse button), change the same BorderWidth to '4'. This gives a nice kind of 3D effect: run the experiment to see how the border is affected by moving the mouse over the button or clicking it.

If it does, then notice that clicking the button doesn't do anything. Why, E-Prime, that was the only possible option! It appears we still need to add the button as a response option.

Step 2

Go to the KeepInMindRecog general properties, Duration/Input tab, and remove the Keyboard. Instead, add a Button, keep the Allowable as is '{*}', but change the Correct property to '{PictureInMind}'.

Now it works, but it's a bit easy if the answer is always given.

Step 3

Selecting the button, press CTRL+C to copy it, then press CTRL+V to paste it. From the drop-down list (between the global properties icon and the subobject properties icon) click on Button1 and go to its subobject properties. Keep everything the same as for the PictureInMind button, except the X (in the Frame tab): this should be '50%', and the image, which should be 'img/[irr1].jpg'.

Step 4

Repeat step 3, but make the new button irr2 with X being '70%' and the image filename 'img/[irr2].jpg'. Repeat this for irr3, irr4 and irr5, but with their particular filenames, and Y values being 70%, and X values successively 30%, 50%, and 70%.

Now if you run the experiment, it should show something like Figure 3.9. This is otherwise pretty OK, but note the weird instructions (still referring to the old, single stimulus screen). Fix this by editing the text field to something reasonable. More problematic is the fact that the correct image is always in the same position. That makes it easy for our participants to cheat.

Figure 3.9

Step 5

To change the location of the correct button, create different slidestates by clicking on the drop-down menu below the green plus and selecting Clone SlideState (see Figure 3.10). Notice that this creates a new Tab to the bottom of the working area saying State2. Within this state, select the first button (PictureInMind), move it to the right (give it coordinates 50%, 30%), then select the second button (irr1), and move it to the left (30%, 30%). Now the positioning in state 1 has the correct image as the first image, and state 2 has it as the second image.

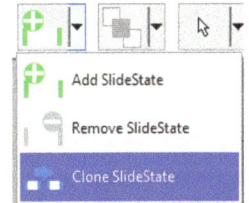

Step 6

Keep repeating step 5, for State3, State4, State5, and State6, such that the PictureInMind can appear at any position. This can involve a lot of clicking

Figure 3.10

and inspecting properties, so it might be easiest to drag the buttons around directly and then fixing the properties inside the Properties Inspector (go to View > Properties if this menu isn't already showing).

Now you should end up with State1, State2, State3, State4, State5, and State6, but if you try to run it, you'll see that E-Prime will keep showing always the same State1 – the relevant picture in the top left. If you look inside the global properties of the KeepInMindRecog slide, you can see why: under the General tab, the Active State is set to State1.

Step 7

Fix this by going into the TrialList and typing 'RecogState' under the Nested column. Accept the question as to whether you want to create this new list. Enter the same value ('RecogState') for every cell under the Nested column.

A nested list does not run trials like normal lists, but merely 'runs along' with the present one. This is very useful when we want to randomise something independently of the present list. Let's see it in action with the final step.

Step 8

Double-click on RecogState, add a single attribute called 'State' and add 5 rows. Make the first value in State 'State1', the second 'State2', and so on until 'State6'. Set the list to randomise the 6 values. Then edit the global properties of KeepInMindRecog and set the Active State to '[State]'. Run the experiment and see if it works.

If so, the experiment is done.

ADDING TRIGGERS

Hold on before you start testing, though. It's all well and good to show all sorts of stimuli to your subjects and have them do the experiment, but afterwards you need to know exactly when things occurred. Otherwise, your data will look a bit like Figure 3.11:

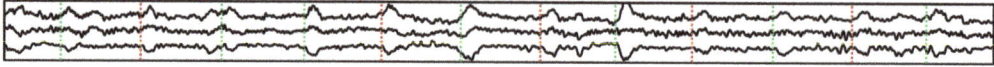

Figure 3.11 Three EEG channels and triggers. Note the positivity in the first channel and negativity in the third that recurs right after dotted lines.

There's a bump in the EEG above, right after each dotted line. Would you have noticed this recurring pattern without the vertical line? Maybe so, but probably not if I hadn't made them about 20 times larger than the rest of the EEG. This is an important point when it comes to EEG: when we know *when* something happened, we might see an effect on the EEG, but we are rarely if ever able to guess mental thing X happened just by inspecting raw EEG. For this reason, (1) by inserting into the EEG these kinds of lines, as in the figure, we know *when* something happened, and (2) by further defining the lines, we know exactly *what* happened, for example using the red and green above. In biosignal analysis (of EEG, but also EMG, EDA, or even fMRI), we do that using **triggers** (sometimes called markers or labels) that act like little flags so we can later understand what is going on. For example, in our experiment, we would expect something like Figure 3.12:

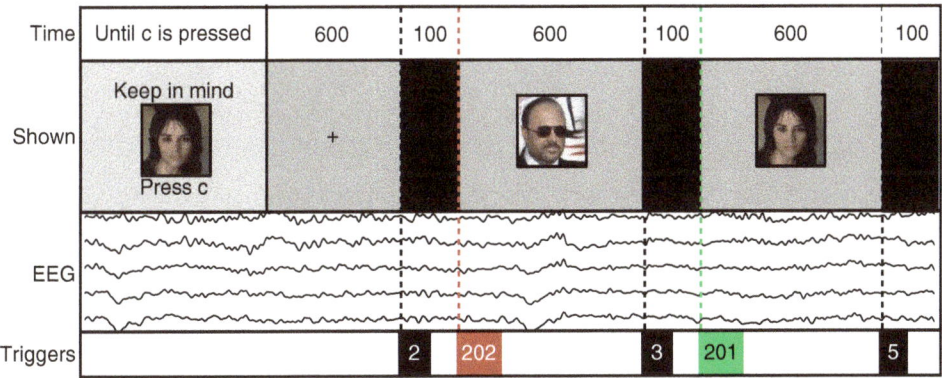

Time	Until c is pressed	600	100	600	100	600	100

Figure 3.12   EEG is recorded synchronised with known events. By recording triggers, we will know afterwards whether an image was relevant (201) or irrelevant (202), and which specific image it concerned.

In the above figure, I marked down our planned timing of showing pictures as per the Procedure from Chapter 2. After recording, however, we only see the EEG part. By putting in the little numbered flags, we can create a kind of logbook for ourselves by using codes that say 'if there's a trigger saying 202, an image was shown that was irrelevant' and 'trigger 201 means a relevant image was shown here'.

Sending out triggers can occur in various ways, depending on the EEG equipment you use. I will discuss the most common way here, as various manufacturers use complicated hardware to allow communication between the experiment computer and the EEG recording, but these differ across brands. However, the companies will normally be familiar with E-Prime and have software, tutorials, and a helpdesk in place to help you out. Psychology Software Tools is also aware of the peculiarities of most common EEG system vendors and may give you advice on how to implement EEG synchronisation. I will further discuss various EEG systems in Chapter 5, but consider this as a general rule:

If there's a flat, very wide (really old looking 'printer cable' type), cable sticking out of your computer, connecting it to the big EEG box (the amplifier, further discussed in Chapter 5), you are good to go. If not, you might want to ask around before proceeding with this chapter.

To make sure the onset of your stimulus is integrated with your data with high temporal precision (**trigger synchronisation**), E-Prime needs to send out a trigger exactly synchronised with the event that is occurring. Here, that event is the onset of the presentation of the face pictures. The most standard way of doing this is by sending out a very short pulse exactly when the stimulus is shown to the EEG amplifier (see Chapter 5) via parallel port. Parallel port communication is a common type of messaging between computers and equipment (usually printers before USB became the standard). It is very old, and has low bandwidth (i.e. it can send very little information in a set amount of time), but it has extremely low latency (it sends the information without much delay). For this reason, it is still in use in psychophysiology and associated fields.

Setting up E-Prime for communication with the EEG amplifier

Given that you have a parallel port and given that somebody already used it to record EEG, you should be able to find out the specific communication port address – the **COM port** – of this parallel port. The COM port specifies the particular location of the parallel port, which is necessary information: much like your average postal worker, E-Prime cannot know where you want a package to be delivered if you do not write down the address.

To find out what the address is, in Windows press start, type Device manager and click to open the device manager. In the Device Manager, you'll likely see a section Ports, with one that is called a Printer Port (for me, that is the ECP Printer Port LPT1). Our port of interest is thus situated at LPT1. Sometimes, E-Prime can be a bit more specific and request the hexadecimal value of the port. This can be found by double-clicking on the Port in Device Manager and looking under the Resources tab, where you should find the address. For me, it's set to go from 0378–037F (we'll just use the first number). Write down the LPT (LPT1) and port address (0378) for use later on.

There are multiple ways to send triggers from E-Prime, but for convenience I'll cover two. Steps 1 to 4 explain how to use Task Events, Step 5 will explain WritePort commands.

Step 1

Task Events can be used in E-Prime to send triggers at the same time as some sort of additional procedure is carried out. For example, at the time a picture is presented (its onset time), E-Prime does its best to send out a trigger via the parallel port by doing the following:

Go to Edit > Experiment > Devices and Add a ParallelPort. Make sure the LPT port is set to be the same as what you wrote down above (for me 1).

Step 2

Go to the Stim slide of the SingleStim procedure, Task Events tab, and Add an event to the Stim.OnsetTime. Select the Stim.OnsetTime, now under the Events menu, and click, under

Task, on the small button next to Name: (Choose a Task). Click on Parallel Port so that the Stim.OnsetTime event triggers an Action for the ParallelPort. Choose WriteByte next to action, change Data Type to 'Byte'. The result should look like Figure 3.13.

Now what this does is to send a byte value (a byte is a number between 0 and 255) to the parallel port at the onset of Stimulus. For simplicity, let us assume a trigger channel can only have two values: 0 (down), and 1 (up). So, when the first face image appears, E-Prime is set to send out the trigger, making the trigger channel's value go up, as shown in Figure 3.14.

This works fine, but the problem comes with the next image. If that image also sends 1, it will be undetected, as the trigger channel is already up. That is, the PC does not automatically reset the parallel port back to its down state so the 'up' of the second face is never received. This is not the case with every single EEG amplifier, but when it is, it causes much confusion. To fix this, we need to make sure our parallel port returns to 0, as in the lowest row of Figure 3.14 above.

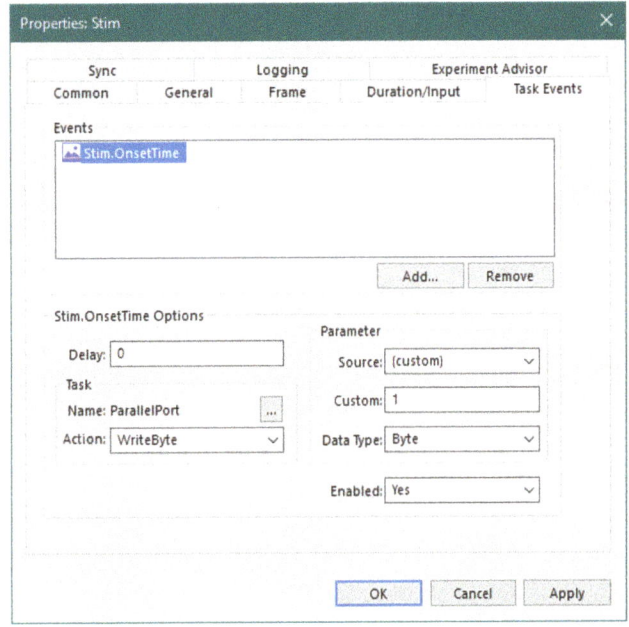

Figure 3.13

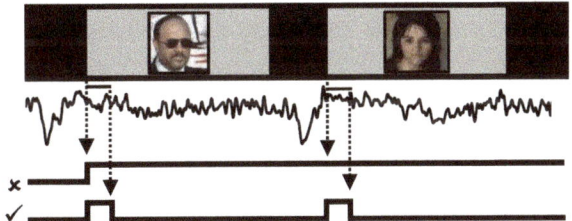

Figure 3.14 Stimuli shown, 1 channel of EEG, and how to set up triggers incorrectly and correctly. The latter sends two triggers for each face: one at the onset of an image, another with a very short delay to reset the trigger to its default state.

Step 3

Again, go to the Stim slide's global properties, and add another event. Make this also based on Stim.OnsetTime with a ParallelPort action sending a Byte using WriteByte. However, make sure the Custom value is 0, and set the delay to 20. This way a 0 is sent out 20 ms after the 1, re-enabling the use of the 1 for the next image.

But then again, we don't want to send out the value 1 because it's uninformative. Better would be to send out a specific trigger so we know which stimulus is (1) relevant, (2) irrelevant, or (3) odd (pet). The parallel port allows the values 1 to 255, so there's limited availability for creativity here. We will use the triggers 201, 202, and 203.

Step 4

Add an attribute to the RSVPList, call it 'ConditionTrigger'. Give the value '201' to the top row (which has Picture [RelPicture]), '202' to the next 5 rows ([irr1] to [irr5]) and '203' to the last two rows (with 101, 102). Then, go to the Stim slide's global properties, yet again, and change the first event (the one without delay, and with Custom value 1) so that its Custom value is now '[ConditionTrigger]'.

Step 5

The **WritePort**-method is a relatively simple way to send triggers using code, rather than the graphical user interface. This is not the preferred way to do things when timing is critical, such as with the onset of the stimulus, but it makes an amount of sense to do this when it comes to various other events in your experiments, which you do want to remember in terms of the EEG data collection. For example, you might want to remember what the beginning and end of the entire trial was.

Add an inline just after TrialStart, call it 'SendTrialStartTrigger', and write down the following:

WritePort &H378, 240

But replace the 378 above with the value you looked up in your system, keeping the &H. This bit of code simply sends the number 240 to the parallel port.

Hexadecimal and binary

The &H means that E-Prime understands the value as referring to a **hexadecimal** one, which means based on a numbering system in which 10-based digits (from 0 to 9) are replaced by 16-valued hexadecimals (from 0 to 9, then A, B, C, D, E and F). Hexadecimal 378 is thus converted to decimal number 888 ($3*16^2 + 7*16^1 + 8$). The parallel communication converts the 240 into binary, however, which means that the 240 is converted to 8 **bits** (0s and 1s) of data, which is why the reset trigger is still necessary. For example, 240 is in binary 11110000, so if after this image 1 appears (1 hex in binary equals 00000001), so the amplifier will mistakenly record a value of 11110001, which converts back to decimal as 241. To get a sense of how normal numbers (decimals) convert to binary and hexadecimal, try the Windows calculator and change the Standard view to Programmer.

If the boxed text about binary, decimal and hexadecimal went way over your head, don't worry. The TL;DR is that you should just resend a 0 slightly after you send any value to the parallel port, just to be sure. To do this, change the above line of code to:

```
WritePort &H378, 240

Sleep 20

WritePort &H378, 0
```

This tells E-Prime to wait for 20 ms, and then send the trigger 0, resetting the state of the parallel port.

The experiment is almost ready for the lab. Follow the essential exercises to complete it and the additional exercises to make it really shine.

ESSENTIAL EXERCISES

Essential exercise 1: More cats

In the design as yet implemented, the two cats are always the same. Adjust the design by employing the same technique as explained in Advanced stimulus generation in this chapter, by adding two new attributes to the TrialList, and referencing these in the RSVPList so that different pet pictures are shown between trials.

Essential exercise 2: Image number

Currently, we are sending a specific trigger to mark the condition type, but afterwards, it will be difficult to figure out which stimuli are presented. Perhaps you want to know whether a specific image gives a bigger EEG response than another one? Perhaps you want to know what the effect of seeing yourself is? For such reasons, it makes an amount of sense to send a trigger out that also codes the exact image. Luckily, the images were named with numbers, so that makes things rather easy. Send a trigger at the onset time of the ISI in the SingleStim procedure so as to send a trigger coding the specific image (see Figure 3.12), irrespective of the relevance condition.

By obtaining a picture of each participant, and renaming that image to '1.jpg' just before the start of an experiment, you will later on be able to figure out what the effect of seeing yourself is.

Additional exercise 1: Better randomisation

There's still a problem with the design, which makes for a bit more complication. Notice that pictures 1 and 2 always share a trial, but 1 and 7 never do. Is this a problem? Well, let's say 1 and 2 are pictures of Alice and Bob, respectively, but Bob happens to be so stunning a lad that he completely masks the effect of Alice (who somehow doesn't make any sort of mental impact), then we might not be able to measure the effect of keeping Alice in mind if she happens to co-occur with Bob, even though she may safely co-occur with Frédérique (picture 7) – who's less interesting than her name suggests. What we need, then, is to randomise better, more randomly somehow. One way of doing this is by disconnecting the picture number from the file name: anybody can be Alice!

1. Add a new nested list '(People)' to the BlockList. Give it a single attribute, 'Pic', and give it values from '1' (on the first row) to the number of pictures in your group, on that row. So, if my group has 30 people, I keep on numbering until row 30, with value '30'. Set the List to randomise.

This will be the stimulus set (for people). The convenient thing is that if we at any point in time want to change a number, we can always do that within this List: a new addition in our group is easily done. More importantly, if our TrialList now references a *random* row in the People List, number 1 will be a random person. To do this, picture 1 (in TrialList) needs to reference a single picture in the People List (say, person 25), such that picture 2 in TrialList is not person 25. One can do this in E-Prime using **level references**, which works as follows:

2. In TrialList, change the first row of relpicture to '[pic:0]'. This references now the People List (the one that has the attribute called pic), specifically to the first level of this attribute, which can be anyone, depending on the randomisation. The 0 refers to the row (inconveniently, this first level is 0, the second 1). So, give the other rows in relpicture values from '[pic:1]' to the last, minus 1 (so if you have 25 rows, that would be '[pic:24]'. Now, give the same treatment for irr1 to irr5: for the first row, given relpicture being '[pic:0]', irr1 to irr5 should be '[pic:1]' to '[pic:5]'. Repeat this for all the next rows.

3. Notice it may now be harder to use the participant's own picture in the experiment. Change the People and TrialList such that every experiment will *also* use 1.jpg (see Essential exercise 2).

Finally, apply the same way of working, but then also for the pets. Start by typing under Block-List's nested column 'People, Cats' and create a new attribute 'CPic' in the new list.

Additional exercise 2: Error trigger

Previously, you sent a trigger at the beginning of the trial. Now sometimes, trials are repeated – if the participant makes an error. It would be useful to send a trigger in the event that such an error occurs (for example, we might remove this trial from analysis). Edit the previous CheckErrors inline script to send a trigger: if the participant responds inaccurately to KeepInMindRecog, send the trigger 241. If they respond accurately, send the trigger 242. Make sure to reset the state of the parallel port after writing the trigger.

EXPLORE: FURTHER E-PRIME MATERIAL

If you want to continue exploring the capabilities of E-Prime, some more material can be found here:

Psychology software tools provide on their website (pstnet.com) official support, as well unofficial support with the E-Prime forum (support.pstnet.com/), and a bunch of pre-programmed studies previously known from the System for Teaching Experimental Psychology (support.pstnet.com/hc/en-us/categories/360003738794-STEP-Experiments).

The E-Prime googlegroup mailinglist (groups.google.com/g/e-prime) is a useful resource for asking questions and finding answers, particularly if you're stuck with programming an experiment or have a technical issue

And of course, Spapé, Verdonschot & van Steenbergen (2019), *The E-Primer: An Introduction to Creating Psychological Experiments in E-Prime. Second edition updated for E-Prime 3*, is your ultimate textbook for all things E-Prime.

4

DEVELOP YOUR EXPERIMENT IN OPENSESAME

IN THIS CHAPTER, YOU WILL LEARN ABOUT:

- Developing experiments in OpenSesame
- Creating an RSVP study
- Synchronisation with external equipment

The landscape of software used to be pretty simple. We used Windows on our PCs, Office for documents, and SPSS for scaring students with statistics. E-Prime, which I demonstrated in the previous chapter, was a native to this land (as were alternatives such as Neurobehavioral Systems Inc., **Presentation**), even though the company behind it is hardly like a giant corporate machine such as Microsoft or IBM (of SPSS). Still, it is a commercial product, which means that development and support depends on sales, which in turn entails costly, closed-source user-licences.

As elsewhere, the open-source movement has disrupted this convention, with various small (e.g. Python, at least initially) and big (Linux, Google) players providing licence and opening source-code, making money in various other ways, or without any profit motive. In the land of experimental design, a multitude of free, open-source alternative experiment development software has sprung up. Some of these are browser-based (e.g. **PsyToolkit**) running on many different operating systems. Another popular type runs in an otherwise proprietary platform such as MAT-LAB (e.g. **Psychtoolbox**), or Qualtrics (see **QRTEngine**, by Barnhoorn et al., 2015). Finally, there are various programs that use Python, a programming language which itself is quickly becoming the most popular across sciences. These include the original, Jonathan Peirce's **PsychoPy** (Peirce, 2007; see Peirce & MacAskill, 2018, for a great introduction), which is magnificent, and **Expyriment** (Krause & Lindemann, 2014).

Given the many alternatives, I had a hard time to decide which to discuss and promote for this book, so I took some time learning the various software packages to determine which ones were powerful enough for present purposes, while remaining relatively easy. However, what *easy* means depends on your background. To someone who already knows Python, the use of well-documented application programming interface (APIs, a kind of building block of computer code), writing solely code to console screens, comes naturally and is much easier than searching for the right icon to click in an extensive graphic user interface (GUI). For such people who would rather code than click, I would recommend Expyriment or PsychoPy (although PsychoPy has a GUI, the Builder, it seems more a starting point for doing the actual coding). For everyone else, I recommend **OpenSesame** (osdoc.cogsci.nl/; Mathôt, Schreij, & Theeuwes, 2012), which is open source and free, uses Python and has a pretty solid graphic user interface that makes sense to me.

That said, I will not pretend that I do more than scratch the surface of OpenSesame or Python. OpenSesame is in constant, rapid development, which means that what I write today may no longer be current tomorrow. Also, while OpenSesame's GUI is quite capable, our experiment is actually not all that simple, so in order to keep my description within a reasonable number of pages, I found myself having to resort to writing quite a bit of code anyway. Luckily, my Python skills are pretty basic, so what I write hopefully follows the mindset of someone who has just started exploring it.

Before you begin: in this chapter, and the last, I explain how to develop the experiment to run in your lab, either using OpenSesame (this chapter) or E-Prime (Chapter 3). I would recommend you choose one based on availability and skill: OpenSesame is open source, free, but may require some understanding of programming.

GETTING STARTED IN OPENSESAME

The great thing about OpenSesame is that it is incredibly flexible. Not only does it run 'on top of' PsychoPy and Expyriment, it is also built as a bridge between many different environments and abilities. That is to say, it can run on OSX, Windows, Linux, and even Android (so you might run experiments on your mobile phone); it can run on normal displays, eye-trackers, and even some head-mounted displays (for virtual reality); it can run on the web or at home in your lab. There is also a veritable cornucopia of experimental building blocks that one can integrate in experimental design by dragging and dropping elements: joysticks, forms, and even synths. The use of such a toolbox seems inspired by E-Prime, and this makes it much easier for a dedicated E-Primer to make the jump.

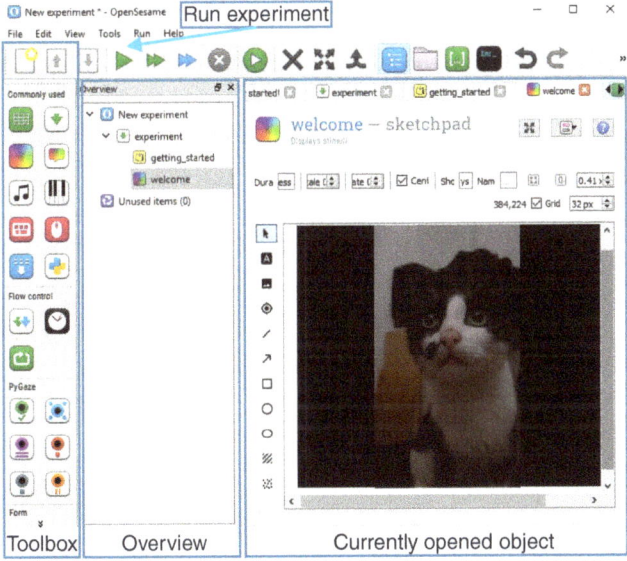

Figure 4.1   Starting OpenSesame

For this chapter, I will assume you run OpenSesame 3.3.8 *Lentiform Loewenfeld* on Windows, although it should not particularly matter. While some of the figures might look a bit different on later versions of OpenSesame, the core functionality tends to remain in place.

Setting up

Make sure you have the following set up and ready to go:

- *OpenSesame has been installed on your computer.* You can freely download it from https://osdoc.cogsci.nl/. The installation is relatively easy, and does not even require one to download Python separately. However, for purposes of learning, it may be useful to download and install Python itself and a **debugger** that can test and run bits of code. OpenSesame includes its own debugger, but other options are Spyder, (www.spyder-ide.org/) or Visual Studio Code (code.visualstudio.com). These make coding easier, for example by showing you information about the code as you type (code completion, IntelliSense).

- *Everything that is also mentioned in Chapter 3: setting up* (see p. 39). In particular, make sure all images are numbered (1 to the number you have available for people, and the oddball images (pets) the same plus 100).

Welcome to OpenSesame

One of the best parts of OpenSesame is that it runs right out of the box. It is not uncommon for academic software to immediately give you warnings due to your computer being in some obscure way different from the one used by the main developer, which inevitably means one has to trawl through all sorts of user forums, be told to *RTFM*, install various patches in specific orders, until only the most dedicated fanatics survive. That rarely includes me. Not so for OpenSesame. I ran the self-installer and it immediately works.

Step 1

Starting OpenSesame (Start › OpenSesame) and you'll be greeted by a getting started screen within an empty experiment. Save it in your ExperimentDirectory (see p. 39) as something reasonable so you won't tear your hair out when you forget to save it later on.

Step 2

In the Overview (Figure 4.1), click on New Experiment. Then, in the currently opened object (i.e. the right side of the screen), change the name of 'New Experiment' by clicking on it and make it something reasonable like 'Oddball'. By default, the experiment is set to run at 1024 × 768 pixels, which is perfectly nice for a good old CRT monitor, but most of us who are not into vision science use wide-screen LCD monitors. Check what your own resolution is (in Windows: Start > type 'Resolution' > Display settings; scroll down, find Display resolution) and fill that in here. Additional information on the difference between CRT and LCD information can be found under Chapter 3, page 47.

Step 3

There are a few other settings as well, such as the default background and foreground colour. Change these to grey in HTML: #7d7d7d and black (#000000) respectively. The default font, Mono, is fine, as is the back-end. The back-end relates to the underlying code structures used by OpenSesame, such as Expyriment and PsychoPy. Presumably, OpenSesame's developers have a good idea what the default should be, so I left it at that. These back-ends also have a few global settings, which can be adjusted by clicking on Back-end settings, but these mainly concern sound (which we're not using here), and a few advanced video aspects related to the display.

Run the experiment by clicking on the biggest PLAY button on the screen (see Figure 4.1) and see if it works. It should provide a welcome screen ('OpenSesame version number, version name'), and pressing any button on the keyboard ends the experiment.

Note on the left, in the Overview, that a new experiment by default provides the following items: A sequence called experiment, a notepad called getting_started, and a sketchpad called welcome. The sequence constitutes a serial ordering of the items within the sequence running from the top to the bottom, so currently just the notepad (which is just to document the experiment) and then the welcome screen.

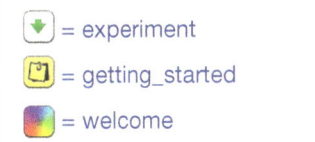

Step 4

Click on 'welcome' and edit the contents of the sketchpad. Double-click on the text and change it to '<i>Welcome to the experiment</i>
Press space to start!' It should now look like Figure 4.2.

Note that the parts between greater than/smaller than signs disappeared as these merely gave instructions on HTML style **markup**, tags that denote how text should be displayed. That is, <i> *italicises everything until the closure* </i>, and <br> introduces a line-break, which is pointless here as putting in

Figure 4.2

a new line by pressing enter after 'experiment' would be interpreted as a line-break as well, but this makes it explicit. However, the bigger point here is that if you know HTML, you can apply it in OpenSesame for making things look a bit more stylish; or if you don't, there are zillions of tutorials on the internet that will help you (it being the Hyper Text Markup Language, which to old farts like myself is like the asphalt of the *information superhighway of the world wide web!*

Step 5

In the top right of the currently opened object (still 'welcome'), note the middle of the three buttons, and change it to Split View. As you do this, a little window with script opens, saying:

```
draw textline center=1 color=white font_bold=no font_family=serif font_italic=no font_
size=32 html=yes show_if=always text='<i>Welcome to the experiment</i><br />Press space
to start!' x=0 y=0 z_index=0
```

…But hang on! That's more or less what you wrote before! Indeed, it turns out that OpenSesame just tricked you into writing little bits of code without your even knowing about it. While I cannot approve of non-consensual coding, it does make an amount of sense to look under the hood and see what OpenSesame is up to. Also, the split screen is pretty useful as (1) it allows you to quickly copy parts of the experiment, (2) to adjust parts that are hard to do manually, such as placing text exactly in the middle (note the x=0 y=0 stand for the centre of the screen), and finally (3) to tweak more advanced parts directly using Python. For now, however, I will mainly concentrate on using the GUI.

Step 6

Drag two additional SketchPads to the experiment, after welcome, renaming the first 'Instructions', and the last 'Goodbye'. Edit the text of Instructions to make clear to the subject what is expected: they are to keep images in mind while looking at series of images. It is also a good idea to remind them of what is generally expected during EEG experiments (Chapter 6). Similarly, edit the text of Goodbye to make clear the experiment is now complete and participants may soon be free again. For Welcome and Instructions, make sure you explain to the subject what to press to continue with the next part (i.e. press a button), but for the Goodbye, change the duration to '4000'. Work on the properties of Instructions and Goodbye to give a satisfactory look and feel. Run the experiment and keep changing the properties until you're happy with the result.

Basic structure in OpenSesame

Now you have the beginning and ending of the experiment, but no actual experiment. OpenSesame (and experimental design in general) works with a hierarchical kind of structure. One experiment is a single sequence, as is already the case: there's a welcome, an instruction screen, and a goodbye. Between the beginning and end is the meat of the matter, a number of *blocks*, which can mean a training part and a testing part (or, say, a questionnaire part and an experiment part). Each **block** has a sequence with a beginning and ending (e.g. 'training begins', 'training ends'), with many different **trials**. In most experiments, the trial is the smallest, most frequently occurring sequence that is described.

   Or think of it the other way around. For example, an emotion processing experiment testing whether we can localise emotional stimuli (such as snakes) faster than neutral stimuli (like chairs) could show a fixation, followed by a stimulus (say, a snake), that requires a single response (depending on where this picture was presented, press LEFT or RIGHT). Given two pictures and two locations, we would have four possible combinations to create what is called a fully **orthogonal design**:

Picture	Location	Emotionality
Snake	Left	Emotional
Snake	Right	Emotional
Chair	Left	Neutral
Chair	Right	Neutral

(Note: this would be a poor experiment design as the *picture* is completely confounded with the *condition*, as described in Chapter 2)

   I call variables like Picture and Location **manifest variables** while Emotionality is here the **latent variable**. That is, we wish to measure the effect of emotionality, but sadly OpenSesame doesn't know emotions, so we have to explicitly tell it what we mean: the different pictures. In OpenSesame we commonly define conditions like these as part of a **Loop** (like a List in E-Prime), in which the same sequence is run multiple times, but with different pictures and locations. Let's see how this works in our experiment.

Step 1

Drag a Loop from the Toolbar to the space just between the previous Instructions and Goodbye Sketchpads (Figure 4.3). Rename it to 'BlockList'.

At the moment, the BlockList does not do anything, because it does not currently 'loop' anything. To have it repeat something, we will need to move a sequence inside the BlockList.

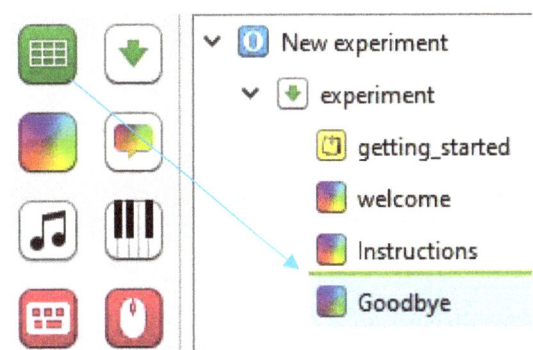

Figure 4.3

Step 2

Do so by dragging a sequence from the Toolbar into the BlockList in the Overview and call it 'BlockProc'. Repeat these two steps to create another level in the hierarchy: Add a Loop to the BlockProc ('TrialList'), and a Sequence to the TrialList ('TrialProc'), such that your Overview will resemble the image in Figure 4.4.

Now we have the skeleton of the experiment set up: the BlockList loop may have a training and testing part (see additional exercise), and the TrialList will be showing rapid serial visual presentation (RSVP) trials. It doesn't yet, as TrialProc is suspiciously empty. The TrialProc will be showing (1) a picture to keep in mind; (2) the series of flashing images; (3) a little test to see whether someone actually kept the picture in mind. But, *which* picture exactly?

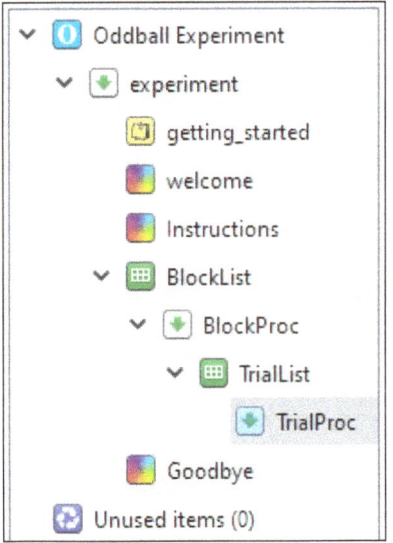

Figure 4.4

Step 3

Click on TrialList, and change the first column header to 'relpicture'. Enter, under the column header, the numbers 1, 2, 3, 4, 5, and 6 in different rows. Then add two SketchPad items from the Toolbar to the TrialProc, naming them 'KeepInMindPresent' and 'KeepInMindRecog' (Figure 4.5).

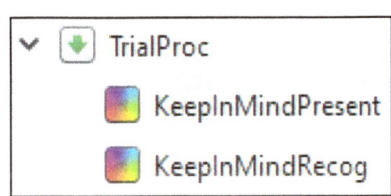

Figure 4.5

Step 4

Click on KeepInMindPresent, and add an ImageElement. Find the ExperimentDirectory that contains all the pictures used in the experiment (see p. 39) and select one. I thought this guy was particularly awesome in a sort of Bruce Willis kind of way (Figure 4.6).

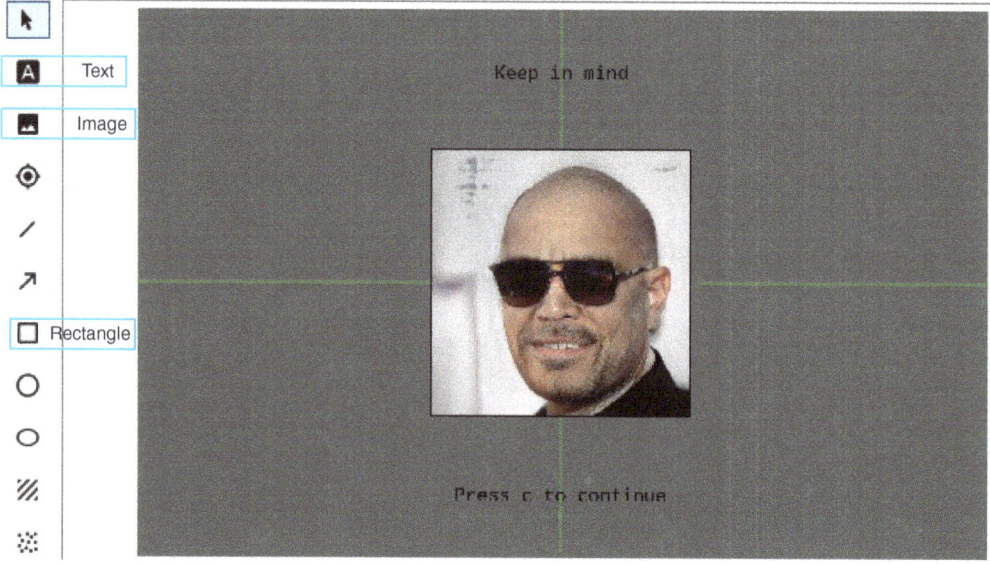

Figure 4.6

Step 5

Change the rest of your KeepInMindPresent to look like mine (with or without Bruce Willis) using the tools I have outlined in red in Figure 4.6 above. Some tips: (1) I used a rectangle with PenWidth 4 around Bruce to create the outline for the image; (2) it helps to use the split view, paying especial attention to having the X of both Text items set to 0; note that the Text uses a centre origin (X = 0 means the horizontal middle), but that the rectangle's origin is the top-left. Given that all images are 512 × 512 pixels, the top-left corner of the image is –256, –256, while it has a width (w) and height (h) of 512. In other words, I just produced the following code:

```
draw image center=1 file='58.jpg' scale=1 show_if=always x=0 y=0 z_index=0

draw rect color='#000000' fill=0 h=512 penwidth=4 show_if=always w=512 x=-256 y=-256
z_index=0

draw textline center=1 color='#000000' font_bold=no font_family=mono font_italic=no
font_size=36 html=yes show_if=always text='Keep in mind' x=0.0 y=-400.0 z_index=0

draw textline center=1 color='#000000' font_bold=no font_family=mono font_italic=no
font_size=36 html=yes show_if=always text='Press c to continue' x=0.0 y=400.0 z_index=0
```

Ouch, that looks pretty unfriendly and heavy on code, but it's not too hard if you take it one command at a time. The image, for example, is drawn in the centre (center = 1), is picture 58 (58.jpg), is drawn at its normal scale, with an offset from centre at x and y 0 (i.e. no offset, thus in the centre).

The z_index refers to its ordering from top to bottom, such that figures in the background have a higher z_index than those in the foreground, and are overlapped by them.

Now, as should be clear from the picture, we are asking the participant to press C to continue, but OpenSesame will still allow any option. Unfortunately, the easiest options are for Open Sesame to accept ANY key, or some duration. There are two ways to allow only C. In step 6, I will explain the first strategy for adding response buttons, which is to put an explicit keyboard-response item into the experiment.

Step 6

Drag a keyboard_response item from the Toolbar to just after KeepInMind-Present. Click on it, call it 'CtoContinue' and set its allowed response to c. If you would keep the experiment as it is and run it (try it!), you would  see OpenSesame first shows KeepInMindPresent until you press ANY button, and then until you *also* press C, because KeepInMindPresent is set to show until a button is pressed – the CtoContinue comes after that. So to change this, set KeepInMindPresent's duration to 0. It will still show, but the experiment will only continue after CtoContinue catches the C.

Why C to continue anyway?

Well, I am always worried that participants will be randomly pushing buttons, and never mind what is on the screen, or accidentally press a button (say, space) twice. It makes sense, in such cases to occasionally switch whatever you designate as the 'next' button.

Step 7

Repeat step 5, but apply it to KeepInMindRecog. It should work more or less the same, but ask the participant 'Were you able to keep the picture below in mind?', and 'Press Y or N'. Obviously, *two* response choices are now available, so instead of the 'c' you had before, let's change the KeepIn-MindRecog's particular keyboard challenge to use the *second strategy for adding response buttons*. Use the Split view, and enter the following code in a line just after the set duration keypress:

```
set allowed_responses 'y;n'
```

How did I know that? Well, if you would look at the script underlying the CtoContinue code, you would notice that the only unique aspect of it is the set allowed_responses bit, so I just copied that part to the KeepInMindRecog slide.

OK, now run the experiment to see if it actually works, that is, if it does not crash. If it does, it would be a good idea to go back and see what went wrong before continuing.

Implementing variables

While seeing Bruce is good fun for a while, at some point you have to say 'Mr Willis, maybe it's time for the next generation of actors to take the limelight?' At the moment, there's still the one picture, presented twice, and although we can change the second picture to Ryan Gosling, we will still need to do something about the direction of this experiment. Like any good *romcom*, you just have to ask, when and where is the RSVP? And this recognition scene, isn't that a bit naff? Anyone could see that coming from miles off! And as my analogy is ageing much faster than Bruce Will is, I'll just say, what we need is surprise, variety, and *variables with a Vengeance*!

In OpenSesame, the least coding intensive way to implement variables is through the table as implemented in loops. Remember you added, in the TrialList, the column 'relpicture'? We can use, or *reference*, the data within the table by referring to its name in square brackets (like E-Prime attributes). Let's see how that works:

Step 1

Change the KeepInMindPresent by changing the text Keep in mind to say 'Keep in mind: [relpicture]'. Set the TrialList from its default random to sequential. Now run the experiment, and notice that the first time the picture is shown it will say 'Keep in mind: 1', the next 'Keep in mind: 2'. Exit the experiment by pressing Escape.

OpenSesame does not suddenly do a trial counting operation. Instead, think of it like this: Each loop in OpenSesame is a cycle in which a variable – relpicture – is set to a different number. In cycle 0 (the first one), the relpicture is '1'. Within the sequence defined in the cycle, KeepInMindPresent is found, in which the brackets [relpicture] denote a reference to the variable, which thus equals 1.

Step 2

This is a bit abstract, so you might find it helpful to change a few numbers in the TrialList under relpicture: change the third, for example, to everyone's favourite psychologist, 'William James'. Run the experiment and observe the result.

Now, this is, of course, a bit pointless: we need no trial counter, and although we will always need William (♡), we would prefer to see the actual pictures to be kept in mind.

Step 3

On the KeepInMindPresent, remove the trial counter in the text, but change the picture (e.g. double-click on it or look in the script) such that the file is changed to 'img/[relpicture].jpg'. Again, note that this refers to where your pictures *should* be (keeping in mind the instructions of Chapter 3 – setting up!).

This may look cryptic, but it is in no way different from what you did with the trial counter. Run it to see if it works. If it does, but crashes on the third trial: excellent! Remember, on the first trial, the image will be the first level of [relpicture], therefore, img/[relpicture].jpg becomes 'img/1.jpg', which points OpenSesame directly to a picture in the 'img' sub-directory you created. On the second, it will be 'img/2.jpg'. On the third, it will be 'img/William James.jpg', which I doubt you have placed in that directory (although he does deserve it). So, go back to the TrialList, and fix the third trial to just be '3': now it should work.

Step 4

Now try to extrapolate this thinking towards KeepInMindRecog. Add a column to the TrialList, call it 'recogpicture'. Enter, on the 1st and 2nd row '1', on the 3rd and 4th row '3', and on the 5th and 6th row '5'. Then change KeepInMindRecog so that it will use the recogpicture attribute to change the picture. Change the instructions to say something like 'Was this the picture you were to keep in mind?'

Step 5

But, how do we know if our participant isn't just pressing buttons randomly, and how do we keep a record of their performance? To do this, add yet another column to the TrialList, calling it 'correct'. Now change it such that when the picture to be kept in mind is the same as the picture that is to be recognised, the correct response would be 'y', and otherwise 'n'. To use the variable, change the script of the KeepInMindRecog by adding another line:

```
set correct_response '[correct]'
```

Or alternatively, if you prefer the first strategy for adding response buttons, change the KeepInMindRecog duration to '0', add a keyboard_response item after KeepInMindRecog, and set it to allow 'y;n', but make correct '[correct]'.

However, OpenSesame doesn't automatically log things, so there's no way to know whether your participant was just randomly hammering those keys. To do so, drag a **logger** item from the Toolbar to the end of the TrialProc. Call it 'TrialLogger'.

Loggers log, and by default, they log everything. Run the experiment and after finishing it, inspect the resulting comma delimited CSV file (e.g. 'subject-1.csv'), for example using Excel. You should find a column saying correct_KeepInMindRecog, which is set to 1 if the participant was paying any attention.

= logger item

Step 6

Finally, have a look at the TrialList again. Make sure that the order is set to random, and have it repeat the cycles 2 times, such that the experiment has 12 trials in total.

The RSVP

At the moment, it is of course rather missing the main part of the experiment: we were going to look at a sequence of images (the Rapid Serial Visual Presentation, or RSVP) while keeping an image in mind. Designing this repeats many of the steps I just explained, so you should not have trouble with this part. If you do, you might want to read back. We start a short fixation screen:

Step 1

Drag a Sketchpad to the place in between CtoContinue and KeepInMindRecog, call it 'Fixation', and give it a duration of 695 ms (see blue box). Make it so it will show a '+' symbol in the exact middle, in Courier font and a point size of 30.

Step 2

Create a new Loop, 'RSVPList', in between Fixation and KeepInMindRecog with a new Sequence, 'SingleStim', and let RSVPList have a column header 'Picture' with the values of 1, 2, 3, 4, 5, and 6 below it. Let it randomise, and make cycle 10 times.

Step 3

Drag a SketchPad (call it 'ISI') to the new SingleStim procedure. Add a filled, black rectangle of 512×512 pixels in the middle with a penwidth of 4 and a duration of 95 ms.

As you will remember from the procedure (Chapter 2, p. 31), the fixation was supposed to be 700 ms, and the flash in between pictures was supposed to be 100 ms, so why do I write 695 and 95? Displays update at a constant rate – usually at about 60 Hz or every 16.67 ms – which means that presenting an image in between two refreshes (say, halfway between 16.67 and 33.33 ms) could mean that for a brief moment, half a picture shows and half does not. To avoid this, OpenSesame (and E-Prime, not to mention many computer games) synchronise presentation such that the display receives its information *just before* the new refresh. This means that even if I tell OpenSesame that my Sketchpad *should* have a duration of 95 ms, it will only show up just before the next refresh occurs – which happens at 100 ms. Imagine what would happen if a tiny timing error occurs (which happens often enough!) such that the instruction 'show this on the screen' is sent 4 ms late: now the stimulus is shown at 99 ms – which rounded up to the next refresh recycle is still at 100 ms. If, on the other hand, I entered a (hoped for) duration of 100, and a 2 ms timing occurs, then the next stimulus will suddenly occur one refresh later, or at 116.67 ms. So it's good practice for timing critical stimuli to have a duration that is slightly shorter than their prescribed duration, keeping in mind the refresh rates. For example, a refresh rate of 70 (refresh every 1000/70 = 14.2857 ms) could have an ISI of 86 (14.2857 × 6 = 85.71), or 100 (14.2857 × 7 = 100), but not 90. So, entering a duration of 90 will mean OpenSesame shows the next image at 100 anyway.

Step 4

Drag a SketchPad (call it 'Stim') just after the ISI. Let it be 512 × 512 pixels, a 4-point black border, a duration of 595 ms, and let it *reference* the attribute you made in step 2, using the same strategy you used in Implementing Variables step 3 (p. 72).

Run the experiment and see for yourself how you can improve it. As such, it should basically work, but it is really the minimum of what I would expect from a student. What can we do better?

- There are only 6 images in use, always the same ones.

- It is still extremely easy to guess the answer, and there's no penalty for doing so.

- There are no cats.

IMPROVING THE EXPERIMENT

This is the age of the internet, which means that any experiment worth its salt needs to have cats. Let's start with that then.

Two cats

Whether or not you use pictures of cats as pets (I have nothing against dogs, horses or hamsters), all *odd* pictures described in Chapter 2 had 100+ numbers, such that cat 1 is 101, and pet 2, a tortoise, 102. In the procedure, I describe 62 images being shown, 10 of each person, and 2 oddballs. That gets difficult with our list, which randomises 6 images 10 times. The easiest way to sort this is by doing the following:

Go to the RSVPList and change the number of repetitions back to 1. Then, copy and paste the first 6 rows 10 times, such that row 1–60 has the numbers 1–6 ten times. In row 61 and 62, change the value to 101 and 102.

Why? Previously, we mixed a stew of 6 ingredients (images) 10 times, meaning that after eating a dish in some random order (e.g. 132465), the ingredients were remixed. What we now have is a pot of 62 ingredients, that is mixed just the once, with some ingredients (1–6) 10 times as much in the pot as two others (101–102). This means that sequences may feel less random than they were: a sequence like 1112144 was impossible, but isn't anymore. This may be problematic. After noticing something relevant once, you might not have the same evoked potential upon repetition. Some experimental designs therefore try to restrict randomisation.

For example, one might create 4 RSVP lists with 4 cycles, and just the people (1–6), *then* include the list with cat 1, then another 4 RSVP lists with 4 cycles, and the single list with cat 2. A further possibility is to look ahead inside the list and kind of randomly replace a human with a pet. Each possibility has its own issues, though, so I will keep it like it is here, as what we have now is much simpler to implement and less restrictive for the randomisation.

But there are still just 2 cats being shown (I have 4 cats myself) and 6 humans (I do not have 6 humans) throughout the experiment. That is insufficient.

Advanced stimulus randomisation

The problem is that we have many images but we are only showing 6 different ones (and 2 cats) within a single trial. This is part of the design: If we always show *all* images within a trial, it would be much easier. However, this would severely impact the degree to which we control for differences in stimuli, and how well we can generalise towards other stimuli (see Chapter 2). Also, if we have, say, 25 different images, and 1 of them is relevant, then only 1 in 25 stimuli would show the image to be kept in mind. In other words, participants would mainly be spending time on watching images that were *not* kept in mind. Of course, we could add the relevant image to the pool four times, so it would occur in 4/25 stimuli, which is still about once every 6 stimuli. This is not ideal either, however, as the stimulus that is kept in mind would then be showing much more often in a run than any other picture. In other words, it would now confound *frequency* with *relevance*, making it impossible to say afterwards which of these would cause the observed effect.

Ideally, we therefore pick 1 random picture that is the relevant one, and 5 irrelevant ones each trial, then continue using only these throughout a single RSVP. There are various ways to accomplish this in OpenSesame, but to keep the amount of coding to a minimum, let's start with the following:

Step 1

Change the TrialList to include 5 extra attributes, named 'irr1' to 'irr5' (see Figure 4.7, left list). Whenever the relpicture has a number, fill the irr1 to irr5 attributes with the pictures that could be irrelevant. Continue the list so that each picture can be relevant: I went until 12 below (because I'm lazy), but if you have 20 people in your group, then you should continue until 20. No pictures should be irrelevant if they are not also at some point relevant, so if the relpicture is 16 (and there are 20 people in your group), for example, the irrelevant pictures should be 17, 18, 19, 20, and 1. This type of design is sometimes referred to as a **Latin square** design, and is popular in textbooks on experiment design.

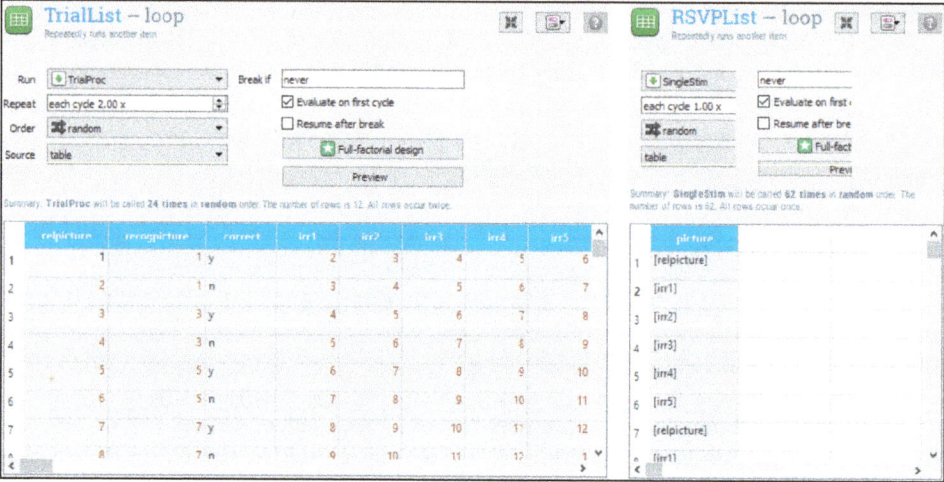

Figure 4.7

Step 2

Change the RSVPList so as to match the TrialList (see Figure 4.7, but note RSVPList has 62 rows). Notice that instead of absolute numbers, the values below picture now show variable references. Thus, in the third trial, if we weren't using randomisation, the Pictures should retrieve the value 3 for the first row of the RSVPList, then 4, 5, 6, 7, and 8 for the irrelevant pictures, and still 101, 102 for the pets.

Feedback and avoiding guesses

As it remains extremely easy to guess the answer, a less cooperative participant might just press a button, then dig out their phone and start doing something entirely different instead. They won't be told off for wasting your time and they're free to carry on guessing. So, let's make it a bit harder to guess the answer before we start with the real cruelty and punish wrong answers in the harshest way I can think of: by making them do the trial again. Let's start with the friendly reminder for participants that allows them to learn from their mistakes.

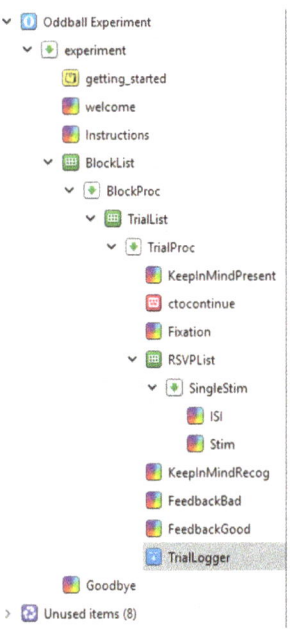

Figure 4.8

Step 1

Drag two Sketchpads from the Toolbar to the end of the TrialProc but just before the Trial-Logger and call them FeedbackBad and FeedbackGood. Your overview should now look like Figure 4.8.

There are many ways of giving feedback to participants, but my experience is that participants generally prefer to leave sooner rather than later. That means that positive feedback is only useful to the point that it doesn't add too much extra time, which brings us to Step 2.

Step 2

Edit the FeedbackGood to have a duration of 1000 and to show a single word (I often use a single :) in plain text). Also edit the FeedbackBad to say something sufficiently threatening like :(. Finally, make it last until the participant presses the spacebar (inform them by adding instructions). To make it respond only to the spacebar, add 'set allowed_responses space' to the script. Requiring a specific button takes a bit of time, adding to the experience. Run the experiment to see what it looks like and whether you're impressed with your own reinforcements!

Oh, but now it appears to show both the good *and* the bad, no matter what the participant does. To do something about this, run the experiment for more than one trial and open the resulting .csv file (e.g. in spreadsheet software like Excel). It should give a column heading somewhere saying 'correct_KeepInMindRecog'. Inspecting the values below it, you should notice whether these reflect your accuracy (if not, and there is more than 1 row in the csv file, see 'Basic structure in OpenSesame', pp. 68–72). From that, we know that the variable 'correct_KeepInMindRecog' holds the performance, and what we want to do is show FeedbackGood if the correct_KeepInMindRecog equals 1 and FeedbackBad if the correct_KeepInMindRecog equals 0.

Step 3

To do so, click on TrialProc in the overview and notice all the various items that are run within the sequence. More importantly, the second column mentions 'Run if', with all items stating 'always'. Double-click on the always next to FeedbackBad and write the following: '[correct_KeepInMind Recog] = 0'. Now the FeedbackBad is 'run if' the variable correct_KeepInMindRecog equals 0. Using this logic, also change the run if for FeedbackGood such that it is only shown if the participant is responding correctly.

That should work (do check this). However, on a human level, it is clear that a sad smiley is not nearly as effective in punishing behaviour as the threat of having to stay in an EEG lab until Judgement day. A particularly evil subject who is just into it for the money or cinema vouchers[1] will merely laugh hysterically as they keep on hammering random buttons, knowing full well the puzzlement we experience as we look at their EEG later on (see p. 126 for detecting maniacal laughter artefacts). But what if we can make bad behaviour result in *more data*, while simultaneously getting our subjects to behave?

Step 4

Move a repeat_cycle item from the Toolbar to the end of the TrialProc, just after the TrialLogger, and call it 'RepeatBad'.

 = repeat_cycle

Currently, it doesn't do anything. Selecting it and looking into it, you may notice it has few parameters (i.e. changeable aspects of a function), other than 'Repeat if', which currently says 'never'.

Step 5

Using the same logic described in step 3, change it such that it repeats for incorrect responses. Make sure you also change the FeedbackBad to mention to the subject that the trial will be repeated.

The experiment is now almost ready.

ADDING TRIGGERS

Hold on before you start testing, though. At the moment, the experiment is working more or less as expected, but all this information will be lost after recording the experiment. Of course, if you are able to start the experiment and the recording at exactly the same time, you may be able to puzzle out at what point in time exactly which image was shown. However, because participants may take longer or shorter with certain parts, you will need some device that pauses and starts recording each trial, which will be very hard to design. More importantly, the internal clocks of the EEG and of the PC run at very similar but not *exactly* the same speed, so the longer the recording lasts, the more unsure we will be about the timing. For this reason, we need to **synchronise** our EEG recording

---

[1]It is common in Finland, where I work, to give cinema vouchers for participation rather than cash. That means that the film industry in general are kept running with the financial support of scientific research. Maybe you can try this argument when the next person asks what your research is really good for!

with the stimulus computer, so we know later on what part of the recording concerns which stimulus. In EEG and other biosignal measurements, we use **triggers** to do this, annotating the data with information about what is going on. If you are not familiar with the idea of triggers, see page 54, Figure 3.11 and the text up to 'Sending out triggers' for a general introduction.

Sending out triggers can occur in various ways, depending on the EEG equipment you use. The common way is to use the parallel port, although some companies (like EGI) have their own idiosyncratic ways. They think it's probably better for timing or helping you with difficult coding or something, but it makes it hard for me to write a book, and for you to transfer your skills from the one EEG lab to the other. The only reasonable thing to do is to pick up our pitchfork and storm their company headquarters, *viva la révolution*! But as we are still waiting for the movement to reach critical mass, I would recommend sending a polite email to their customer service department asking them how to send triggers. They will have heard this question before and are likely to send you some information sheets written by the engineer who is despairing over the fact that all the precious work that went into their Completely Custom Super Easy System is not being appreciated.

Anyway, while I will further discuss various EEG systems in Chapter 5, to figure out whether parallel port communication should work, use this general rule of thumb:

If there's a wide and rather flat (really old looking 'printer cable' type), probably grey, cable sticking out of your computer, connecting it to the big EEG box (the amplifier, further discussed in Chapter 5), you are good to go. If not, you might want to ask around before proceeding with this chapter.

To make sure the onset of your stimulus is integrated with your data with high temporal precision (**trigger synchronisation**), OpenSesame needs to send out a trigger exactly synchronised with the event that is occurring. Here, that is the onset of the presentation of the face pictures. The most standard way of doing this is by sending out a very short pulse *just after* the stimulus is shown to the EEG amplifier (see Chapter 5) via parallel port. Parallel port communication is a very old standard of messaging between computers and equipment (usually printers before USB became the standard). It has low bandwidth (i.e. it can send very little information in a set amount of time), but it has extremely low latency (it sends the information without much delay). For this reason, it is still a common standard in psychophysiology and associated fields.

Just after? Why?

It makes intuitively little sense to do something *after* the picture is shown, unless you think of it from a computer's point of view (if computers had a POV). Common sense might say at 0 ms, the computer loads the image (done perhaps after 20 ms), beams it to the screen (at lightspeed!), then after 600 ms (at 620 ms)

(Continued)

stops beaming it and performs the next instruction, such as showing the black ISI box. That would be terribly inefficient. Instead, the image is loaded (which can occur much earlier), does some video preparing instructions (rendering), and is shown (possibly still at 20 ms). That means in the next 600 ms, the computer has time to put its feet up or party as it doesn't need to re-render the picture over and over. Of course, we could also tell it to send a trigger straight after rendering is complete (i.e. at 20 ms), which is *much closer* to the actual onset of the image than sending a trigger just before requesting the computer to do the preparation.

Setting up OpenSesame for communication with the EEG amplifier

Given that you have a parallel port, and given that somebody has already used it to record EEG, you should be able to find out the specific communication port address – the **COM port** – of this parallel port. The COM port specifies the particular location of the parallel port, which is necessary information for running OpenSesame on Windows (as opposed to Linux). So, if you use Windows, this is how you can find out what the COM port address is:

Press start, type Device manager and click to open the device manager:

In the Device Manager, hopefully, you will see a section Ports, with one that is called a Printer Port (for me, that is the ECP Printer Port LPT1). Our port of interest is thus situated at LPT1. However, what we need is the hexadecimal value of the port. This can be found by double-clicking on the Port in Device Manager and looking under the Resources tab, where you should

 find the address. For me, it's set to go from 0378–037F (we'll just use the first number). Write down the LPT (LPT1) and port address (0378) for use later on.

Note: 378 is not three-hundred-and-seventy-eight, but hexadecimal 378, each value being one that goes from zero to 9 and then onwards to a, b, c, d, e, and f, so 16 numbers. The 7 in 378 thus stands for 7 × 16 (112), and the 3 for 3 × 16 × 16 = 768, so 378 hexadecimal equals our normal number system *decimal* 8 + 112 + 768 = 888. The fact that there's nothing inherently superior to thinking in *base 10* (i.e. the decimal system) gives me a sense of abstract joy; perhaps because I'm a nerd, or perhaps because it means I can still claim I am in my *cough*hexadecimal*cough* twenties.

There are currently two ways to send triggers from OpenSesame: using a third-party plugin or with Python inline script. In my tests, I found the former to require less hassle in terms of both installation and code (but see p. 84 **using inline**), so I will assume the reader uses the plugin.

Using the Parallel Port Trigger plugin

Currently, OpenSesame does not by default come with a conveniently integrated way of firing triggers to your average EEG amplifier. However, there is a very useful plugin by Bob Rosdag, available from github.com/dev-jam/opensesame_plugin_-_parallel_port_trigger (or Google 'OpenSesame parallel port trigger'), which provides such functionality. On the github website, click on the clone or download button, and download a zip archive containing the plugin. Move the two folders parallel_port_trigger_init and parallel_port_trigger_send to the OpenSesame plugin directory (for me, that is C:\Program Files (x86)\OpenSesame\share\opensesame_plugins). If you now restart OpenSesame and expand the Toolbar, it should now show two new items as in Figure 4.9.

Step 1

Drag the Parallel Port Init item to the start of the experiment, just before the Getting Started item (this is also a good moment to remove that Getting Started notepad item!), call it InitParallelPort (Figure 4.10).

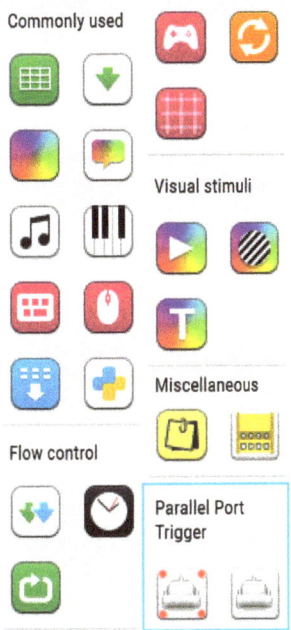

Figure 4.9

InitParallelPort — parallel port trigger init
Parallel Port Trigger: initializes the parallel port device.

☐ Dummy Mode
☐ Verbose Mode
Port Adress   0x378   Adress of the parallel port, value is a hexadecimal number (Windows) or path (Linu

Figure 4.10

The Init item prepares OpenSesame for sending triggers to the parallel port with the parallel port defined as per the options you see in the item. These are:

• Dummy mode: For use when you are developing an experiment on a computer without parallel port, such that you can simulate sending triggers without actually doing so. However, Windows by and large has no idea whether a trigger is successfully sent to a non-functioning COM port, in which case the only additional value of having a dummy is that it may (if you are as flaky as I am) cause you to forget to turn it on by the time you actually want to send triggers.

• Turning on verbose mode makes the plugin slightly chatty, informing you what is going on in the Debug window (if you don't see the Debug window: right-click anywhere in the Toolbar and enable Debug window). This is useful if you don't understand the plugin, so I would recommend turning it on the first time you run the experiment with triggers, but then turning it off to keep the Debug window free for more important information.

So now we have the parallel port ready for action. Before we insert the triggers, however, let's think for a moment of what we would like to see afterwards, in our data. That is, if we directly follow the advice above – of sending triggers immediately after showing images – the computation will look a bit like this:

1. Send the picture to the video interface (which takes a few milliseconds)

2. After the image is shown, wait for the duration (current 595 ms)

3. Send the parallel port trigger

4. Wait until the next refresh (presumably at 600 ms)

5. Send the ISI image to the video

Which would mean the parallel port trigger intended for the image onset would be sent a bit *before the start* of the ISI. That would be useless.

Instead, the ideal would be to go for the following:

1. Send the picture to the video interface (which may take a millisecond or two)

2. As soon as it is done sending, send the parallel port trigger

3. Now wait for the rest of the duration the image is to be shown (e.g. 575 ms)

4. Send the ISI image to the video

If this strategy is clear, you can start to implement it:

Step 2

Drag a parallel port trigger send item from the Toolbar to just after Stim. Call it 'StimTrigger', and give it a value of '[ConditionTrigger]'. Turn Enable duration on, and make the Duration (ms) 10. The 10 ms duration means that, after sending the trigger, the plugin will wait for 10 ms, and then send a value 0, resetting the parallel port.

Why do we need to reset the parallel port?

This is due to the parallel port as a communication device in which values are transmitted as binary code from 8 pins. Much like the previously mentioned hexadecimal, binary code is a form of writing numbers in which the base value is not 10 (such as with normal decimals, from 0–9), or 16 (hexadecimal,

0–f), but 2 (0–1). With 8 pins, and each pin readable as 'on' or 'off', that means there are 256 different values possible (2^8). A value like 201 (irrelevant image, with the triggers in the picture above) thus translates to binary 1111 1011, which means all but 1 pin will be powered. By default, they will be powered until someone tells them they shouldn't be, so if again the number 201 is sent, then the EEG amplifier will not notice at all, as there is no change in voltage. And it gets worse: what happens if I send the trigger 3? This translates to binary 0000 0011, so 5 pins should drop in value and 3 pins should remain the same. The result of this **crosstalk** of binary codes is that the EEG amplifier reads completely wrong values, which is a nightmare later on. To pre-empt such an issue, it is best practice to reset the parallel port after a short duration such that any new trigger is received against 0000 0000, resulting in a clean reading.

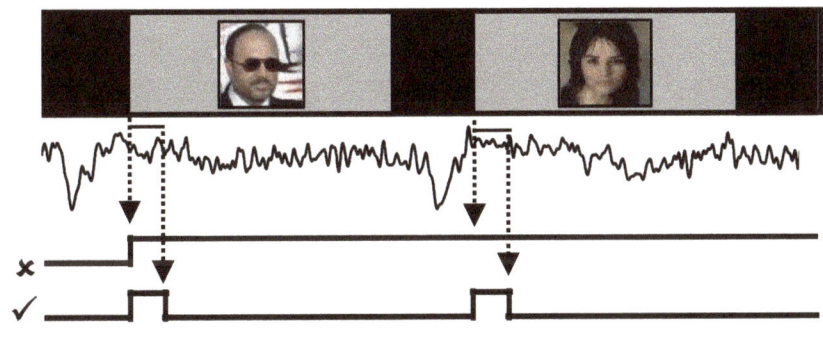

Figure 4.11

Now we need to make sure that StimTrigger uses a variable ConditionTrigger when sending the specific trigger. These can be any value between 1 and 255, as long as we use them consistently. For this book I made the executive decision to send a 201 for any relevant image, 202 for irrelevant images, and 203 for pets. To use these values, however, we need to edit the list. Let's see how that works.

Step 3

Click on RSVP list, create an extra column, calling it ConditionTrigger. For every row that has [RelPicture], make the ConditionTrigger 201, for [irr1] to [irr5] 202, and the last two rows should be 203.

This should work, but it has made a mess of the timing of the experiment: if you already saw what was coming, the images now show for a super short time, or the trigger is still only sent long *after* the image was shown. We will fix this in Step 4.

Step 4

Set the Stim duration to 0, then drag an advanced_delay from the Toolbar to just after the Stim-Trigger and call it StimDuration. Give it a 580 ms duration (notice also that there is a useful 'Jitter' option, which allows randomisation of the delay; we are not using it here, though).

In theory, these timing changes should still work out, as your experiment should now look like Figure 4.12.

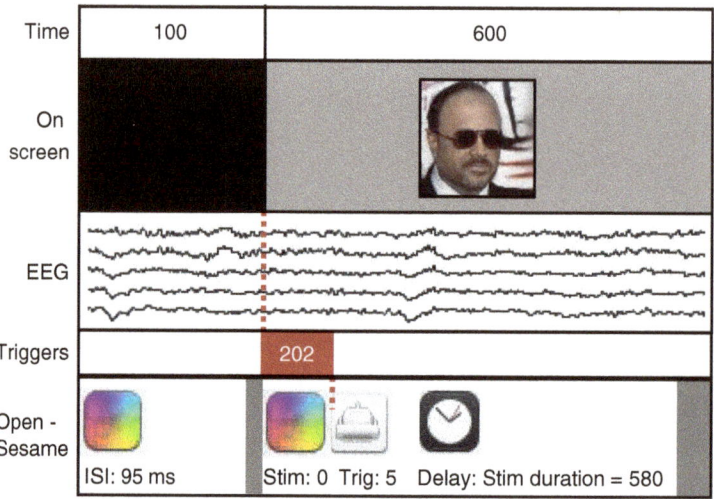

Figure 4.12

The top of Figure 4.12 shows the time, as in the experiment procedure. In OpenSesame, we present an ISI of 95, but the next image (Stim) should only show up with the next screen refresh (see p. 74), which is generally – if your screen refreshes at 60 Hz – at 100 ms. The image is shown, but despite its 0 ms duration, it will not actually be wiped from the screen again. Thus, the trigger for the irrelevant image is sent as near to the image being shown as possible – still at 100 ms, hopefully. After 10 ms, the triggers are reset (by sending a 0). The trigger delay does not add to the duration of the stimulus, but the command itself might take a millisecond or two, and the first refresh of the stimulus itself adds some. Therefore, in theory, we should hope that the minimal stimulus duration is 2 + 16.67 + 580, which rounds up to 600 ms, but we'll have to check this.

Run your experiment and see if it works.

Using inline script

It is also possible to use inline script instead of the third-party plugin. In my tests, however, I found it relied on a set of drivers packaged in a file (dlportio.dll) that can be downloaded from the internet. It is not generally recommended to download DLL files from unknown sources, and although I trust the guy from OpenSesame enough to count him as a known source, I still cannot immediately see the advantage of using code instead of the graphic user interface of Parallel Port Trigger. If you detest pointing and clicking, though, I recommend having a look at the Explore section at the end of this chapter for online resources.

ESSENTIAL EXERCISES

Essential exercise 1: More cats

In the design as yet implemented, the two cats are always the same. Adjust the design by employing the same technique as explained in Advanced stimulus generation in this chapter, by adding two new attributes to the TrialList, c1 and c2, and referencing these in the RSVPList so that the two pets are different between trials.

Essential exercise 2: Image number

Currently, we are sending a specific trigger to mark the condition type, but afterwards, it will be difficult to figure out which stimuli are presented. Perhaps you want to know whether a specific image gives a bigger EEG response than another one? Perhaps you want to know what the effect of seeing yourself is? For such reasons, it makes an amount of sense to send a trigger out that also codes the exact image. Luckily, the images were named with numbers, so that makes things rather easy. Send a trigger at the onset time of the ISI in the SingleStim procedure that sends a trigger coding the specific image, irrespective of the relevance condition. Make sure you add the appropriate delays so that the timing will still work. By getting a picture of each participant, and having that image always named 1.jpg, you will be able to analyse what the effect of seeing yourself is.

Essential exercise 3: Check the timing

As I wrote previously, the timing as proposed should work. But the hesitance suggested by my use of 'in theory' probably gave away the fact that this was hopeful thinking. I do not know how fast the computer at your disposal is, but my own was *pathetic*. Let's first test this in the classic way:

- Use a stopwatch (or phone app) to time the duration of a single trial of 62 images, starting with pressing the C to begin, and ending with the recognition screen. For me it took 1 min 26 seconds, while my work laptop ran the bunch in 58 seconds. Which is better, but still pretty awful: given that the blank screen plus image should take 700 ms, a trial should last less than 44 seconds. Clearly, this isn't great.

What to do? A new computer would be nice, but we should first see if we can find out where the problem is.

Step 1

 = inline_script

Drag an inline_script from the Toolbar to the SingleStim sequence, just before ISI. Inline scripts let you add bits of code directly into the experiment. Write the following in the Run tab of the script:

```
print ('1: Before ISI ', clock.time())
```

Inline scripts in OpenSesame have a Prepare tab and a Run tab. The idea is to do as much code as early as possible in the sequence, so that all the loading and other timing-costly parts of a sequence are done as early as possible and do not interfere with the parts of the experiment that need to be timed accurately. For example, if you want to show a 'subconscious prime' for exactly 17 ms and not a millisecond longer, you would want to make sure that after these 17 ms, it doesn't take 50 ms to load the subsequent stimulus, as that would mean your prime would actually be on the screen for 67 ms. Thus, the prepare parts of Inline scripts likewise are called earlier on (commands like 'load the stimulus from disk'), while the run part can be done at runtime (commands like 'show the loaded stimulus now').

 = run in window

The script, at runtime, executes a simple command: to display (print) to the Debug window a bit of text '1: Before ISI', along with the current Time in milliseconds, clock.time(), as measured from the beginning of the experiment. Run your experiment with the green forward button, so it shows in a window. Move the window around a bit so you can keep your eye on the Debug window to the bottom of the screen (if you don't see the debug window, right click anywhere in the Toolbar).

But if you try to run the experiment now, it may[2] crash, and if it does,a lot of indecipherable stuff is shown in the Debug (Console) window. For example, it says 'TypeError: cannot concatenate "str" and "float" objects'. What does it mean? It refers to the fact that Python is trying to interpret why we want to add a number (a 'float' variable, being a number that can have decimals), being the time, to text (a 'str', or string, variable), being the characters '1: Before ISI'. That is because programming languages are dumb: since text and numbers are both represented as binary information in computer language, the code interpreter fails to understand whether you mean (A) convert '1: Before ISI' to a number and add that to time, or (B) convert time (e.g. 12437) to a bunch of characters (e.g. '12437') and add those to the already existing text. So we have to make this explicit.

Step 2

Edit the Python script to convert clock.time() to text:

```
print ('1: Before ISI ', str(clock.time()))
```

It should now work and output timestamps in your Debug window as you run the experiment. This is what I saw:

---

[2]It does not necessarily. In fact, every time I rerun this part, Python gives me a new answer. If it does not crash for you, be thankful, but continue doing the explicit converting of Step 2 anyway.

```
1: Before ISI 5313.75
1: Before ISI 6053.76
```

So a single stimulus + ISI is taking 740 ms rather than 700 for me. But what is the culprit?

Step 3

Right-click on the Inline Script in the Experiment Overview and select copy (unlinked), then right-click in the same position and select paste, twice. Notice that two new Inline scripts have shown up, each with different names (making linked copies creates inline scripts with the same name that cannot be individually edited).

Step 4

Move one inline script just before the Stim Sketchpad and another to the end of the sequence. Edit their contents so that each proclaims its position (e.g. the second could say 2: Before Stim).

Now for the real exercise: check the timestamps and find out where exactly things go wrong. Then, change the durations of the objects so that you start approaching the desired timing.

Additional exercise 1: Better randomisation

There's still a problem with the design, which is perhaps not immediately obvious. Notice that pictures 1 and 2 always share a trial, but 1 and 7 never do. Is this a problem? Well, let's say 1 and 2 are pictures of Alice and Bob, respectively, but Bob happens to be so stunning a lad that he completely masks the effect of Alice (who doesn't make any sort of mental impact, somehow), then we might not be able to measure the effect of keeping Alice in mind if she happens to co-occur with Bob, even though she may safely co-occur with Frédérique (picture 7) – who's less interesting than her name suggests. What we need, then, is to randomise better, more randomly somehow. One way of doing this is by disconnecting the picture number from the file name: anybody can be Alice!

What we'll try to do, then, is to use the numbers we have previously entered into TrialList to reference a variable, grabbing a random image from a list of pictures. There are various ways to do this very elegantly in Python, but it is useful to know how things work before being clever about things. In the box on page 89, I mention the more elegant feature of randomising columns, but let's first have a look at how to do it with simple steps.

We will start by defining a list of all our people.

Step 1

Drag an inline_script to the beginning of the experiment and, in the Prepare tab, enter the following code:

```
People = [1, 2, 3, 4, 5, 6, 7, 8, 9, 10, 11, 12]
print (People[0])
```

This is now a list of all the normal images we will be using in the experiment (change the list accordingly if you have more, or fewer, pictures) stored in a variable with multiple positions, which in programming terms is called an array. Arrays commonly, and confusingly, are 0-indexed, such that the first datapoint in the array (i.e. person '1') is at position 0. The second line of code shows this in action by displaying People number 0 to the Debug window. Run the experiment to see if it works.

If anything crashes, notice that Python cares about case: all people are equal, but people aren't People! Likewise, you can't Print people, but you can print them.

Step 2

Edit the inline_script to randomise the variable, shuffling it like a deck of cards:

```
People = [1, 2, 3, 4, 5, 6, 7, 8, 9, 10, 11, 12]
print ('Before randomisation: ', People[0])
import random
random.shuffle(People)
print ('After randomisation: ', People[0])
```

Now looking into the output, it should say that after randomisation there's a new first (0th) number. Except, of course, that there's a 1 in 12 chance that after shuffling the deck of 12 cards, the 1 lands again in the same position.

To make further use of variables, we can make them available within OpenSesame's GUI by creating explicit versions of the variables. These can then be read by the list.

Step 3

Replace the last print People[0] statement with the following code:

```
var.p1 = People[0]
var.p2 = People[1]
var.p3 = People[2]
```

But continue it for as many people as are in there (i.e. 12 for me). Once this is done, any of the *var.p*s is globally available in your experiment and lists by accessing it with square brackets: [p1] = var.p1. In other words, we have shuffled a sample of numbers (e.g. 1–12) and assigned these numbers to various specific individuals who will now be different people for every single subject.

Step 4

To apply this, edit your TrialList and exchange all the people numbers 1 to as many as you have, to [p1], [p2], and so on.

relpicture	recogpicture	correct	irr1	irr2	irr3
[p1]	[p1]	y	[p2]	[p3]	[p4]
[p2]	[p3]	n	[p3]	[p4]	[p5]
[p3]	[p3]	y	[p4]	[p5]	[p6]
[p4]	[p5]	n	[p5]	[p6]	[p7]

Figure 4.13

Run the experiment to see if it works and apply the same strategy for the pets.

Note: now that you have this working, you can also use more images in the experiment. For example, if you have a group of 100 people, and images 1–100, the 'People =' statement can easily be adjusted to go from 1 to 100. Or, using slightly more elegant Python, it could use this statement to create a list (a type of array) using the range function, which returns the entire range between its two parameters (1 and 100):

```
People = list(range(1,100))
```

Now, even though there are 100 images used in the experiment, participant 1 will almost certainly see a set of images that is different from participant 2. However, you will have to think about how to make sure the picture of the participant themselves is always included in the range.

Another way of doing the same is by using the *shuffle_horiz* function. To do this (or if you get stuck with the above), edit the BlockList and create as many columns as you have items to randomise. For example, if I had three pictures, Alice.jpg, Bob.jpg, and Carol.jpg, and wanted to randomly assign p1, p2, and p3 to these people, I could create three columns, p1, p2, and p3 and write down their names below them. Then, editing the script of the BlockList, I would insert just below the last setcycle line (setcycle 0 p3 Carol.jpg) the following command: `shuffle_horiz`. The magic word then causes OpenSesame to shuffle the contents of the three columns (by explicitly specifying the column names, you can also shuffle a subset of columns), so that for subject 1, p1, p2, and p3 may be successively Alice, Carol, and Bob, while for subject 2, they could be Carol, Bob, and Alice. Elsewhere, for example in my TrialList, referring to [p1] will always drag up the same randomised person (i.e. Alice for subject 1, but Carol for subject 2).

Additional exercise 2: Error trigger

Previously, you sent a trigger at the beginning of the trial. Now sometimes, trials are repeated – if the participant makes an error. It would be useful to send a trigger in the event that such an error occurs (later we might remove this trial from the analysis). Edit the TrialProc loop and make it send a trigger after KeepInMindRecog based on the participant's response: If the participant responds inaccurately, send 241, otherwise send 242. Make sure to reset the state of the parallel port after writing the trigger.

EXPLORE: FURTHER OPENSESAME MATERIAL

If you want to continue to explore the capabilities of E-Prime, some more material can be found here:

The OpenSesame website (https://osdoc.cogsci.nl/) has the latest versions of the software for Windows, Mac, Linux, and Android, information on using eye-trackers, auxiliary hardware, and how to run online, as well as extensive tutorials and peer support forums.

If you want to get better at OpenSesame, however, it is recommended to improve your Python skills. The above website also links to some fun and friendly tutorials on how to get started with that (https://python.cogsci.nl/), although there's a lot of competition due to the popularity of Python and you'll find videos and written tutorials all over the internet.

As far as I am aware, there's no real book for OpenSesame yet, but there is one for its next of kin, *Building Experiments in PsychoPy* (Peirce & MacAskill, 2018), as well as just *Python: Programming Experiments in Python* (Peirce, Gray & MacAskill, forthcoming).

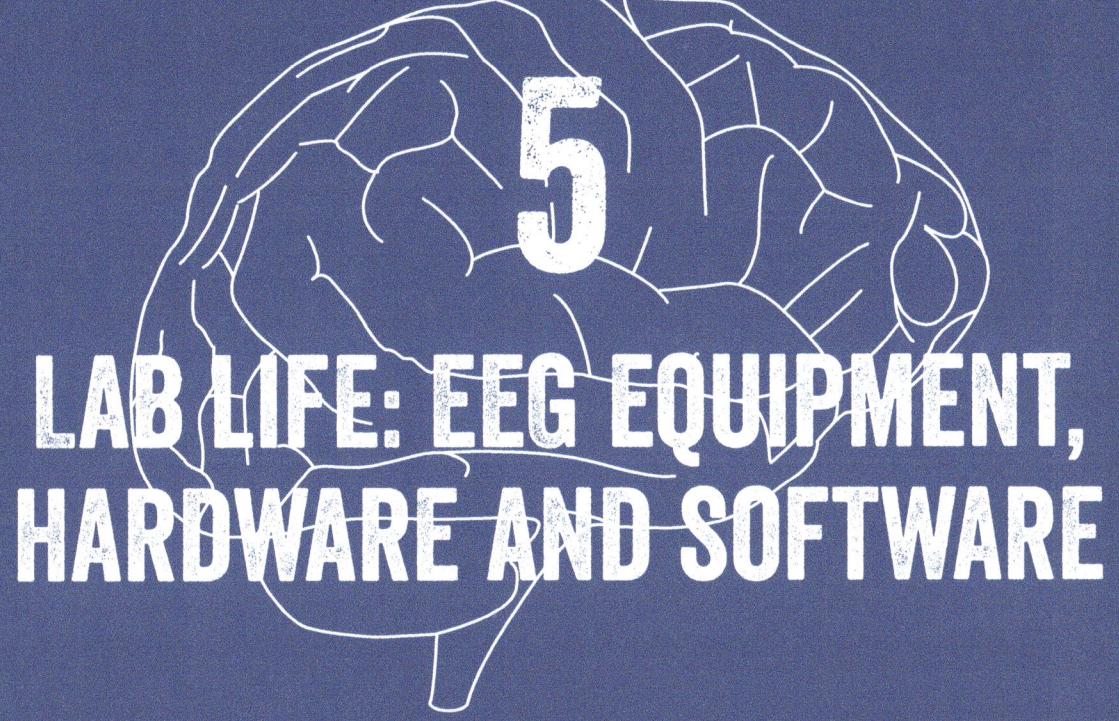

5

LAB LIFE: EEG EQUIPMENT, HARDWARE AND SOFTWARE

IN THIS CHAPTER, YOU WILL LEARN ABOUT:

- What kind of EEG equipment is on the market and how much it costs
- What an EEG lab should look like
- How to set up EEG recording software

When I first entered the psychophysiology lab, I felt rather overwhelmed. So much equipment, so many cables and control panels! It's like going to a hospital and being surrounded by the arcane equipment or glimpsing the myriad dials, levers, and buttons inside the cockpit of a plane. It seemed very much a 1980s sci-fi throwback and the opposite of modern touchscreens. Few things immediately made sense, so the glowing red button on the EEG amplifier might as well have shot the spring-loaded participant's jump seat straight into the ceiling.[1]

The only way to deal with this is by staring in awe at the EEG engineer as they spout their technical jargon and gradually feel stress rising as you swiftly accumulate mental notes to the tune of *look up what on earth they mean*. I remember having this same sense when I failed to live up to my overestimated engineering prowess the moment I managed to turn a slight plumbing issue into a floodgate disaster. The Sunday emergency plumber arrived, naturally assuming that I have the innate ability, as a male, to understand what he's talking about as he rambles on about float valves and gaskets, '*you know what I mean?*'. As he used this exact request for confirmation of understanding about every 10 seconds, I felt myself turning purple as the mental notes started to escape from my ears, while my nodding went from seemingly sincere to tiringly vigorous and desperately manic as the explanation continues on and on.

I thought I would spare you that and I've written this chapter starting from the point at which you have absolutely no idea what an EEG lab might contain but are interested anyway. A practical example seemed to me someone who was interested in getting their own EEG lab – perhaps like an eccentric lord or lady for their mansion, or maybe a dean for the faculty. EEG equipment does not look very fancy – indeed, often enough it looks archaic to the point of Frankenstein's time (hence the lord or lady) – but it can be surprisingly expensive. I explain this, and many other aspects of life in the lab, along a kind of tour through the lab and describe how everything is connected: Faraday cage, dentist chair, TP9, inion-to-nasion. *You know what I mean?*

EEG HARDWARE: STARTING YOUR EEG LAB

If you have been reading this book getting more and more excited about doing mind-reading with EEG research, but becoming a little bit worried about whether the lack of any EEG equipment will become a problem, then yes, you have hit a little snag. But, depending on the size of your pockets, I will help you overcome this issue. We will just need to (1) buy some kit, (2) pop them in their right place, and (3) connect them. I will deal with each of these in turn so that even if you already have an EEG system set up, you should be able to learn something.

Shopping for EEG equipment

The first question on my mind when I want to make a big investment is *where can I get something decent given my budget?* Maybe that vulgar preoccupation is just me being Dutch, because that little detail, price, seems to be missing from most EEG amplifier manufacturers' websites, apart from Biosemi, a Dutch company. Most other companies seem decidedly hesitant about prices,

---

[1]My engineering expertise is revealed by my confusing jump seats, as used by flight attendants during take-off and landing, for ejection seats. The internet tells me the latter are not actually a feature of regular planes.

even failing to give me a rough approximate price when I badger them during conferences. It depends, they say, on the particular distributor in the location, which is probably true, but surely that doesn't mean they can't give a basic guess? I get that a Samsung Galaxy 20 phone may have a different price in Telia Oslo than in Friend Mobile Yongsan, but it's not going to be ten times as much, is it? The lack of transparency comes across as reflecting practices that were common around the time the websites of these companies were designed. Surely this is not intended, but shopping for an EEG amplifier feels a bit like browsing fashion in an insanely upmarket *haute couture* shop: it's very hard to understand the wares being offered, and there are no price tags. In fact, you feel a bit embarrassed about even asking. Strangely, though, I found out you can even try bartering: explain your budget and ask the distributor whether there are demo models available. I mean for EEG, not fashion; I have never dared enter an *haute couture* shop in my life and would probably be laughed out.

What should you look for when you're shopping? This is not an easy question as the information provided tends to be either far too complex for the average user to understand (e.g. 'sensitivity <60 nV (LSB) +− 250 mV'), or far too nonsensical to be of use ('A prosumer EEG used by engaged individuals seeking better understanding of their brains'). So what should be on the top and bottom of the list of features you demand?

- How many channels? More channels is great: it increases signal to noise, allows you to interpolate more missing channels (see Chapter 7), improves average referencing (see Chapter 7), and enhances source estimation (see Chapter 11). Whereas the use of 32 channels used to be normal, advocates of high density (128–256 electrodes) or ultra-high density (700–800, Petrov et al., 2014) have steadily increased the minimal norm. On the other hand, having more electrodes generally takes longer to set up (see Chapter 6), thus reducing the time allowed for the experiment itself, which limits the number of trials you can record and therefore the reliability of your study. It is also more expensive to use many electrodes, both to initially acquire and in operation, as electrodes need replacement every now and then. So, for a normal ERP type setup without grand ambitions on source localisation or identification of EEG micro-networks, I still recommend **64 channels** as hitting the sweet spot of strongly improving signal-to-noise over 32 channels while not adding much time to setup.

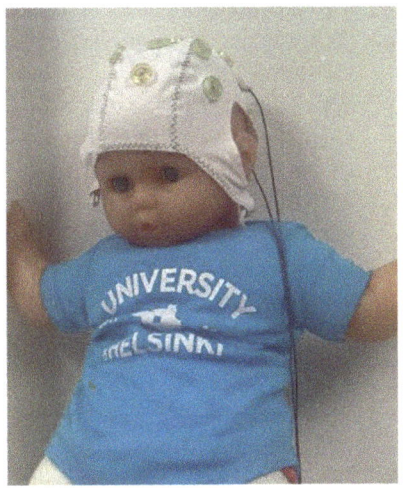

Figure 5.1 Having fewer (e.g. 16) channels is normal in research in infants and children as they have smaller heads and cannot be expected to stay still for the duration of a lengthy setup

- Wet or dry electrodes? Ideally, electrodes should be flush against the skin, but as the head is not a metal cube, they won't be – which is a problem as air is not very conductive. To deal with this, most EEG solutions put some conductive liquid (usually salty or *salty-ish* gel) between the scalp and the electrodes, which works very well, but takes effort (see Chapter 6), and makes the participant feel like they had an unfortunate encounter with a seagull – at least until they can take a shower. Dry electrodes avoid the problem with different electrode shapes (e.g. looking like blunt pins) more reliably pushing against the head. Although this feels not particularly pleasant for the duration of the experiment, the setup is inarguably faster. However, more problematically, the signal to noise remains lower than with traditional, wet electrodes, so the consensus seems to be that **wet** electrodes are still preferable for research purposes.

- Wireless (mobile) or wired (stationary)? This is a question that manufacturers seem very busy with for some reason. Sure, it looks cool if you can publish something about the EEG of somebody going about their own business in real life, or the EEG of someone during a parachute jump, but such research is almost inevitably meaningless. That is, in life (not to mention in freefall), we tend to move a lot, and moving produces EEG artefacts. You may say I'm a grumpy old traditionalist for being against taking super-sensitive equipment out of the lab, but I'm also the one who has to review papers that use mobile EEG in museum visitors and claim that brain activity 'detects aesthetic interest'. No, dear authors, if something is not interesting, you tend to *walk away*, which produces a ton of noise. So, in case my view wasn't abundantly clear, the answer is **wired**. It's cheaper (by channel), probably more accurately timed, and the wires conveniently act as a leash so people don't move.

- Active or passive electrodes? In general, EEG electrodes use a covering of Ag/AgCl deposits (silver and silver-chloride ink), which reduces their contact impedance (Zhou et al., 1994), and therefore noise. Noise defined in hardware terms is any outside, non-brain-related activity that is captured at any stage between the source of activity, its analogue recording (by means of the electrode), and its digitisation. In passive electrodes, interference acts upon the signal all the way as it travels from the silver-chloride to the amplifier. The basic idea of active electrodes is to move the digitisation up the signal chain as close to the electrodes as possible. Then, most of the wire transmits merely the 0s and 1s, and is no longer impacted by impedance and electric room interference (Metting van Rijn et al., 1996). Therefore, as **active electrode** systems improve signal quality, they are often preferable. They are often bundled wire systems that are faster to set up. However, active electrode bundles are generally more expensive in acquisition and operation, as single passive electrodes are relatively cheap and easy to replace.

- How to communicate with external equipment? As mentioned in the previous two chapters, it is necessary for EEG amplifiers to record markers for the occurrence of specific events, i.e. triggers. It is easiest by far if an amplifier can record triggers using a **parallel port input**, as discussed in Chapters 3 and 4. Take care also to find an EEG amplifier with a good sample rate, as this determines the temporal precision that can be achieved (p. 100): a sample rate of 1024 Hz means that data are recorded about every (1/1024)th of a second, and is a reasonable minimum to look for.

- What about consumer products? Consumer products refers to 'normal' consumers, or civilians, as EEG researchers might say, if they were working in the army. The differences with research products are cost – consumer EEG is in the hundreds rather than tens of thousands of pounds (or dollars / euros) – and background knowledge – consumer EEG seems to assume the customer has bigger dreams than brains. That is, such products generally use very few channels of low-quality electrodes that are awfully positioned, record with low sample rate, precision and sensitivity. While they are cheap and easy to use, any of these gains are offset by the waste of time involved in getting good data out of them. In other words, my recommendation for the lab is **just no**. Of course, if the goal is to buy a fun toy for the lab, then by all means, and I recommend the *Necomimi Brainwave Cat Ears*.

Figure 5.2   Nothing spells serious EEG researcher better than their wearing Necomimi Brainwave Cat Ears

EEG companies and cost

How big a wallet do you need if you want more than Necomimi Brainwave Cat Ears then? To find out, I took a sample of some of the biggest companies out there and got them to send me quotes. Now, to make it comparable, I asked all of them what it would cost to acquire a *Starter Pack* kind of system with amplifier, EEG caps, around 64 electrodes, and basic acquisition software, to be delivered in Finland and not accounting for VAT. I did mention to all of them I was writing a book about EEG, but somehow I seemed to have given the impression that I might be interested in buying such a system. The result of this little bit of investigative journalism is that I now have a pack of hungry medical technology company representatives hounding me to this day. I hope the effort was worth it, though, if only so you know just in how much trouble you are if you break things in the lab.

Company	Device
Biopac	Biopac (www.biopac.com) is famous in the field for providing biosignal equipment with a particular focus on what we in EEG call 'peripheral' physiology – EDA, EMG, ECG, respiration, and so on. Accordingly, they are well represented in labs that have a somewhat more social (e.g. research on shame and guilt) or clinical (e.g. studies on anxiety) focus. However, if there already is such a lab, and you want to keep using Biopac equipment, then you might be interested to know that they do also provide EEG options. Their quote involved a 32 active EEG electrode 'Mobita' system with two caps, wireless transmission to the PC, and acquisition system.
Brain-Products	Brain Products (www.brainproducts.com) produces the actiCHamp plus amplifier enabling one to use both active and passive electrodes in a configuration that can be easily adapted to increase channel counts from 32 (which keeps it in the price range of the other systems in this table) to 160. It also allows impedance measurements – which is often difficult for active electrode systems – and even sponge net configurations (see EGI below). They also sell ultramobile solutions like the CGX systems Quick-30 EEG system, which is a 30-channel dry headset that looks like a tentacular biking helmet with amp, electrodes, and recording integrated into its design – no cables at all. As mentioned, this has its own problems if you need temporal accuracy, so you would need a wireless time marker system to follow the experiment in this book, after which it sits neatly in the range of the other setups in this table.

(Continued)

(Continued)

Company	Device
BioSemi	BioSemi (www.biosemi.com) produces the popular ActiveTwo EEG amplifier, featuring the first *active electrode* system available, of which they tell me it is the only commercially available system that uses active electrodes, battery powered front-end, and fibre optic data transfer from front-end to acquisition computer. They furthermore underline that the combination of these three features is essential for the best possible prevention and suppression of interference and optimal subject safety. At this price range, you get a very decent 64 system setup, which can be extended as each box can fits 8 cables of 32 electrodes. It can communicate with the PC through its USB receiver, which is also able to receive trigger input along the lines of a parallel port input – although this requires some additional engineering (ask the supplier) – or the USB trigger interface (a USB to serial convertor). Personally, I have found it a smooth experience, but I'm a bit biased: I learnt EEG on one, so it will always remain as precious to me as the VW Beetle of my driving school days.
EGI / Philips	EGI's (magstimegi.com) geodesign nets are remarkable in the sense that they don't use caps. Instead, the nets are head-shaped, very easy to put on, and feature electrodes, wiring, and wet solution all in one. The electrodes are placed behind little sponges, which suck in the salty water and press against the skin during the experiment. The upshot is that this makes setup quick and avoids the horror of the remnants of gel in the hair, but my experience is that the signal degrades more quickly than normal as water evaporates. Trigger synchronisation does not work with basic parallel port but uses custom National Instruments gear. The indicated price includes the EEG recording Mac computer and EGI's data acquisition software (Netstation), but not the ERP package, which adds another $11,000 for E-Prime 2 licences, stimulus computer, validated experiments, response boxes, and more.
G-Tec	G-Tec (www.gtec.at) is an Austrian company that is famous for marketing the first commercial BCI system. However, they also cater towards more Psychology-oriented people such as myself, and suggest their g.Hlamp-Research 80 channel set (with 64 active EEG electrodes) for ERP research. Their quote featured numerous items, without me having to push for details, but make sure you request a trigger cable to enable compatibility with E-Prime or OpenSesame.
Neuroscan	Neuroscan (compumedicsneuroscan.com/) is one of the largest companies in providing EEG hardware and software technologies for research and clinical applications. Their representative suggested a SynampsRT EEG system, featuring 64 channels, recording PC, active or passive (cheaper) electrodes, and parallel port input possibilities.
Emotiv	Emotiv Epoc+ has 14 channels, it does not allow for an easy or precise way to send triggers, and seems better at recording muscle activity than EEG. Still, I have seen it measure alpha, and it has many more channels than the competition (even compared to Emotiv's newer products, which seem designed for sci-fi *looks* as opposed to usefulness). Essential reading on getting reasonable data from this headset are described by Debener et al. (2012), if you really must take your EEG out for a walk. A special 'flex kit' version is now also available, which at more than double the price mentioned below increases channel count to up to 32 with actual silver-chloride electrodes that are positioned onto a more traditional cap (note: requires 'pro-license' for access to raw data at a monthly cost).

After pondering the question of how to make the information worthwhile but not litigable, I cowardly decided to summarise the prices for you rather than dropping the actual quotes at your feet. Prices do change, of course, by country of import (here: Finland), supplier, and time (quotes were obtained between 2019 and 2020). Consequently, the table above in general tells you what you might be able to get if you had a budget of €30,000–40,000 in 2019 moneys, except with the last product, the Emotiv, which was at the time of writing about $849 for the research edition.

Laying out the EEG lab

Best Christmas morning ever: Santa stuffed an entire EEG starter kit down your lab's chimney! You will naturally want to play the rest of the day with your new toys, but where do you put everything? In the past, I have asked HR this, but was told this was not allowed. So I decided to do it *Breaking Bad* style, basically squatting in one of the special communal spaces that was commonly ignored by the community even though it was still functional in the HR books. This worked. However, if you find yourself in a less ghetto-like position, for example if you were to be handed a bag of money and told to design your dream EEG lab, then you should consider the following building blocks.

Provided you have the possibility, an EEG lab should have access to **basic washing facilities**. Usually that means putting it not too far away from the closest bathroom, although posh universities have a sink inside the lab. This is because you will need to clean EEG electrodes after every participant as they collect what people in the know call gunk and goo (dirt from the participant's scalp and left-over gel), not to mention bacteria that can travel from person to person if you don't wash. I will discuss basic gunk, goo and hygiene in Chapter 6, but for now take it as a given that it's good to have a sink. Ideally, the sink can have a nearby space for electrodes to dry, and a cupboard for electrode caps, hats, cleaning detergents, towels, and the EEG toothbrush (see Chapter 6). While on the topic of these more basic facilities, it is also worthwhile to make sure the lab features good climate control. It is best to keep temperatures relatively cool, so as to avoid participant perspiration (which affects conductivity), and electrode gel drying. Finally, a word to UK residents: avoid carpets in your lab – or in any other wet space! I dearly hope this is common sense anywhere else.

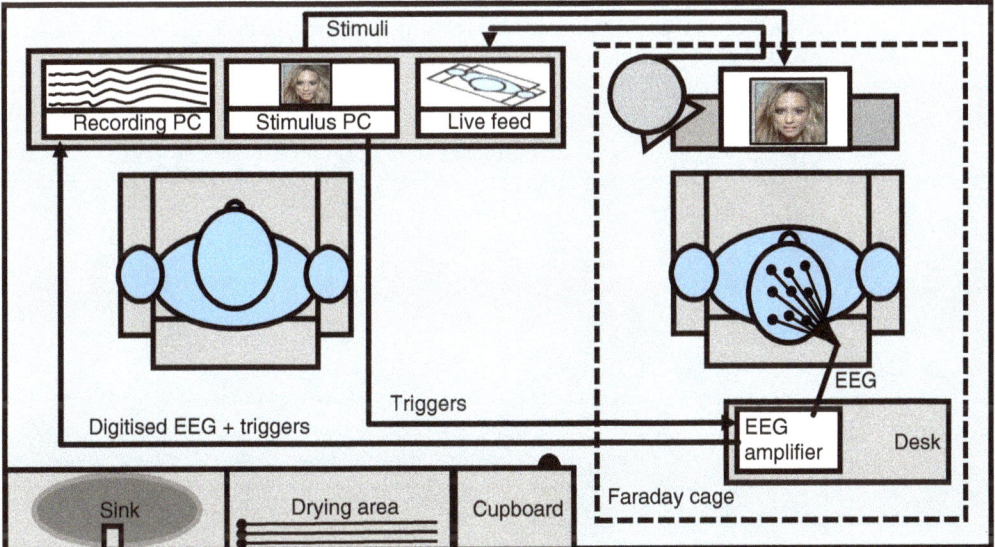

Figure 5.3 The EEG lab layout

Most labs will feature two spaces: one for the experimenter (or lab assistant), and one for the participant. Ideally, the participant (on the right in Figure 5.3) is seated within a large metal bunker, the **Faraday cage**. This is a mesh or continuous covering of conductive material, which creates a shield against electromagnetic interference from outside the cage. For the past few weeks, I've been breaking my head trying to create a great and really simple explanation of just why this works, but without going into physics, I will just stick with the following: It's basically what stops your microwave oven from frying you. Similarly, then, most of the electric interference from surrounding apparatus, the 'mains hum' or 'line noise' which you normally pick up in your EEG recordings (see Chapter 8), is avoided. Despite these significant benefits, a Faraday cage is, in the end, a massive metal bunker that is not cheap to acquire or install. Nor is it convenient, given that it means that all the wiring has to go through a few points (see Figure 5.3). It also makes it hard to see or hear the participant, so if you go down this route, then you should also consider installing a camera/microphone live-feed system so the experimenter knows if the participant has a seizure, or, more likely, attempts to call their friend about how boring the experiment is. Many researchers therefore don't bother with the Faraday cage and just assign half the lab-space as the participant's side of things, perhaps separated by a thin board so your participant will think you are busy doing science rather than looking at listicles of best cat photobombs.

Besides the camera, the Faraday cage should contain the participant on a good chair, looking at the screen while their EEG is recorded using an amplifier located nearby, preferably behind them. A good chair can take a little bit of thought. Make it too comfortable and people fall asleep, but if it is too uncomfortable it elicits movements. The chair should *not* have wheels (again, this invites movements). It should have no, or very limited amounts of, head-rest, as subjects may inadvertently bump the back of their head against it, causing the EEG equivalent of someone tapping a microphone with their hand just as you're trying to record the sound of an ant tapdancing. Arm- and leg-rests are useful, however, as they may help to avoid extraneous movements: some participants tap their feet during tasks, innocently assuming these are task-irrelevant (Q: which part of the body controls feet tapping?). If you look at the overall picture, then, it becomes clear just why discarded **dentist chairs** are coveted pieces of furniture by all right-thinking EEG researchers.

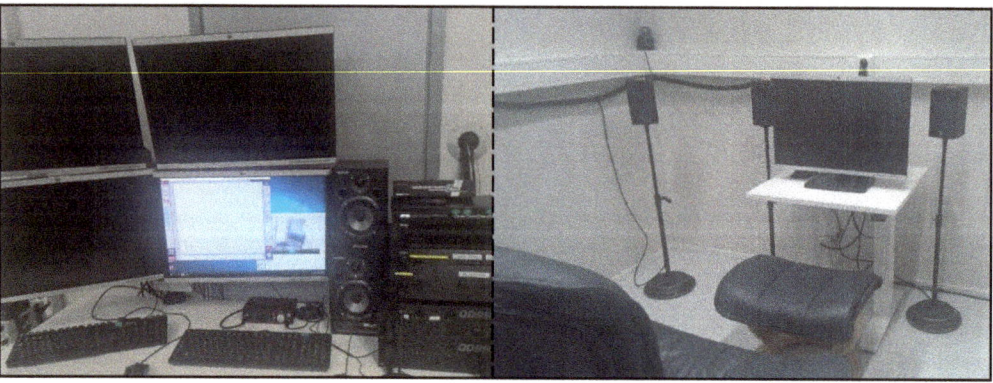

Figure 5.4 Experimenter (left) and participant (right) areas

In a configuration such as that shown in Figure 5.4, the participant's screen is directed from the **stimulus PC** in the experimenter area, which can be cloned (duplicated) towards one of the experimenter's screen so they know what the participant is looking at. Of course, in a configuration without a Faraday cage, one can simply look over the participant's shoulder. The participant's screen should ideally be of high quality, with a good contrast, high refresh rate, support high screen resolution, and be large enough to be comfortable to see even from a metre away. Traditionally, EEG researchers have stuck with CRT screens (the massive, non-flat screens of ages past) as they have long remained the weapon of choice when it comes to high refresh rates (basic flat screen monitors tend to run at only 60 Hz) and swift changes from white to black or vice versa.

Why are slow responses and low refresh rates problematic?

Imagine you want to accurately time a P1 EEG potential observed 80 ms after the onset of a flash (Mangun, 1995). If a 'flash of light' means luminance increases gradually over 20 milliseconds, we cannot say at what time the retina was sufficiently hit to cause a neural chain reaction: Should you report this as 80 ms after onset, 60 ms after maximal intensity, or somewhere in between? While LCD screens have gradually improved over the years, cheap consumer screens can still suffer from these problems, so it pays to know what exactly you are buying if you are planning to set up a lab. In this case, see Matsumoto et al. (2014), who compare performance between various types of screen in terms of timing and electrophysiological response (on the retina).

The stimulus PC is by far the most important computer in the EEG lab as it controls the all-important presentation timing. It should be fast in common ways: having a good video card (look for gamer reviews), professional audio interface (if you plan to do audio experiments), and good SSD hard disk (for fast loading of images from the disk). Furthermore, it should have a parallel port ('old printer port', or more technically, a DB-25 port) – which is by now starting to become challenging – as this is still the easiest way to communicate with EEG amplifiers. Finally, it should be optimised for timing performance, which means that it should have as few as possible unnecessary programs running in the background, doing nothing related to the experiment. This can be difficult, as students may sometimes install random software (for some reason I've found messaging software on a fairly large number). Usually, however, they are unable to do this since IT departments tend to stop people from tampering for security reasons. That too comes with its own problem, though, as they will also very much insist that you install virus scanners. Virus scanners should never be installed on lab PCs as they will try to work out whether files that are being loaded are potentially malicious, which obviously impacts the timing at which files are loaded and executed. If your local IT department insists you need a virus scanner, then try to convince them that the system will be completely disconnected from both your Institute's network and the internet (this is generally a good idea: having no internet improves timing and avoids participants checking social media). Failing that, I would spend £200 on a second-hand PC and install a clean operating system on that – a good, clean system with known slowness is much preferred to a fancy, rich 'ecosystem' of who-knows-what that randomly starts important system updates in the middle of an experiment.

Now let us consider the **recording PC**. While the stimulus PC presents the participant with *stuff to do*, and the EEG amplifier with the notifications of what exactly the participant is doing (i.e. the triggers), the EEG amplifier records the participant's scalp voltages (i.e. the brain waves) at a very fast pace. Instead of streaming this information to the PC as it comes in,

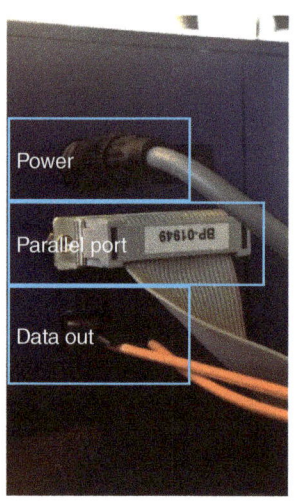

an EEG system tends to involve a buffer, a chunk of data of a relatively short duration, which is then also aligned and time-stamped with incoming triggers. This buffer is then read out by software running on the recording PC, which stores the incoming data at a set **sample rate**, defined as the amount of data-points per second. I will discuss typical data acquisition software used in the EEG lab later in this chapter. For now, however, I should mention that the above-described process means that the timing accuracy of the recording PC is not nearly as important as that of the stimulus PC, since the former only needs to keep up with reading the incoming EEG data before more buffer is created than is read (**buffer overflow**). This is fairly rare, unless the PC enters sleep mode (which I have seen happening!) – likely a condition it contracted from the lab assistant. Critically, however, the recording PC should have a lot of space on the hard-disk, which means that students need to clean up their recorded data after they are done. I would invest in big hard disks rather than expect the latter.

Figure 5.5

Once you have the layout set up as shown in Figure 5.3, with or without Faraday cage, make sure that the following are in place: (1) the stimulus PC is connected to the participant's screen and response devices (mouse, keyboard); (2) the stimulus PC is connected to the amplifier for sending triggers, typically using its parallel port cable; (3) the amplifier should furthermore be connected to the recording PC via USB, optical cables (Figure 5.5), or other. It may also have a separate power supply (Figure 5.5) or use batteries (as with Biosemi amplifiers). We conclude the tour of the EEG lab by switching the amplifier on.

EEG SOFTWARE: GETTING READY FOR RECORDING DATA

Now you have turned on your EEG amplifier, the question arises how you might see if it all works. Initially, you might think the best way to go about doing this is by sacrificing yourself. Indeed, I tried this multiple times either setting up the EEG on my own head with a mirror (not recommended!) or asking a colleague to help out. The problem, however, is that as soon as you need to move from the participant's side of the room to the recording side, you'll find that there was a point of having a mobile EEG after all: you're literally stuck. That means you won't be able to do what you set out to do, nor get any hardware out from the other side of the room, nor get more coffee to get through this wasted day. It is therefore best to call up a friend or colleague and ask them to sit there and basically do nothing while you tweak hardware and software until everything is satisfactory. Sure, it's not a very rewarding job, but at least they'll get some coffee.

What you need is a way to **monitor** that the recording PC is picking up reasonable stuff, so that once you start **recording**, you can be safe in the knowledge that it's going to be fine: the EEG data and triggers are recorded in good order. EEG amplifiers typically come with software that show the data to the experimenter. In this section, I will discuss how to inspect the data and make sure everything is OK given that you do not have a subject in there. In the next chapter, I will discuss monitoring once it involves actually having subjects there: making sure you have good EEG signals.

Setting up software

As discussed, EEG amplifiers generally keep an amount of EEG and triggers within a short duration buffer, which is then retrieved using the recording PC, and saved to disk. This is the most essential aspect of EEG recording software, as data processing (such as concerns almost the entire next half of this book) can generally be performed at any later point. Of course, if you want to use EEG for creating a brain computer interface (see p. 284), time is much more critical: If you want to spell BRAIN by concentrating on flashing letters, you probably don't want to come back the next day to identify the first letter in the word. Apart from that, the most useful functionality of EEG recording software is that which allows us to look at the quality of the data. That means that even though I will discuss BrainVision Recorder (default with BrainProducts systems) and ActiView (default with Biosemi systems), you should be able to extrapolate to different EEG recording programs, such as NetStation (with Philips/EGI) and ACQKnowledge (with Biopac).

The first step to set up your EEG recording software involves defining the location of the electrodes. That is, all EEG electrodes you will be using are the same, and interchangeable, so there is nothing in the electrode that differentiates a frontal electrode from a parietal one (see Figure 5.6). Therefore, you will need to make sure that the software on your system 'knows' where the EEG comes from. The simplest system is to use an elastic EEG cap in which electrodes are placed within holders (see p. 117, Figure 6.2). An elastic cap will make sure that, as long as (1) the cap goes from inion to nasion and from left tragus to right tragus, (2) isn't rotated, and (3) has the Cz right on top of the head in the midpoints of the measurements of 1, then everything should align correctly. An electrode over C3 for your cap's electrode 11 (in the 23 electrodes plus ground setup below), should then be similar to my electrode 29 in a 64-electrode setup. So, if you have no cap at all yet and just new caps, number them similarly to the system I show in Figure 5.6 (in 'rows' from anterior to posterior and left to right). Of course, if your caps are already named, then we'll just have to deal with a pre-defined system. In every case, however, you need to find out from your cap's electrode holders how each number (Figure 5.6 left) corresponds to each of the 10/10 system (Figure 5.6 right). To do this, make a note and find out with your measuring tape. Location 1 in Figure 5.6, for example, might refer to Fp1 and 9, to me, looks like FT6. Likely enough, however, your cap's manufacturer will be able to tell you.

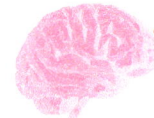

Electrode positioning and the 10/10 system

Amplifiers and electrode caps may permit many configurations, but numbered electrodes are conventionally standardised using the 10/20 system, later revised as the 10/10 system. This internationally used system uses four landmark positions on the head to identify positions on the scalp from left to right and front (anterior) to back (posterior). Use a roll-up measuring tape and put one end to the bit of cartilage (the **tragus**) just anterior to the left ear's passage, or the **tragus** as shown in Figure 5.6, and measure the distance between this one and the right ear's one. Take the middle and let this be 0% (zero, or **z**). Now, starting from there and moving 20% to the right over the tape, we get z, 2, 4, 6, 8, and 10 (100%). The convention in EEG is that these **even electrode numbers are on the right side**. The higher the number, the more lateral, or here, the more to the right. Applying this convention then to the left, we have the same relative positions, but odd numbers are taken, we call them z, 1, 3, 5, 7, and 9 (100%). Thus, the electrodes P7 and P8 are in mirrored positions, P7 is just far left, P8 far right. Now take the measuring tape and put it instead from the part where the bridge of the nose turns into the forehead (often there is a small groove here), the **nasion**, and measure the distance from here to the **inion**, which is the bump just above the neck on the back of the head (fun fact: according to phrenology, this is where your *philoprogenitiveness* is located). Again, there are 11 points from 0–100% and back to front, but now they are described using letters: Inion, Occipital, Parietal-Occipital (between O and P), Parietal, between C and P, Central, between F and C, Frontal, between Anterior-Frontal, and Pre-Frontal. However, given all these names describing brain-lobes, further to the left and right are not actually over the central lobe anymore, but rather placed over the temporal lobe (T). Thus, any C higher than 6 becomes a T: FC7+, C7+, and CP7+, are called FT7+, T7+, and TP7+. The whole system takes some getting used to, so print a diagram such as Figure 5.6 and hang it on the wall.

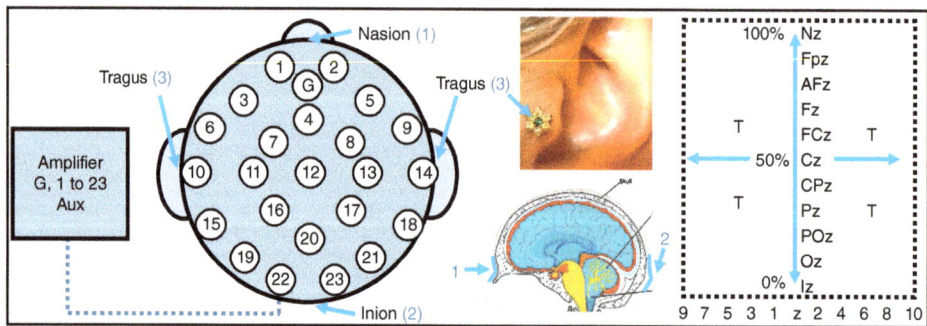

Figure 5.6

Using the recording PC, start up your recording software. These tend to be very similar, but normally they allow you to edit what kind of setup you have: Do you have 24, 32, or 256 electrodes? What kind of **auxiliary signals** (EMG, EDA, eye-movements) are you recording? Usually, your EEG system will provide you with a particular default configuration that goes along with your EEG cap. If it has 64 holders, the locations on the cap are known so chances are that a default configuration can be loaded up. I will explain what to do with ActiView and BrainVision Recorder, but you should be able to follow the same steps if you use another program.

In Biosemi's ActiView, this is under the About or TCP server tab, with a big, maroon button saying load config file. By default, there are .cfg files in the ActiView directory, named for the specific configuration you have in the lab. Load the one you have, then press the Start monitoring button of the monopolar display tab (see Figure 5.7), which should update the labels for EEG to something we can understand. If the EEG amplifier is on and everything works, things should look a bit like they are with the left screenshot in Figure 5.7 – except for the lack of EEG signals, given that you have no subject yet. However, you can still see the triggers coming in (to the bottom) if the experiment is running, which is what we will be looking at later in essential exercise 2. Furthermore, you can check the status of the battery, and adjust the sample rate (stop monitoring first). ActiView uses a *decimation* algorithm, which means that rather than storing all datapoints polled from the amplifier, it keeps 1 out of n points. For example, if decimation is 1/4, and the default sample rate is 2048, then ActiView is storing 1 sample per 4, out of 2048 per second, which means it stores data at an actual sample rate of 512 Hz. Of course you lose data this way, but there is very little EEG

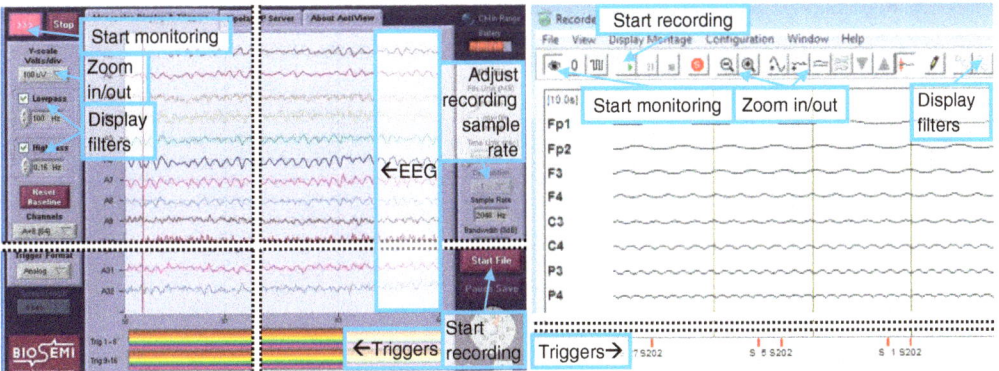

Figure 5.7 Principle monitoring view of ActiView (left) and Vision Recorder (right). In ActiView (left), triggers show up as lovely rainbow stripes showing which of the 16 bits are on and off, which can be changed in the Trigger Format menu, to the left of this panel, for convenience. In Vision Recorder, labels show up as numbers with a red marker that indicates the onset. By looking back to how we implemented triggers (in E-Prime, p. 54; in OpenSesame, p. 78), remember that triggers denoted the onset of pictures, where 201 were relevant pictures, 202 were irrelevant ones, and 203 odd ones, while the trigger just before each image was the image file. So Vision Recorder saw the onset of 3 irrelevant pictures (successively numbered 27, 5, and 1).

information above 100 Hz anyway. On the other hand, trigger information will also be stored at the actual sample rate, so to keep the timing accuracy at millisecond level, my advice is to store data at close to 1000 Hz (1 sample per millisecond); so with a default sample rate of 2048, you could enter a decimation of 1/2.

In BrainProducts Vision Recorder, there are a few more options. Start Vision Recorder and, if this is the first time you run it, go to Configuration > Select Amplifier, and select your amplifier. Configurations that are specific to your setup, independent from other users of the lab (they will presumably all use the same amplifier!), are defined as part of a 'workspace'. Go to File > New Workspace, change the data files so that the default saving directory is some place you will later be able to find, and press Next. The amplifier settings screen will show up. Assuming that your amplifier is connected and turned on, then pressing Scan for Amplifiers should populate the dialogue with channel numbers and names. Here, make sure that each numbered channel is connected to each physical channel number, and make sure that the names properly identify their locations, as explained in Figure 5.6 (p. 102). Furthermore, set the sampling rate to a value near 1000, as I have just explained in more detail with the *decimation* setting within ActiView (p. 103). The same principle applies, but without the simple calculus. Pressing next will move you to the menu with the data filters. I will further discuss filters when we actually have some incoming data (Chapter 7), but for now it suffices to say that they can 'clean up' the data. **Low-cut** (high-pass) filters remove drift (so that the signals don't go so far as to run off the screen), **high-cut** filters smooth out the fuzzy sort of noise that makes your EEG look like a startled cat's tail (what might be called 'hairiness'). **Bandpass filters** combine low-cut and high-cut, *passing* only a select band, while **notch filters** do the reverse, *cutting* only a select band.

Back to the data filters of Vision Recorder, the first tab, Raw Data Saving Filters, is mainly important (see blue box below) as anything you adjust there will affect what is saved, so if you filter out something, you will never regain it. For this reason, I advise caution and only set a low-cut filter of about 0.01 Hz to remove very slow drift (occurring over c. 1/0.01 Hz = 100s) from your incoming data signal, so they will not go above the minimum floor or maximum ceiling level. Some labs set a high-cut filter to remove noise due to aliasing (resulting from sampling a continuous signal, sometimes set to 1/3th or 1/4th sample rate; however, specific amps might do this already internally) or high frequency noise (e.g. above 80 Hz), or a notch filter to remove mains hum. However, unless specifically recommended by engineers, I would avoid these operations as you can always apply them after recording. But what if you want to look at your data without having to stare at ugly line noise and such? This is where **Display filters** come in. It uses a very similar menu, but does not affect the data that are stored; it just looks better while you are monitoring it. This is useful if you want to check the data, see if you can produce alpha, or show it to the subject, and you do not have to worry about filtering. It is also a good exercise to adjust the display filters and look at what happens to your incoming data. As a starting point, you might set a low-cut of 0.2 Hz, a notch filter at line frequency

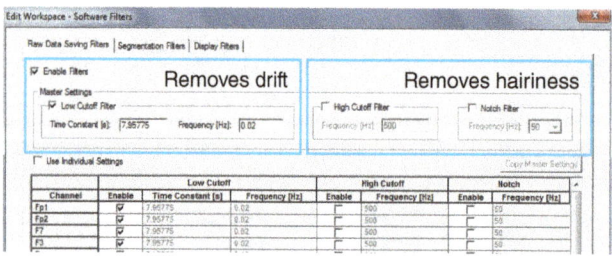

Figure 5.8

(50 Hz in the UK/European Union, 60 Hz in the US), and a high-cut filter of 80 Hz. Once you are done with the filters, close the dialogue and start monitoring (see Figure 5.8).

The dangers of raw data saving filters

I started writing this book because I had the impression that anything that could go wrong with EEG, I had already done wrongly, so perhaps I could act as a public service announcement. And then I managed to mistake the Low cut (Hz) in raw data saving for the Time constant (s). Specifically, I inserted a very conservative measure to counter drift (remove all data below 0.01 Hz) as a time constant. That sounds innocent enough, especially if (like me, evidently!), you find it hard to keep everything clear when it comes to double negatives. Low-cut filtering means we want to remove data that are slower than 0.01 Hz, which is equivalent to saying we want to *keep* data that are higher than 0.01 Hz (high-pass), so anything that is *faster than about 1/0.01=100 seconds* is kept, which is pretty much everything, but the very, very slow changes due to sweatiness or such are removed. The 100 seconds is related to the time constant[2]. Embarrassingly, however, I filled in that I want to keep a *time constant of 0.1 s*, which meant any waveform longer than 100 ms was removed. This means that pretty much all EEG (most of the P3 I planned to record!) was removed, and most of the noise was kept. Adding to the embarrassment, I blamed my seemingly solid experiment and kept tweaking its design and recording new participants, so it took many hours of work from me and 8 participants to notice the problem. They had virtually no ERPs because the accidental switch-up between frequency and time-constant meant most of the signal was filtered out by the ridiculously strong raw data saving filter.

Now that your recording PC is connected to your amplifier (or an error would appear when you tried to start monitoring), it's time to see if the stimulus PC is also working. Turn the stimulus PC on, make sure that all necessary experiment files, including images, are available on the local hard disk, and run the experiment. If you are with at least one other person, ask them to act like a participant and do the experiment, while you check the monitoring. If not,

---

[2]I oversimplify the relation to make the calculus more straightforward. The time constant refers to the time it takes for a signal to decay to 63.2% of its strength.

start the experiment to the point where the RSVP is shown (the rapid 'stream' of images), run back to the recording PC, and check the monitoring. The triggers should be clearly shown, for example as pretty rainbow flags in ActiView, or straight lines with labels in Vision Recorder (see the bottom of Figure 5.7, p. 103).

Once the triggers are shown in the recording software, you can be certain that everything is connected. Which means it's time to think about getting some real participants.

ESSENTIAL EXERCISES

Essential exercise 1: Set up recording software

Produce all necessary steps to set up your recording software to record all necessary data. Ideally, this should include at least 32 channels of EEG data, plus auxiliary channels from the eyes. Make sure that the locations are known in the software (if possible), and draw a diagram showing all used electrode numbers along with their 10/10 position names. This will make it easy, later on, to find out quickly whether blinks are shown in frontal areas.

Essential exercise 2: Dry run of the study

Test the experiment by setting the recording software to start recording, then starting the experiment on the stimulus PC and run the experiment on yourself. A **dry run** is when we run the EEG study without actually recording EEG data, but just to make sure everything is OK. Take a piece of paper and write at the top the present time as you start. Write down anything you note that can be improved. When you finish, make a note of the total amount of time the experiment took.

Additional exercise 1: Time the triggers

Once you get the EEG recording software to monitor and record, start the experiment and inspect the triggers. Is the onset of the pictures actually in accordance with the trial procedure as outlined in Chapter 2?

> RSVPs comprised 62 sequentially displayed stimuli, each with a duration of 600 ms [3], and a inter stimulus interval (ISI) of 100 ms.

Try to find out from the triggers as they appear in your recording software whether they conform to this description. If they do not – which is likely – tweak the E-Prime or OpenSesame files until they do.

Additional exercise 2: Present breaks

Edit the E-Prime or OpenSesame code to present a break in the middle of the experiment. This will require some additional study to find out how to (1) keep track of the current trial number; (2) check to see if the current trial number matches the midpoint of trials; and (3) present, when this happens, a custom screen showing a 'Please take a break' for at least 60 s.

6

LAB LIFE: EEG EQUIPMENT AND DEALING WITH PEOPLE

IN THIS CHAPTER, YOU WILL LEARN:

- How to handle the ethics of EEG
- Screening and recruitment procedure
- How to set up the equipment
- Why you need a toothbrush, and several other EEG utilities

If this wasn't already abundantly clear, I would confess that I am not really a people person. As a cognitive psychology student, I could simply wait until a subject arrived at the lab, tell them to sit in front of the PC, and have them follow the instructions as they were presented in E-Prime. If you see a circle, press left; star, right. Done? OK, bye. Not exactly a challenging social transaction. I get some reaction time data, the subject course credits, and everybody leaves happy in the lack of appreciation of the other's full, human existence.

EEG research requires one to not only deal with people by talking with them; establishing some kind of rapport is actually necessary. That is because EEG data are much more costly than survey data or data from behavioural experiments. They require a significant amount of time to obtain, and to analyse, so if you find out long after the fact that the participant had been pressing right instead of left for circles, say, you will quickly start to see stars yourself. It is therefore more convenient if participants explain when they don't understand something and, I found, they will not do so if they do not trust you. Also, if they feel uncomfortable during the process of setting up – which involves needles – then, ideally, they tell you this during the study rather than make official complaints to the board long after the fact. Likewise, by getting participants on your side, they might follow your exaggerated, minute instructions on staying still and concentrating rather than pursue some extra-curricular but surely innocuous form of activity like leg-swinging, gum-chewing, mind-wandering, and social media checking. Strange as it may seem, all these activities, even commenting on social media, do *involve the brain* and likely mean the data will be useless.

So, it's not like you need to *be* a people person, but for your own sake, try to pretend for the duration of this chapter.

PREPARING FOR PARTICIPANTS

In dealing with real participants, I normally take the following approach. First, I ask the person who wants to learn EEG to be the subject and observe and take notes as I set up. This allows them to get some inside experience, allowing them to empathise with being a subject. It might also teach them how to set up, but it is usually too much to remember. They find out this with the next participant, who is someone they recruit, and who is someone they are comfortable with (e.g. their partner), and I help setting up. Essentially, this is when the EEG trainee notices just how many steps there are and how many are already forgotten, so a good EEG trainer patiently explains the same things as with the first subject, but makes the EEG trainee do as much as possible on their own. Having a friend for this is very useful at this point, as they are more likely to forgive the sheer amount of boredom on their side, and chaos on yours. Think of it like this: it's a great way to learn just how forgiving your friend is! It's also good to have someone with you who is unrestrained when it comes to telling you when something is uncomfortable (which will be covered in the next section), instead of suing you. It is a pretty awkward experience, but by the end you should have a better grasp and maybe a stronger friendship, although I wouldn't vouch for this. I normally tell trainees to then get a third subject, during which I will be present but stay as silent as possible while the student will hopefully have figured it out by then.

Figure 6.1 My (now) wife on our first date, in the EEG lab back in 2008

But that is still in the future. Before you have subjects, you should think about ethics, who to recruit, how to recruit, and finally setting up the lab materials to ready yourself for the first subject. And of course: when do you get to wear that lab coat?

Ethics

I am assuming that you already know something about ethics; if not, please inform your local police department. But surely you know that any research study involving human participants normally requires a statement from your institution's ethical board approving the proposed study before you run the experiment. In some locations and departments, this might not be required for simple behavioural studies and social media surveys but, by and large, the trend seems to go in the direction that any study needs approval by an ethics committee. Also, when it comes to the division of studies into those that are ethically neutral (e.g. a linguistic corpus analysis of newspapers in the 1880s) and those that are ethically difficult (a randomised clinical trial studying the effects of lobotomy on driving behaviour in densely populated areas), studies involving actual brain measurements are more like the latter than the former.

So, consult the ethical board and ask for the forms. They vary widely across institutions, but typically discuss the principles of safety, transparency, voluntary participation, and data privacy. A good, informed consent form gives participants information on all these principles.

1. Safety

Safety, or freedom from harm, means that we're not going to inflict pain (unless you study pain), and that there are *minimal risks* involved in participation. Why minimal risks? You will need to think about and discuss the very minor risks that may occur, *and how you mitigate these risks*: What if your participant is allergic to your EEG gel (we use hypo-allergenic gels)? What if you use alcohol prep pads and you rub a little too hard and now the participant's skin is a bit red (it will usually be fine in an hour or so, but otherwise see a doctor)? What if your material contact causes bacterial or viral infections (whenever possible, we use single-use materials and disinfect)? Finally, what if your emotional stimuli are a bit shocking (we inform the participants and they are free to not participate – see principles 2 and 3). Most labs I know do not mitigate against all the risks I can think of, which is a problem. Another point is that it is probably a good idea to soothe your participants against groundless worries that may exist in the general population: the EEG equipment cannot apply shocks (EEG involves brains and electricity, but not electro-compulsive therapy) and the needle that is used will not break the skin (see p. 121).

Safety and pandemics

As I am writing this during a lockdown in 2020–2021, the topic of viral infection remains a strong concern. Several labs are still operational, but it is a challenging question that is put to every researcher whether the information gained from their

(Continued)

proposed EEG study is worth the risk to the participant and lab personnel. For more information and protocols that are designed to mitigate the risks, see Luck and Kappenman (2020), but in general it boils down to avoiding contact as much as possible, wearing masks, face-shields, and gloves, cleaning extensively, and making sure it is really worth it.

2. Transparency

Transparency usually involves (not) applying deception in psychology studies. Deception involves misleading participants or telling them lies: for example, telling your subject they will be giving a public presentation after 'these simple tasks' to evoke stress, while the simple tasks are of course the real experiments and the presentation never takes place. It does *not* mean that all possible information is essential and should be provided to avoid deception. Not telling participants what is the exact order of a Stroop task is not being nasty, it's just allowing the randomness to do its work. Similarly, not informing participants about your exact predictions isn't being opaque, it's how experiments (should) work. So, be transparent, but don't shoot yourself in the foot. In terms of ethics, spend time thinking about providing information on potentially harmful aspects (e.g. provide 'trigger warnings' for strong emotional stimuli), or data sensitivity issues, so participants may give an informed choice on their participation.

3. Voluntary participation

Voluntary participation means that people should not feel forced to participate in terms of pressure or due to some misinformation , and that they should feel free to end their participation at any point in time without fear of negative consequences. In terms of pressure, I have noticed that many institutions require students to do a minimum of participation in experiments 'in exchange for course credits'. Is that pressure? Alternatively, some give money, but if you are really in need of money, does offering participants money in return for helping count as pressure? It seems a bit much to me, but I know at least one institution that required the term 'inconvenience allowance' to avoid suggesting that participants were 'paid volunteers', which was felt to be an oxymoron. I think there is something to be said for that, but changing the terms will hardly affect the underlying reality! Finally, the part that goes 'free to end their participation at any point in time without fear of negative consequences' is one that is in pretty much every single psychology publication. This is an official part of the Declaration of Helsinki (which I also mention in most of my publications, since I live there), a set of 37 *Ethical principles for medical research involving human subjects*, declared by the World Medical Association which convened here, around the corner, back in 1964. In the current version (wma.net, 2013), principle 26 mentions it as follows: *The potential subject must be informed of the right to refuse to participate in the study or to withdraw consent to participate at any time without reprisal*. I would recommend reading the full set of guidelines, as they provide a much more comprehensive introduction to research ethics than I can currently provide, but in the meantime, inform your participants of their rights to withdraw, and don't yell at them if they want to get out (even if you'll probably want to).

What if they withdraw?

So, withdrawing is OK. If people do end their participation midway through an experiment, obviously you smile and say it's OK, even as you realise half a day just went down the drain, but do you still give them the full amount of money (or credits)? (I suggest yes, though it feels unfair from our side of things) What do you do with their half-finished data? (Ask them if they want to withdraw these as well). Notice that according to some organisations that your department may be part of, such as the British Psychological Society, people can also withdraw their participation *at any point afterwards*. This is nice in principle, but what do you do if you are so busy protecting the anonymity of your participants (see principle 4) that you haven't got a clue who is who? How do you delete the data of Bob Smith while protecting their sensitive data? One possibility is to give participants a key (such as the last four numbers of your telephone number), which codes the data, such that at a later point they can contact you and say they have thought about it and for some unknown reason they want to withdraw their data. That should be OK, provided you haven't already published an article about said data, in which case contact the ethics board to help you. But that means your data are now *pseudonymous*, as a code linking the personal data and sensitive data will exist, which then starts to involve legal obligations, such as the European Union's General Data Protection Regulation (GDPR) (European Council, 2016). All this needs to be considered for ethical applications, never mind the fact that to date, I've never had a single subject interested in withdrawing their data.

4. Sensitive data

Sensitive data in EEG is a difficult subject. People perceive EEG data as extremely sensitive – after all, it does measure the brain – but anyone with a bit of EEG knowledge knows it does not identify people well. Indeed, the same person having different measurements between two days, or two EEG systems, will have radically different patterns (but see p. 228). The ethics board won't necessarily know this, however, so be open about it and explain that you are collecting physiological data. If possible, anonymise the data (however, see the previous blue box) after the experiment is complete: the largest threat to sensitive data comes from it leaking, passing on to those who can cross-reference it with other data (e.g. your social media information). In that sense, I like to keep a minimum of personal data, and avoid recording names or birthdays, and other personally identifying data, such as handedness and neurological history. Of course, this principle can conflict with the need to screen for exactly such data, so for a publication, it may be necessary to keep a record describing merely the sample pool, and not specific datasets. Finally, notice that the need to be sensitive about data, and data-protection, can be difficult to balance out with the principle of **open data**. Open-access is popular, especially with funders (the EU,

UK Research Council, etc.), and does present ethical benefits over closing off your data solely for your own purposes. However, uploading to open archives such as Zenodo (free, CERN and EU-funded) means that anyone – including Mark Zuckerberg – can access and use your data, at which point you will need to be transparent to your participant about which risks exist, and which do not, and how you aim to avoid such risks.

So, as you may notice, research ethics are a minefield and, likely enough, you cannot make everyone happy. In my communication with ethics boards, I have found that the more reasonable ones understand this, and appreciate it when they see you have given proper thought about ethics, rather than write a minimum of information and say it's all good. Approach an ethics board like a group of real people and they may be reasonable and lenient; trying to browbeat them into just accepting this pretty safe study can quickly lead to a very tedious process. This is sound advice with any committee, really.

Screening

While the ethical application is underway, it is now a good time to think about participants. How many do we typically need? The proper answer is *as many as constitutes a representative sample that provides enough power*. Undoubtedly, you have found out that much of what we know about people is in reality based on samples of first-year psychology students. These tend to be white, female (at least in the last few decades), and middle-class, and so on, which leaves out a significant part of the world. But psychology isn't sociology, and fundamentally, people are pretty similar, at least in our nervous systems. The eyes of a man have rods and cones in their retinas just like a woman does, and not swords and stones, and so too do they have slower reaction times to GREEN than GREEN, and a P3 to relevant faces. And as such, normal samples back in the golden age of ERPs (1960s) had around six subjects of unknown gender identity, but I will hazard a guess they were male. These six needed to be naturally similar, as they were generally expected to show the evoked potential of interest based on their physiology, not due to their group identity, so variation that might possibly affect the brains was generally avoided. This meant that the six people were generally *right-handed* and *healthy*.

Right-handed, because as you know, left-handed people have their language centre on the right side, which is why they are so creative. Oops, it turns out that language isn't processed solely on the left, even in right-handers (Lindell, 2006; Neininger & Pulvermüller, 2003). Most left-handed individuals also have language left-hemisphere dominance (Carey & Johnstone, 2014), and the left-brained versus right-brained people is an infuriatingly stupid myth. Nevertheless, it's possible that the brain structure and connectivity of left-handed people is different from that of right-handed ones, and thus it remains common even today to report that only right-handed people were recruited (having done some EEG studies on motor control with left- and right-handed participants, I found that this is mainly due to left-handers being harder to find!; Serrien & Spapé, 2013), or, when you didn't think of it, for a reviewer to ask whether all participants were right-handed. In which case, it would be good to give an answer along the lines of '17 out of 20 participants self-reported being right-handed'. Either you know this because you literally asked people this, verbally or within the experiment, or because you had them fill out the default questionnaire for this, the Edinburgh Handedness Inventory (Oldfield, 1971), which is way cooler, #OldfieldNeverOldHat.

Healthy pertains to the suggestion that people with neurological issues may have structurally atypical brains, in contrast to what is now sometimes called 'neurotypical' ones. In the past, this was somewhat obvious: if I want to say something about memory using EEG methods, I probably want to avoid people with Alzheimer's disorder, and in general we want 'normal, healthy' brains in our research that concerns normal brain-functioning – unless the interest is in the brains of people with Alzheimer's, but then you want to avoid them having comorbid issues as well. However, while things were easier with a strict medical view of the world, the question of what neurologically healthy means has become a somewhat tendentious issue. Most psychological disorders are now seen as brain-related, while the behavioural manifestations of interest for psychological diagnosis are continually expanding (e.g. DSM-I had 145 pages, but DSM-V has 970). So, are people who have ADHD (Attention Deficit Hyperactivity Disorder) or ASD (Autism Spectrum Disorder) 'healthy'? I would say yes. But neurotypical? Maybe not.

How do we cope? My advice is to undertake screening during recruitment, so that people at any point in time can stop communicating without feeling they are being pushed into something they aren't comfortable with or pressured into providing sensitive information without consent. This can be easily implemented during a staged approach towards recruitment, which I will now discuss.

Recruitment

We all have different ways to recruit a representative sample: asking friends, flyering, using a course credit computer system, posting an ad online – all have their merits. However, by and large, EEG studies are uncomfortable, boring, and messy, so the competition can be fierce (paying more helps). Furthermore, the amount of preparation involved in an EEG study is such that a person not showing up is a significant waste of time, so you'll want to avoid this. I train my lab assistants and PhD students to use the following three-staged approach to recruitment.

In stage 1, we create flyers (hand out in cafeterias) that do not outright lie about the study, but do make it sound perhaps a bit more exciting than it is. A 'fun and easy experiment', which takes 'only two hours of your time' while 'paying you pretty well', and let you walk away with the warm glow of charity, 'for the sake of science', not to mention deep insight – 'see your brain activity!'. 'Neurologically healthy', interested people between the age of 18 and 40 may contact [email-address] for further information. A good short ad as a flyer or mailing list message will get more responses than a lengthy explanation of what the study is about.

In stage 2, send individual replies to recruits from stage 1 – copy–paste much of it from a template, but people tend to remain more interested if they don't think you are a robot, so at least address them individually. Send a quick explanation of the study (two sentences), and its purpose (but not specific hypotheses), and what is required of them (one sentence, like 'keep a person in mind while you look at pictures flashing by'). For transparency, you may also send the full instruction sheet (essential exercise 1) and informed consent form, but make sure it's not required reading and they know they don't have to do anything with such attachments – no one likes homework! Here, it's a good time to inform them just what you mean with neurologically healthy. I've previously used the phrase: 'Not using any psychopharmaceutical (e.g. lithium, Ritalin) or neurological (sodium valproate) medicine, nor having any diagnosis commonly associated with such medicines (bipolar disorder, ADHD, epilepsy). Sometimes aspiring recruits ask whether concussions that happened years ago are reasons not to participate – I tell them they are not, but in reality, we don't know. There are, however, conditions that we do know are problematic. For example, I once forgot to screen for colour blindness in a Stroop study, resulting in a curious lack of difference between GREEN and GREEN stimuli. Finally, you might want to screen for non-medical issues that just interfere with EEG (such as having amazing hair, as explained in the box below) or other biosignals (a big, bushy beard makes it difficult to place EMG electrodes near the muscles around the mouth). Thus, when you write to potential recruits that, at this stage, they can make an appointment to come in or they can stop communicating, the reason for their hesitance is hidden: they might have bipolar disorder (a sensitive issue), or a big, bushy beard (a hipster issue).

In stage 3, finally, you have negotiated a suitable date, and you hope they will come. Will they, though? I sometimes give them my phone number and suggest that if they want, they can give theirs also, in case I need to suddenly cancel. Of course, I won't, but I'm cleverly using 'unconscious priming' to increase the mental representation of the rule of common courtesy that if you can't make it, you let someone know.[1] In any case, it's not enough. At least one day before (unless the experiment is on Monday, then both on Friday and Sunday!), send a reminder. In this reminder, you don't want to sound like you're nagging, so you might as well give some information as a pretext: 'I forgot to say, if you happen to use make up (other than eye-liner), you might want to consider that we need to remove this in order to improve the contact between our electrodes and the skin. We use alcohol swipes for this, which is safe [etc.], but you might want to take a bit of time afterwards to "refresh yourself"'. Indeed, EEG experiments do not generally improve anyone's look. Let me explain the reason for this in the next section.

Hair and EEG

We commonly screen our participants for all sorts of neurological disorders – to make sure they either are, or are not, present depending on our interest – but *hair* is often unexpectedly important. If you know a little about EEG, you might expect bald participants to be ideal: after all, having no hair between the scalp and the electrodes should be great. The problem, however, is that a bit of insulation keeps electrodes from drying out (particularly if you use water-based solutions), so having thinning hair to the point of, but not quite being, bald is best. The general rule is indeed that the better your hair, the worse your EEG, which I tell participants if preparation takes long. Sometimes people have such amazing hair that you can forget about getting any signal – this can be the case with very thick or curly or afro textured hair. Another time I found myself in want of an explanation was when it turned out that even permanent hair-dye, if recently applied, presents a problem. That is, not to the participant, but how do you explain to the owner of a bespoke EEG lab that their precious EEG net of salt-water-and-apparently-hair-dye-soaking sponges are suddenly a not very festive shade of brownish blue?

---

[1]No, I know that this isn't an unconscious social prime, because for one it actually works. Also, with any luck, you now have a phone number to ring in case no one shows up.

Preparing lab materials

Given that EEG measures the electric potential from the scalp using electrodes, and given that air isn't very conductive, we normally try to minimise the amount of air between the skin and the electrode. Dry solutions (discussed on p. 93) use pins and such pressing on the head, which is faster and doesn't leave as much *gunk*, but these tend not to give as clean a signal as wet solutions. Wet solutions either use a salty-water plus sponge system, or an **electrolyte gel**. Whichever you use, try to have them conveniently placed and ready for use before the subject arrives. If you use an **elastic cap** and know the participant's head-size (unlikely but convenient) then you can already plug in the electrodes, a process that is made much easier if the lab has a Styrofoam **head mannikin** (Figure 6.2).

Then there are **eye-electrodes** to record the presence of eye-movements and blinks (see Chapter 8). These are often bipolar, so rather than recording the scalp potential along with the EEG, they record the difference in activity between the one electrode and the other. To capture horizontal eye-gaze movements, they are placed to optimally capture the eyeballs rolling left and right. They are therefore typically 1 cm to the left of the left eye's outer corner (where the upper and lower eyelids meet: the canthus), and 1 cm to the right of the right eye's outer corner, such that a dipole of the two eyes would be in the middle (see p. 6 explaining dipoles). Likewise, vertical eye-movements, but also eye-blinks, are measured with another pair of bipolar electrodes, commonly placed 1 cm above and below the eye. These positions should be close to the eye, but without hurting too much when taking them off. Below the eye is quite sensitive, soft skin, while on the eyebrow, you will appreciate that your participant might like to keep having eyebrows after the experiment. Which eye should you measure? Well, conveniently they tend to operate at the same time, so the choice doesn't particularly matter: take the eye that is nearest to the amplifier for convenience. Whatever choice you make, be consistent about it between participants.

Eye-electrodes, and other auxiliary electrodes that aren't placed on the cap (such as EMG electrodes), tend to work with a more DIY system reminiscent of the early days of EEG. As can be seen in Figure 6.3, they can look rather different and are often placed using stickers. In the images, (1) I placed a sticker on one of the electrodes in the pair, (2) filled it with conductive gel (similar, but not the same as for the normal EEG) and used the sticker-paper to wipe away the extraneous gel such that, as in (3), when I removed the sticker, a nice, flat area without air bubbles remains inside the hollow of the electrode. (1) and (2) can already be done before participants are in the room, speeding up the process (but don't remove the sticker, as the glue might dry up).

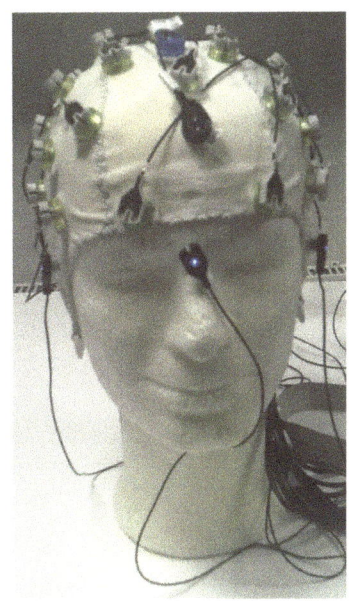

Figure 6.2

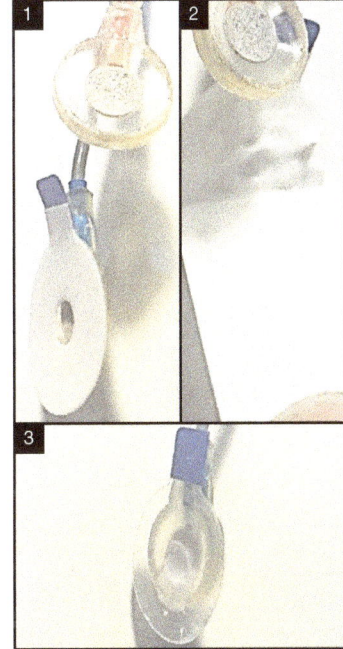

Figure 6.3

What you should have ready before your participants arrive are therefore the following:

- Recording software: already set to monitoring and ready to start recording

- Forms: instructions, consents, questionnaires

- EEG electrodes placed on (freshly washed) elastic cap

- Electrolyte gel or liquid

- Eye-electrodes, prepared, and have more stickers and gel on standby

- Surgical tape in case stickers do not hold

You should also have available, but their use will only become apparent later:

- Single-use hollow needle for applying EEG gel and scrubbing

- Cotton swabs, alcohol prep pads for skin cleaning

- Single use gloves and a lab coat, for hygiene and looking professional

- Snacks for participants during a break

- Water mister to avoid the 'familiar static charges'

- Toothbrush for cleaning electrodes and cap

- Measuring tape for measuring your participant's head and electrode location

- Two towels, one for drying electrodes, another for the participant's hair

RUNNING THE EXPERIMENT

Finally, your advertising, recruiting, and endless nagging have secured a willing participant. You bring them to the recording room, give them the bundle of paper to read, the informed consent to sign, answer their questions, and so on. The next step will require about 30 minutes for someone experienced with an average number of electrodes (32/64), but expect about an hour if you don't have much experience, or if the participant has amazing (i.e. difficult) hair. That means that you have to take a good hour into account during which you will become intimately familiar with the participant, in the sense that you'll be constantly touching their hair and skin. This is the part I meant in the introduction that is very different from the regular psychology of talking therapy, survey questions, and button presses. It is your job to make participants feel comfortable in your presence, because to some extent they hand their physiological responsibility over to you.

When I started, I found that a frightening prospect, being an introvert science nerd who suddenly has to pretend to be a hairdresser, hands in their hair and making casual chit-chat. One strategy is to a have a stock of anecdotes to entertain them with. Here's one.

Does looking and acting professional work?

On the one hand, participants should feel happy, and feel the researcher is approachable enough to ask questions like 'I don't know what to do' or 'this wire I'm sitting on, is that part of the experiment?'. Indeed, in the wise words of a colleague: 'Happy participants produce better data'. But there are some limits: Be too cheerful and nonchalant and participants may feel that *anything goes*, which may provoke inattention at best and behaviour that causes you to bin your data at worst. I had one participant start a phone call during an experiment. So, a professional act may convey the idea that while you are friendly and approachable, the work concerns *important, scientific research*. Maybe a lab coat will help establish this impression.

On the other hand, it is easy to over-shoot with looking professional. Famously, the colleague of the 'happy participant, good data' connection once welcomed a subject at the door, accompanied by the assistant (both of them sounding *foreign*), wearing white lab coats and holding, in their white latex gloves, needles dripping with mysterious green liquid ... the participant promptly fainted.

Another strategy is to get inspiration from hairdressers and use their questions: 'What are your majoring in?' and 'Do you have plans for holiday?'. If, however, you disliked answering these questions so much that you refused to ever see hairdressers again, there's another strategy: 'I'm sorry this is going to take about an hour, feel free to browse the internet/use your phone'. Even so, you will still need to act like a hairdresser in terms of working with people's scalp and hair, so let's have a look at that now.

Setting up

After a participant has signed consent and is comfortably seated in the lab, it's a good time to measure their head size. If you were a Victorian, you might have asked them in advance what their hat size is during the recruitment process, but nowadays you get to be your own tailor and manservant, so the process described previously (p. 102) will have to be performed by yourself. Take your soft measuring tape and measure a full circle horizontally around their head, from over the bump of their eyebrows to the inion on the back of the head (see electrode position system for inion in Figure 5.6). If you have well-labelled elastic EEG caps, this measurement will inform you which hat to take, but keep in mind that a slightly tight fit is better than a slightly loose fit, for example, I have elastic caps of 56, 58, and 60 (cm), but should I have a participant with a hat-size 57, I'd take the 56. While that is clearly against principles of good taste in fashion, (1) we're trying to achieve minimum distance between the skin and the electrodes; (2) nobody ever looks fashionable wearing EEG caps; and (3) it's not you who's wearing it.

Speaking of fashion and hairdressers, Grey Walter, of Chapter 1 fame, advises dampening the hair if it is 'too dry' in order to avoid 'the familiar static charges' (Walter, 1937). You can use the water mister for this, but don't overdo it.

Steps to start recording EEG

Step 1

Put the cap on the participant. Hold the cap from the inside, around your hands, front facing towards you. Ask the participant to face you and keep their head still as you swoop the cap on, stretching the front of the cap first over their forehead, and making sure there are no wrinkles. Help the participant's ears pop out through the holes to the sides. Check your positioning by using the landmarks explained in electrode positioning and finding out whether the cap isn't too far to the left or right, front or back, or rotated. Likely enough, one of the checks fails (often Cz is more anterior than people think). You can either do it again (if the check failed badly) or, in most cases, you'll manage by getting your participant to hold their head still, like a hairdresser would, and then reposition the cap by sliding it over their hair.

Use a measuring tape to check the cap positioning with three checks:

1. Check the length from inion to nasion (mine is 40 cm). Cz should be at half this length (20 cm).

2. Check the length from left tragus to right tragus (mine is again 40), again check if Cz is halfway.

3. Look from the front (into the participant's eyes) and from the top to see that the hat isn't rotated.

I have often wondered why, but many people *apologise* for their head being big. This is not true to begin with: the average participant has an average-sized head (except Dutch participants: our heads are obscenely big). It must be a mad hatter culture thing, generally designing hats to be too small and rather than admitting failure, they suggest customers should feel ashamed. Again, very similar to fashion in general.

Step 2

Clean the scalp directly under the electrode holders. Passive electrode systems in particular benefit from an initial cleaning session using cotton swabs dipped in a bit of mildly abrasive, conductive skin-gel that is especially designed for EEG (e.g. Nuprep). I was taught that even though brochures of active electrode systems say no extra cleaning or scrubbing is necessary, a better signal is obtained through further scrubbing of the skin using the blunt needle. My experience is that this is true, but you can avoid the procedure with more sensitive subjects (or, say, infants). Take a fresh disposable, blunt needle, pop it on top of the syringe, and stick it through the cap's hole until you think it touches the skin, then lightly scratch the scalp around where the electrode is positioned. Make sure you can actually feel the needle scrubbing the surface of the scalp and not merely rotate the syringe such that the needle itself is slowly digging a hole, which is not a pleasant experience for the subject.

Note: With step 2, it starts to become crucial that you have a reached a good rapport with the participant. Inform them every step of the way what you are going to do, why you need to do it, that no part of it should hurt, and to inform you immediately if something does anyway. Needles, in particular, are scary, so I always make sure I show the participants so that they see it is very blunt, that it won't pierce the skin, and give a demonstration on the top of their hand to show what the scrubbing should feel like. During step 3, I also ask them what it feels like, and if everything is well, they will say it is like a scrubbing and kind of pleasant. Importantly, if they say they don't feel anything, this is problematic as well, as you are likely failing to touch the skin.

Step 3

If you use a passive electrode system with ring-shaped electrodes, put them in their holders then fill the space between scalp and electrode with conductive gel; if you use active electrodes with a single touch-point, place the electrodes after the filling. Using the needle is a pretty scary experience if you're not used to holding needles, so let's take that slowly. Depending on the syringe and conductive gel, *suck it* or *squeeze it* into the syringe. The default mode is to stick the point of the syringe without the needle into the conductive material, then pull the plunger of the syringe in a single, smooth motion causing suction to fill the barrel, making sure you get as few air bubbles as possible. If that is somehow impossible, you can likely squeeze the gel directly from the tube through the back of the syringe, then pushing the plunger in, again watching out for air bubbles. Once it's filled, place the needle part onto the syringe, and push out a little bit of the gel, *as seen in the movies*, though mainly to get a feel for the amount of force required to get the gel out – often more than expected. Hold the syringe with your dominant hand and place two fingers of your other hand in a V shape around the electrode hole in the cap you are about to fill (preferably the ground electrode), pushing the cap flush against the scalp. Now softly press the blunt needle through the hole, making very sure you touch the skin (this operation can also be combined with the needle scrubbing explained in step 2). Then, push the plunger so the gel flows out of the syringe, while moving the needle just a tiny little bit up so as not to have the skin block the passage. When the gel is visibly coming out of the holder, you can let go with the two fingers and move on to the next electrode. Use the same operation on a few (say, three) other electrodes. *Once you have completed the ground(s) and a very small number of measured electrodes, continue to step 4.* Note: some systems replace the single ground with two electrodes: CMS (common mode sense) and DRL (driven right leg), for technical reasons.

Step 4

Check the data. Have a look at your monitoring: if there's only flatlining and yet your participant still breathes, there's probably something that went wrong in step 3, most likely in the ground. Most EEG amplifiers have some way of indicating whether the ground (or CMS + DRL in Biosemi amplifiers) is not properly connected via LED lights; for example, Biosemi systems show a blinking blue signal if there is no proper connection. If a signal does show, but it goes all over the place, then likely the electrodes are to blame: they are either broken (unlikely) or not touching the skin properly. EEG recording software usually comes with some diagnostic tools visualising *impedance* and *DC offset*, as explained below.

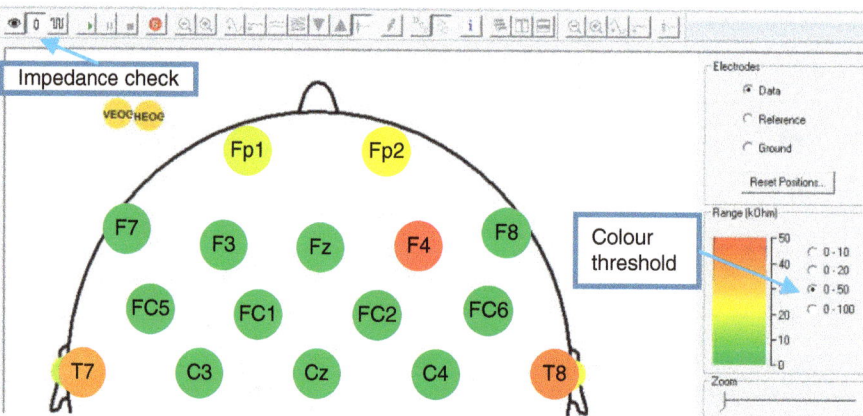

Figure 6.4   Vision Recorder Impedance check screen

Brain Vision Recorder (Figure 6.4) shows the impedance (the opposite of conductivity) of electrodes. Air in particular is not very conductive, so if the gel was placed just above the scalp and in the hair rather than on top of the skin, it will show a high impedance. Skin flakes, grime and other sorts of dirt on the scalp also increase impedance. In Vision Recorder, this is shown in pretty colours using its 'Impedance Check'. Use step 5 below to get each electrode to turn green, then move the colour threshold down from its default 0–100 kOhm range to something like 0–20. Some of the electrodes will jump again to orange and red. Perform the same action until the impedance of all electrodes is below 10. ActiView is not quite as informative, as active electrode systems cannot easily provide impedance measurements. Instead, the DC offset visualises the degree to which good skin contact is achieved. Values over 50 or under -50 µV may indicate an electrode should be checked or, if problems persist, replaced. Use step 5 below to get each electrode with higher than normal offset to go down.

5. Clean individual spots on the scalp. For any electrode you suspect could do with a better contact with the scalp, remove the electrode from its holder (if necessary). For example, in Figure 6.4, the F4 electrode is showing high impedance and therefore might have poor contact. Take a bit of abrasive cleaning gel on a cotton swab, making sure you clean the skin (and not the hair). Do a bit more scrubbing with the needle and fill it up again with gel. Check to see if this reduced impedance (see above), and the signal using the monitoring screen.

How do you spot bad data?

Some researchers out there report impressively low impedance values, but getting good data in is more than just impedance. The best way to get a good signal is by learning to spot it, and this comes with experience. A good exercise is to look at your participant and see if they are sitting still or looking around – chances are they are doing the latter. Ask if they can hold their head still for 10 seconds, and notice the large difference due to the absence of **movement artefacts**. Unless you specifically ask people to keep still, it is very hard to spot any issues with the signal. Next, check if any channel is drifting slowly up through the other recordings: **EEG drift** suggests a bad signal if it's just one line doing this, while if there are multiple going at the same time, then it's either the reference, or some sort of repetitive behaviour: is your subject swaying their legs? Then, see if there are any signals with **sudden jumps**, as this also indicates a bad signal. Again, if *all* electrodes do this in one go, it's likely to be the reference. Finally, ask your participant to frown or look very concentrated and notice how **high frequency noise** starts to show (mainly in frontal electrodes). That problem can be fixed by asking the participant to relax their face, but if you spot high frequency noise in single electrodes, it is likely due to bad electrode placement. Have a look at some of the figures of artefacts in Chapter 7 for a more visual guide: noisy and dead channels are shown in Figure 7.11, page 146, high frequency noise in Figure 8.8, page 175, and drift as well as effects of filtering in Figure 7.18, page 153.

Step 6

Repeat until either you can't improve the signal anymore, or until so much time has passed you run the danger of the gel starting to dry up. This depends a bit on the type of gel used, with water-based solutions being particularly prone to drying out, but in general it's good to keep it under an hour. Indeed, it can take pretty long, so if you don't want to bore your participant to death, you can strike up a conversation. Of course, if you're panicking about the quality of the signal, then that can be difficult, in which case maybe give them back their phone and let them fiddle with that. Once you're done, use some surgical tape to tie the bundle of EEG wires, with a little bit of extra space (so the participant can move their head) to their shoulders. This reduces the chance of wires accidentally getting tangled up and dropping off.

Recording other signals

Most normal EEG amplifiers will allow you to record many different signals in addition to EEG. EEG researchers will sometimes call these 'auxiliary' signals, because in the end they're not the meat of the matter (or the brain of the analysis), but additional channels in your amplifier can provide a wealth of information by collecting electro-oculography, electrocardiography, electromyography, electrodermal activity, EEG from an additional reference sites, and other signals.

   Electro-oculography (EOG) is often useful for the later analysis. As will be discussed in Chapter 8, EEG is hugely affected by eye-movements and blinks. In order to find out when this happens, we usually record EOG. Use the stickers as described earlier (p. 117) and make one pair of electrodes for horizontal eye-movements (lateral to each eye), and one for vertical eye-movements (above and below one eye). Connect the horizontal electrodes to an auxiliary channel on the amplifier you name HEOG in your recording software (for horizontal EOG) and vertical electrodes to VEOG.

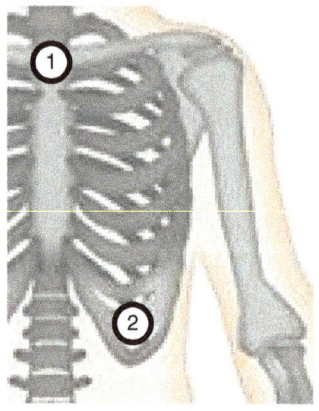

Figure 6.5 Placement of a pair of ECG electrodes over the manubrium (1) and over the second rib counting from the lowest one on left side (2)

   Electrocardiography (ECG) is for some reason often called by its German abbreviation EKG (elektrokardiografie), and uses two or more high-impedance electrodes, generally with the heart – the source dipole it is trying to detect – in the middle. I have good reason to believe you can use basic EEG electrodes as well (see posterior artefacts, p. 128), but single-use electrodes that are clipped on top of a reusable wire are preferred, because re-using electrodes that have been on the middle of the body feels a bit icky. To apply a minimal configuration of one pair of ECG electrodes (more is possible, and typical in clinical settings), use the stickers and place one on top of the flat, bony bit (the **manubrium** of the **sternum**) between and below the two collarbones (the ends of the **clavicles**). Give the other sticker to the participant and let them place it over the second rib on the left side of their chest, counting from the bottom (Figure 6.5).

   Electromyography (EMG) records muscle activity with electrodes placed c. 1 cm from one another. Common reasons are to record hand muscle activity prior to movement, or in response to TMS (transcranial magnetic stimulation), and to record face muscle activity in response to emotions or expressions (the default guidelines are by Fridlund & Cacioppo, 1986). Their application is similar to EOG electrode pairs, but require more scrubbing with alcohol to obtain a good signal, especially if your participant uses make-up.

Electrodermal activity (EDA) records skin conductivity with two electrodes placed on positions with high densities of eccrine sweat glands. The highest densities occur on the palms of the hands and soles of the feet (Shaffer et al., 2016). The feet are, however, somewhat inconvenient and their use feels invasive to participants, while the palm of the non-dominant hand – even though it generally responds strongest to stress and stimuli of psychological interest – can be affected by hand movements. Personally, then, I find finger positions most convenient, with electrodes placed at a similar distance on the middle phalanges of the index and middle finger (on the side of the palm, the **volar** surface) (Figure 6.6). This is convenient as one can just roll surgical tape around the finger and they won't fall off easily.

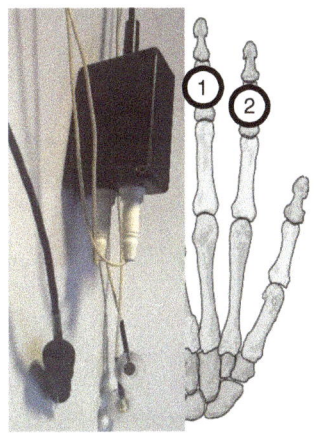

Figure 6.6 Placement of a pair of EDA electrodes over the intermediate phalanges of the non-dominant hand

More EEG can be recorded using the auxiliary inputs. A common strategy is to record, with some additional electrodes and a bit of surgical tape, EEG from **alternative references** other than those involved in the recording. For example, many studies have used the classic *mastoid* references, which is the flat, bony location behind the outer ears, but ear-lobe references and nose-tip references are common as well in some fields. In general, because EEG is always a difference in activity rather than a real activity, the idea used to be that by making the reference as little affected by the brain as possible, the EEG would be 'purer'. Thus, mastoid references, and so on, recorded EEG from what were termed 'silent' sites. While the premise of this (mastoid references pick up lateral activity despite the bone) has long been discounted, the literature with such references remains. To make one's results comparable with this literature, it may be useful to record EEG from these locations and use offline re-referencing, recreating signals 'as if' they were recorded with these references (see p. 147). With mastoid references, make sure to use alcohol swipes for cleaning as there is truth in the old-timer wisdom that people do not clean behind the ears.

Other signals can also be recorded. Apart from other electric signals wired straight into the auxiliary inputs (read the amplifier's manual and be careful!), you can also ask your local distributor about what other sensors they can provide that work for your device. I have worked with all sorts of sensors like accelerometers (for high accuracy movement measurements), respiratory belts (for measuring breathing strength), microphones, and light sensors. The last are particularly useful if you want to make absolutely certain that the triggers you send out are exactly synchronised with the onset of stimuli.

Recording data

With all EEG and auxiliary channels looking reasonably good to you, or as good as you can get it within the amount of time you have, it's good to do a last few checks before you hit record. If at all possible, it would be ideal if you can show the monitoring screen to your participant at this stage, because (1) most people like to see their brainwaves, it's not something they see every day,[2] and (2) it gives you a chance to demonstrate just how finicky EEG is, underlining the point for them to *stop moving*. I like to run the following test:

1. Getting the EEG to calm down: 'Please sit still for a few seconds. Keep your head straight and look to the centre of the screen'. *Explain this to the participant so they understand just how much EEG is affected by movement*. Also, if they want to drink some coffee or such, to please do so in between two trials.

2. The eye tests: 'Don't move your head, but look up… and down. Now look left… and right. Now please blink 3 times'. Make sure you really make the silence (…) a pronounced expectation. Notice how your EEG and auxiliary channels are affected by eye-movements, particularly in frontal channels (but also elsewhere) showing block-type changes in signal, and blinks, showing massive bumps. *Explain this to the participant so they understand to keep their eyes looking straight at the screen.*

3. The muscle test: 'Could you try to look really, really angry?'. Chances are they will try for a bit, immediately followed by embarrassed laughter. Both work to cause high-frequency noise in frontal channels, and in facial EMG (should you record these). Note also that it takes a few seconds for the noise to calm down after the frowning and smiling ended: the muscle remains active a bit longer than people think. *Explain this to the participant, ask them to keep their expression neutral and to wait a bit after any movement before proceeding with the experiment.*

4. The alpha test: 'Now close your eyes and let your mind wonder for a bit. Think of a holiday or whatever comes to mind.' Try to spot *alpha waves*: the big waves with a period of c. 100 ms (illustrated on p. 3 historically and p. 175 contemporarily). They should become bigger as the participant's mind is drifting more. This is bad for two reasons. First, if someone is losing concentration, whatever you're doing with the experiment makes less of an impact. Second, alpha waves are very strong sources of noise that can't be filtered out afterwards. *Explain this to the participant, ask them to take a break when they get tired and feel free to have some coffee.*

With all that explained, it's almost time to start. Before you hit record, though, don't forget to ask the participant to *turn off* and *not just silence* their phone. It might even be helpful to ask them to hand it over to you, and although this request feels almost dictatorial these days, they likely won't complain (for now, anyway – I suspect the Right to Phone will become a universal human right), as it is widely believed that phones will interfere with EEG. This seems unlikely, but it is a very useful myth, as the physical presence of phones causes distractions (e.g. beeps, vibrations – which obviously don't just affect ears and skin, but also the brain), and, as any

---

[2]Default joke: 'Oh wow, good to see my brains work!' 'Yes, especially since you seem to be moving. We're legally obligated to report you to the Zombie response unit if you were not showing any activity.'

in-flight crew-member will tell you, people generally fail to turn their phone off even if you ask them numerous times.

So, hit record on your recording software. Be very sure you give the recording a new file-name, for example the subject number, and be very sure it actually *does* record. The BrainVision recorder requires you to press the 'play' icon in order to record, and will show a red bar once it is recording. ActiView first requires you to start recording, but only does so after you *unpause* the recording. Many people I know (certainly yours truly) have wasted hours by simply failing to record.

Once the recording software is recording, start E-Prime/OpenSesame on the stimulus PC, and run the experiment. I find it best to hover over the participants while they do the first minute or two of the experiment to make sure they not only say they understand it, but also show understanding. Then it's time to wish them luck, while you sit down and relax!

…Not so fast, though. While they're doing the experiment, keep an eye on them. While I cannot seriously expect anyone to scrutinise the EEG as it is being recorded, reading a book and glancing at the screen should be sufficient. Keep an eye out for the following:

1. Periodic artefacts show up as cyclic bursts of noise or signals drifting slowly up and down are often caused by repetitive movements such as leg-swinging, nodding or chewing.

2. A strong increase in noise may be due to coughing, which cannot really be helped, or extremely problematic behaviour. The aforementioned participant who picked up their phone and started to have a chat during the experiment is an example of avoidable problematic behaviour. I was told by a colleague that they managed to record the EEG of a hungover subject vomiting in the lab, which is less avoidable (perhaps at the recruitment stage).

3. Dead signals or flatlining don't normally indicate death, but can result from electrodes disconnecting due to movement or gel drying up. If the effect is found across all electrodes, then it's the reference or ground electrode, which you'll have to clean up again and reconnect. A single channel flatlining is best left alone: During the analysis it's necessary to make a decision to drop a channel or not drop it, so a channel that was useful *some* of the time just complicates things. Finally, if the participant stopped breathing, then panic presents a reasonable course of action.

4. Alpha activity is quite normal in any experiment, but once there is too much, it might be time for a short break. I personally prefer a scheduled one-minute break every 20–30 minutes or so during the experiment, and unscheduled breaks whenever necessary. Ask the participant if they would like caffeinated drinks and biscuits.

In each of these cases, wait until a trial is over and at just the right moment, storm in and shout 'Stop!' – before they carelessly continue with the task while you're in the room talking. Then, politely explain what you would like them to do. Asking people if they want to take a break doesn't always work, because they don't want to be any longer in the lab than they have to be. However, the breaks aren't primarily for their comfort, but to get good data. So if there's too much alpha, just chat with them for a minute.

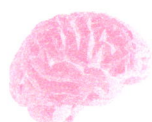

Posterior artefacts

Checking on EEG data tends to require very little concentration. Notice cries for help, requests for information, and avoidable artefacts (e.g. chewing). The rest of the time, barring sudden catastrophes such as heart attacks, fire alarms or epileptic fits, is a good time to read books. Anything too serious is impossible as there will be many interruptions and anything too easy (e.g. social media) causes envy in the subject and irritation in your supervisor, so my suggestion is to spend your time wisely reading literature or popular science. Check every minute or so, because the weirdest things happen. The most mysterious occurrence was when I saw one of the EEG channels give a peculiarly heart-rate type of signal. Not just blood pumping causing an artefact (see Chapter 8), but a veritable, hospital A&E kind of signal. As I looked up the number of the electrode, I found the electrode was nowhere near the heart or neck artery. Besides, the scalp blurs EEG sources, so it's pretty much *impossible* to see any kind of very distinct signal in just one electrode. But then I noticed that C4 was missing on the cap, and as it had dropped off, the participant had slightly repositioned herself and was now sitting right on top of the electrode formerly known as C4. So, accidentally, instead of EEG, we were measuring the difference between the scalp potential and what might be called the buttock's posterior potential, which apparently provides a pretty solid ECG signal.

Washing up

Once the experiment is done, you can stop recording. Most recording software streams the EEG data to disk, so there is no need for an extra 'save' button, but do make sure this is indeed the case. Then close the recording PC, stimulus PC, turn off the amplifier, and release the participant from the EEG. Usually, I find they are very willing to help in any way that can get them out as fast as possible, so they can help getting the auxiliary electrodes off their face. Then take the EEG cap off their head, give them a fresh towel and let them wash their hair and face if necessary. It generally is necessary, to be fair, although I've seen some experienced participants bring their own hat to cover the worst of it.

Meanwhile, it's time to clean up your equipment. It is essential to take your time for cleaning because, formally, we want to avoid bacteria spreading from participant to participant, and, informally, because smearing somebody else's grimy skin flakes onto the next person is disgusting. Some labs use special disinfectant for such reasons of hygiene, but make sure the detergent is gentle for the electrodes. Manufacturers generally hold that good washing with warm water is sufficient, but obviously that won't kill viruses or bacteria. Make sure you wash each individual electrode properly, as some gel is difficult to see when the equipment is wet. A regular, soft toothbrush (I used the head of an electric toothbrush here) we call the **EEG HyperBrush™** is a useful tool for the job (Figure 6.7). Make sure you get the remaining gel out of the auxiliary electrodes also, brushing carefully so as not to wash away the conductive material (e.g. silver-chloride). The elastic cap, and each of the electrode holders, also need washing. If the EEG cap has very small holes (such as with Biosemi electrodes), then a cleaned mascara brush may be more convenient than the HyperBrush. While electrodes generally can take water, make sure you do not let any water enter via the connectors on the other side of the wires, as this can cause serious problems. Finally, the syringe can be reused if cleaned properly, but a proper needle is single-use and may need special care to dispose of.

In general, I find students take about half as much time for cleaning as is required. Washing up takes about 15 minutes of actual, manual labour, so I would suggest checking anyone who does this for the first time. Chances are some sites on the elastic cap still contain goo. Once washing is finished, take a bit of time to get the lab back into a clean state: put the electrodes on another towel to dry, wash off any gel and such from surfaces, remove coffee mugs and cookie crumbs. If you are in

Figure 6.7 The HyperBrush™ is indeed any toothbrush or head of an electric toothbrush you can get your hands on

charge of the EEG lab, expect this advice to be completely ignored and send out a passive-aggressive email every few months reminding people you are not their cleaning staff.

And that brings an end to the recording session. Debrief your participant on the design of the study, if they are interested, and help your participant to find their way out if they are not. Don't forget to copy the new data as sometimes old data will have to be removed from the recording PC.

ESSENTIAL EXERCISES

Essential exercise 1: Informed consent and instruction forms

Any participant recording their data will have to sign an informed consent form, which states they have read and understood the instructions. To make that possible, you will have to write an instruction form, which simply and clearly states both what you expect of them in terms of the experiment (Chapters 2–5) and how you expect them to behave in such a way as to avoid artefacts. Furthermore, make clear what the expected risks are and are not, what you intend to do with the data, and every salient point discussed in the ethics section of the present chapter. And when you write to participants about their rights, please don't turn into Google disclaimer mode. The point is for them to know there's nothing to worry about, not to avoid being sued! However, try not to get sued anyway and, if you do, don't mention me.

Essential exercise 2: Data acquisition

Finally, the moment arrives: Record the data. How many people you want to recruit depends completely on the kind of study you want to run. If you're doing this in the context of a class, then a group of six will be sufficient to do most statistical analysis. If you are in a hurry, you can likely do with fewer, or take a few datasets that are made digitally available as part of this book (https://study.sagepub.com/Spape). More is never a problem. When you run a subject, make sure you obtain a picture of them before the experiment, adjust it so that it fits in with the rest (see Chapter 2, pp. 27–28), and name it 1.jpg, so you will later on be able to analyse with EEG what the effect of seeing yourself is.

Again, as suggested at the beginning of this chapter, make sure you know the first person fairly well, so you get a feel of what it is like to be so physical with people, so to speak. Don't worry, even shy people like myself do get over it eventually! Before you begin, or when you are struggling with the first participant, consider the additional exercise below.

Additional exercise: Write a checklist

Write down everything you need to watch out for and which you are worried you might not remember. Create a checklist that you can later use during data recording, or to jog your memory afterwards while writing your report in case, by then, you have forgotten what you were supposed to be doing at this stage. Make a chronological ordering of steps and checks that need to occur before the experiment, during the experiment, and after the experiment. Add a few additional, empty pages to this checklist and create a **lab diary**, in which you log all strange and wonderful occurrences that happen during recordings and any potential problems you spot.

Don't panic. And good luck.

7

PREPROCESSING EEG: IMPORTING, REFERENCING, FILTERING

IN THIS CHAPTER, YOU WILL LEARN:

- How to get started with MATLAB and EEGLAB
- How to import data
- How to have a first look at your data
- Which data you should get rid of
- Why re-referencing is necessary and how to do it

Have you ever wondered why some of us teachers like popular kids? It would make pedagogical sense if teachers mainly focused on those pupils that need their support most, but this takes effort. Teachers are lazy, cognitive misers, much like actual humans. So instead, most teachers instinctively like geeky, lonely children who enjoy reading and learning or even don't require a teacher. That situation is counterproductively self-perpetuating, since this sample of children make exactly the kind of primordial soup that produces the next generation of teachers. Naturally then, I was the teacher's pet despite not really listening to whatever they had to say, being stuck with my nose in a book.

But there is a certain brand of teacher who has a perverse distrust for my group and instead shower their attention and praise on the type of person who's least likely to be interested in the opinion of a complete nobody. Is it a form of self-loathing or are they honestly feeling that standing in front of a class is a form of teenage idol stardom, getting you to grovel for the attention of a child? A TV psychologist might wonder if there's some childhood trauma or learned helplessness at play here. I, for one, have always found it baffling.

But then I found out about EEG, and I learned the power of popularity.

That is to say, a popular person will be able to rustle up multiple people and say, 'Hey, I want to completely muck up your hair and make-up, after which I'll bore you to death'. You can do that with friends a fair few times before they stop reacting to you and start whispering to their peers that you have fallen on some hard, unpopular times. Of course, you could pay them, but that's like paying for friendship, which is generally frowned upon. Or, more to the point, usually you will want to test out your experiment before you run the real version, and you don't necessarily get funding for such pilots. So the more friends you have, the longer you can sustain your gradual slide into ignominy.

As you might have guessed, my pool of resources[1] was pretty slim to begin with, so I was left to mine some subterranean vaults in order to accrue some data for this book. Eventually, I got one student assistant (paid), one PhD student (cajoled), and one wife (private) to act as participants, resulting in far too little data to reach any sort of statistical significance. So that left me with only one way out: subjects 1, 3, 6, and 7 are all me.

OK, so I know that's not proper statistics, right? Independent observations assumption, yada yada. Give me a break, Andy Field. If you had done the exercises in the last chapter you would know how boring these experiments are. Worse, I had to look at my own, really awful picture for four sessions long. I got so fed up with it, I grew a beard, just to not look like that guy anymore.

Anyway, so here's a pro-tip for EEG teachers and learners alike: when it comes to recruiting subjects, make it the task for the popular kid. EEG: the great equaliser.

Before you begin: Make sure you have raw data to work on. If you have been following the book and have recorded your own data, then that's great: copy it all to a directory you can find easily (such as a special EEG folder on your desktop). However, for tutorial purposes, I would still recommend also downloading my data from the online resources (https://study.sagepub.com/Spape), so that

---

[1]And if you think I'm talking about it in a rather exploitative manner, do remember there's an entire sector out there who *call themselves* HR, for managing human resources, by which they seem to mean paying you.

you can work along with me and get a good idea of what the result of each processing step should be before you start with your own data. The online material is organised by chapter, with each containing the starting files required for the chapter. Download this chapter's material and extract the contents, which will result in a new folder ('Chapter 7'), in turn containing folders for code and data along with a text file providing more details (readme.txt). The latter will have datasets for seven recording sessions, which I rather egocentrically divided into being either me or not me:

Name	umeeg101	umeeg102	umeeg103	umeeg104	umeeg105	umeeg106	umeeg107
Amplifier	BrainProducts	Biosemi	BrainProducts	BrainProducts	Biosemi	Biosemi	Biosemi
Details	Me	Not me	Me right after 102	Not me	Not me	Me	Me weeks after 106

No, I'm not *that* self-centred! The real point is that it allows you to see how much of EEG is due to personality (101, 103, 106 and 107 being the same) and static biological features like skull thickness, how much is due to hardware (contrasting 101/103 with 106/107), and how much due to setting up (101 and 103 have nearly the same set up, as the participant never left the lab, while 106 and 107 are weeks apart. (Contrary to local legend, I do sometimes leave the campus. During harvest moons.)

GETTING STARTED

In this book, I make use of MATLAB, which like E-Prime is neither open-source, nor cheap, although unlike E-Prime, it's developed by a rather large company who are not nearly as prepared to answer individual questions. On the other hand, most universities that have a physics department will likewise already have MATLAB licenses anyway, so the cost is not immediately your problem. Why physics? Well, of all the software out there that can do mathematical wizardry, computer scientists tend to be more often open-source-minded by principle and choose R (or Python), engineers like the schematics provided by LabView, while physics boffins focus on timeless problems that don't require fast work or pretty interfaces and so use MATLAB. Well, with such an introduction, I'm sure you feel the urgent need to start right now, so let's get cracking!

Starting MATLAB

Once you start MATLAB (please, follow along with me and start MATLAB now), the interface in Figure 7.1 should appear.

It does not always show up like this, as it depends a bit on the version and layout. To start with, click on the layout button, and tweak the various options so that most of the windows above show. The **variables window** and **script editor** perhaps do not, but we will get there as I give you this very basic introduction to MATLAB.

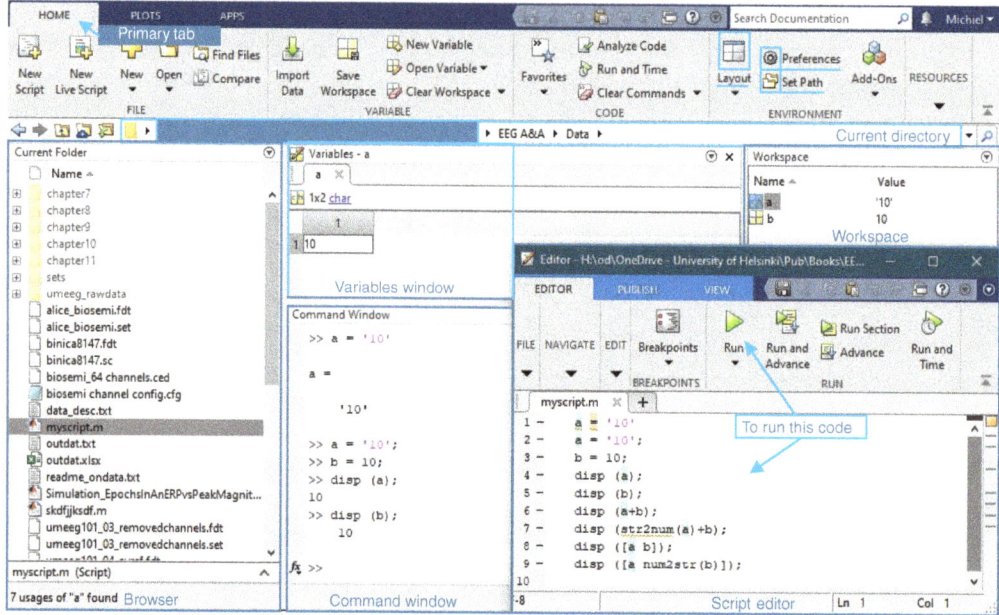

Figure 7.1

Source: Reprinted with permission of The MathWorks, Inc.

In the **command window**, you can write code directly, and MATLAB will execute it as soon as you hit enter. This is often convenient for troubleshooting your code (debugging): if error messages cause your script to crash, you can check out what is happening by taking a closer look at what's happening. That sounds very abstract, so try the following. Whenever I start a line with >>, I mean to convey that you should write what follows directly in the Command window. So:

```
>> a = '10'
```

If you do this, MATLAB will helpfully tell you that a = … and on the next line '10'. MATLAB tells you how variables change whenever you fail to end a line with a semi-colon, so make it:

```
>> a = '10';
```

And MATLAB remains silent. Should you still want to get a read-out of the value, try:

```
>> disp(a)
```

The function disp displays (what in many code languages is called printing) the content of a variable to the command window. Of course, right now this is pretty useless information since we just assigned '10' to the variable a, but if you want to know the size of an ERP, say, you will probably be interested in a read-out. Still, printing the number on the screen is not always necessary, as MATLAB comes with a helpful variable viewer. Notice in the **workspace** a is now shown (name: a, value: '10'). Double-clicking on a opens the **variable viewer**, in which, again, the contents of a ('10') is shown.

Assigning a text value ('10' – notice the single quotes denote text) to a variable is not extremely useful. However, what MATLAB loves more than anything is doing calculus over matrices, which is exactly the sort of thing many students love to hate, so that's a match made in heaven: MATLAB can do what it does and we sit back and enjoy. For example, try:

```
>> a = magic(5)
```

It shows us:

```
17   24    1    8   15
23    5    7   14   16
 4    6   13   20   22
10   12   19   21    3
11   18   25    2    9
```

Which is a magic square of 5 by 5 (entering 10 as parameter, as in magic(10), would give a 10×10 square). What's so magic about it? Try to add all numbers in each row on paper, or you can also just write 17+24+1+8+15 in MATLAB). Now do the same but for each row (17+23+4+10+11, etc.). And now the diagonals... It's all the same number! Why would you make a five-letter **function** (magic) for that?! I don't know! Why am I shouting?! It's so nerdy, I just can't anymore.

OK, one more:

```
>> a = magic(99);
>> imagesc(a);
```

It should produce a pretty figure, a two-dimensional image in which the colour indicates the value at each coordinate within the 99×99 matrix, which you can inspect in the **variable editor**. Why does it have this pattern? I don't know. My mother used to say she studied maths because she likes puzzles and while this reasoning never quite convinced me to really do my homework, I do understand her sense of wonder when I look at this graph.

So, MATLAB is all about matrices, but why is that useful for us? Remember what EEG data looked like as it came into the recording software as a bunch of squiggly lines (channels) of electric potential measured at the scalp. Imagine every channel as a row, and every point in time as a column: it's a matrix. Calculating ERPs is simply chopping this very, very wide matrix into comparable chunks (e.g. defined by the onset and offset of a face on the screen) – smaller matrices – and selectively averaging across these. OK, so it's not super simple, but that's why MATLAB does the nitty-gritty matrix algebra for us, while the clever people who designed a toolset for MATLAB, that is EEGLAB, made it easy for us to apply the correct commands. It's a bit like driving a car: You don't need to know how the engine works (see introduction to Chapter 5) because EEGLAB will give you a steering wheel, horn, pedals, dials, and so on. But you should still try to know where you're going, not use the steering wheel for acceleration, and a little knowledge of mechanics never hurts.

Starting EEGLAB

EEGLAB is hosted by the Swartz Center for Computational Neuroscience, whose main source of traffic is most likely people trying to download EEGLAB. It's found here: sccn.ucsd.edu/eeglab/download.php, although googling EEGLAB download will get you there. It seemingly suggests it will ask you to register for an account, but it won't check such data (I apologise to the developers for having entered a, b, c as respectively name, email, and research area literally hundreds of times). Following the download, extract the archive in a known directory you will be able to find later on, such as on your desktop. Then, in MATLAB, click on Set Path. A dialogue with quite a few directories should show up: these are all the locations on the drive that are instantly findable by MATLAB, so that a function like 'magic' can be called at any point, and MATLAB will know you mean the magic that is found in a specific folder (for me 'C:\Program Files\MATLAB\R2020b\toolbox\matlab\elmat\'). We want EEGLAB to be noticed by MATLAB, so click on Add folder (*not* with subfolders), and find the directory you just extracted. It's important that the right one is added, so browse to where you put it (e.g. the desktop), then double-click on the new folder (for me 'eeglab2021.0'), such that you see the contents as four folders (see Figure 7.2). Click on Select Folder, save, and close the Set Path dialogue.

And then type in the command window:

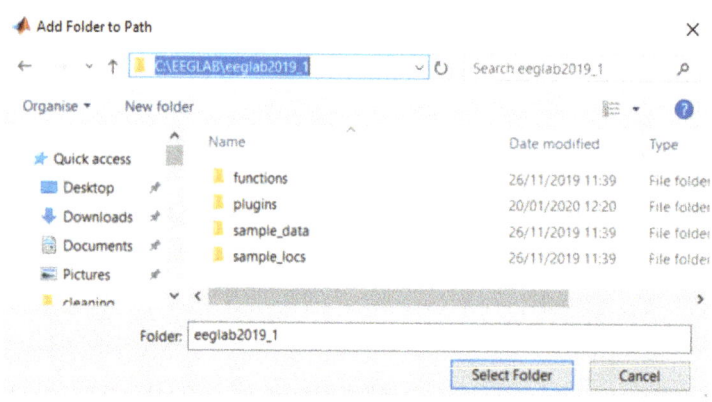

Figure 7.2

Source: Reprinted with permission of The MathWorks, Inc.

```
>> eeglab
```

If this doesn't do anything, then likely something went wrong above. One thing that can happen is that the local IT does not permit you to tweak the Path settings. This is annoying, but not a severe problem. In MATLAB's **Browser**, find the same EEGLAB folder shown in the previous screenshot and continue as below. EEGLAB will always add itself and its used subdirectories to the path, so it's not essential to fix it in MATLAB's settings. However, it will likely mean you will have to go to the EEGLAB folder and run EEGLAB at least once every time you do EEG analysis.

If you see an ugly blue square with some cryptic instructions on how to get started popping up, then yes, that's it: Welcome to EEGLAB (Figure 7.3).

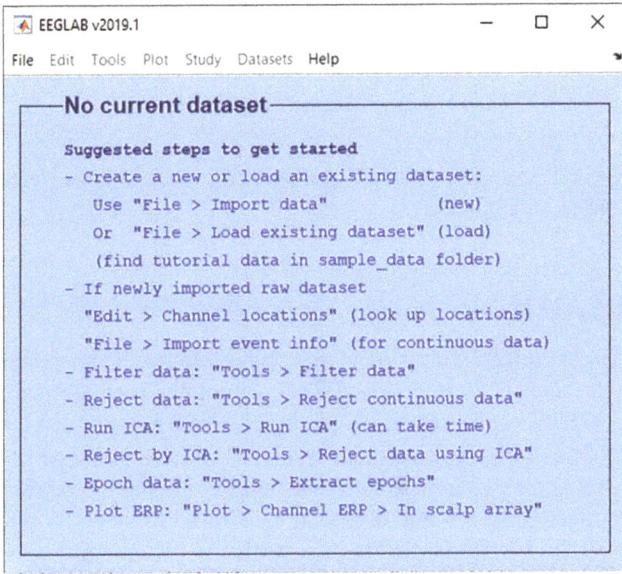

Figure 7.3

Setting up

For the rest of the chapter, I will expect that you have at your disposal:

- *MATLAB, EEGLAB*, running as described above. In EEGLAB, go once to File > Preferences, and enable 'If set, show all menu items from previous EEGLAB versions'.

- *My recorded data files*. These can be downloaded from the online material (https://study. sagepub.com/Spape). I would recommend that you work along with this tutorial first to see if you can replicate my data processing before you start with your own recorded data files.

- *Your recorded data files*. Make sure these include markers, headers, and actual data files. Some EEG systems (BrainProducts) save these as separate files (.vmrk, .vhdr, .eeg) while others (Biosemi) combines all into one. Just copy everything saved as a result of your EEG sessions to your EEG directory and you should be OK.

A FIRST LOOK AT YOUR DATA IN EEGLAB

The central EEGLAB 'box' reveals a stunningly authentic retro design. I'm not entirely sure about the how and why either, but it must be admitted that at least it looks distinct from most other software I have running, so even if you are drowning in floating windows (which you will find out happens if you use EEGLAB), it's easy to find EEGLAB. Sometimes, you might accidentally close it. Don't worry that you forgot to save: the EEGLAB box doesn't contain your data, MATLAB does. The EEGLAB box is just a graphic user interface, or basically just a menu that makes it easier to run a lot of MATLAB algorithms on variables stored in MATLAB. If you can't find the EEGLAB box, just write the following in MATLAB:

```
>> eeglab redraw
```

This opens the EEGLAB box again *while not overwriting your previously stored* data (note: it does if you omit the redraw command).

Importing data

As EEG amplifiers are manufactured by many different, often small, companies, the accompanying recording software tends to store digitised EEG data on the hard disk in many different formats and never the EEGLAB one (AFAIK). Luckily, EEGLAB is used across the world, so many labs have developed their own plugins that sit on top of EEGLAB and permit it to import their type of data. EEGLAB comes with many of those plugins, enabling it to read from Neuroscan, Philips, and Biosemi data, although not, for example, Brain Products. If you don't see the data type you're looking for, such as the Brain Products .vhdr that is required for some of the data files that come with this book, then go to EEGLAB's menu, File > Manage EEGLAB extensions, click on the one you're looking for (e.g. bva-io, which enables reading Brain Products files), and click on Install/Update. Should this process somehow not work, you can also go the traditional route: search the internet (e.g. google bva-io), download the plugin, and extract it within the EEGLAB\ plugins folder. If you are not sure which plugin you need, you can always ask the EEG amplifier manufacturer; they should know what EEGLAB is.

So, let's import some data. I just added the bva-io plugin, and after restarting EEGLAB find it in EEGLAB > File > Import data > Using EEGLAB functions and plugins > From Brain Vis. Rec. vhdr file. I select the header file (*umeeg101.vhdr*), press enter. A dialogue appears that differs across the various plugins, but for bva-io, it asks whether you want to import fewer data points than recorded, or fewer channels. That doesn't make it faster and modern computers don't easily run out of memory, so there's little reason not to import everything. Pressing OK, EEGLAB loads the data into a new 'dataset'. A **dataset** in EEGLAB initially refers to the file we now import, for example '*umeeg101.vhdr*', but whenever you do an operation, such as re-referencing or filtering the data (see later in this chapter), EEGLAB requests whether you will want the result to be a new dataset and/or save either to disk. We'll see how this works shortly, but for now, take note to give a reasonable name any time EEGLAB creates a new dataset. For example, we might call this 'umeeg101_01_raw' (Figure 7.4).

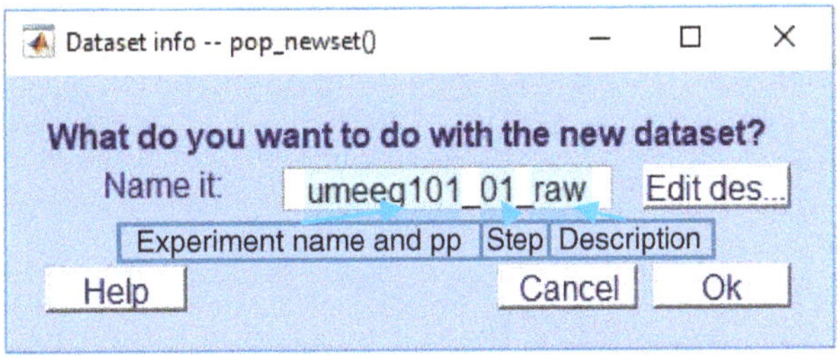

Being a very messy sort of person myself, I easily get lost in my own data, so I try to compensate by using version numbers. Thus, the 01 here refers to it being the first version (or step), and raw tells us what I just did – I imported the raw data. Use your own system if you like, but never, ever name a file 'final version', for as any scientist knows, this brings bad juju. Another file will succeed it and be called 'next final final-most version 3'. It's a law of nature.

If this went well, and the file importing plugin is doing its job, then MATLAB will display comments in the command screen:

Figure 7.4

```
pop_loadbv(): reading header file

pop_loadbv(): reading EEG data

pop_loadbv(): scaling EEG data

pop_loadbv(): reading marker file

Creating a new ALLEEG dataset 1

Done.
```

There are no warnings or errors, so everything is OK. Looking at the EEGLAB box, it now shows some actual information, such as Filename, number of channels, number of epochs (the data are continuous now, meaning in one big chunk, or epoch, which we'll be dividing into many epochs in Chapter 8, p. 192), and number of frames per epoch. Frames are digital **samples** in the time domain, or **datapoints**. Remember that the EEG was recorded at a certain sample rate (Hz), in this case 1000 Hz, so the 2013980 frames reported mean that we have about 2014 seconds of data, or about 33 and a half minutes.

Let's try this again, but now with a different dataset. I do the same operations as above, but now use the 'From Biosemi BDF file' and import '*umeeg102.bdf*'. The Biosemi plugin has some peculiarities, such as that it says it requires one to indicate a *reference*, which will be explained later. For now, it is safe to say that while it may require a (new) reference at some point, that doesn't mean it requires it right now. So again, keep the defaults as is, and ignore the plentiful warnings generated by MATLAB, such as:

```
Warning: line (432: 65440,'dyne s m2 cm−5','Vascular Resistance Index','dyne seconds
square meter per centimetre to the power of 5',#obsolete#) not valid
```

Unless it entirely fails to load. If it doesn't, EEGLAB presents a new dialogue, which will return many times (Figure 7.5):

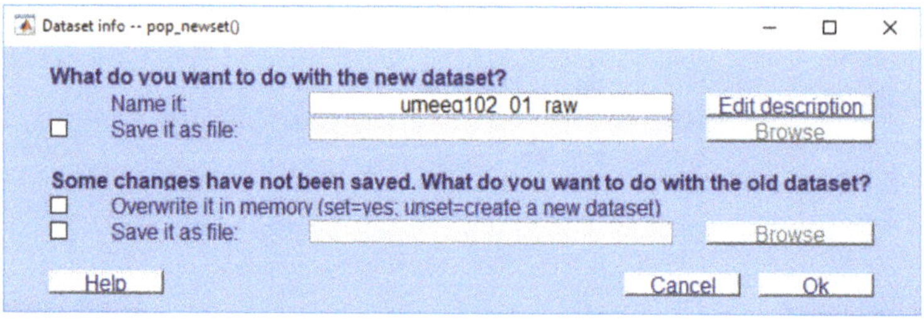

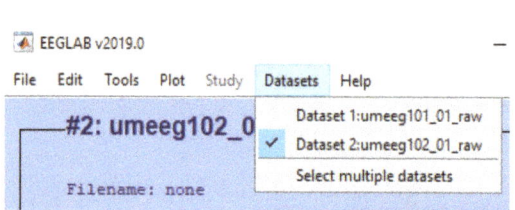

Figure 7.5

This time, since EEGLAB already had one datafile open, it wants to know how you want to continue. The top is the same as previous (notice in the screenshot that this is a different subject, hence 102, but the same step, 01), with the addition that EEGLAB can immediately save the converted file to disk. More importantly, however, it asks what to do with the previous dataset (i.e. *umeeg101_01_raw*). In general, I find it useful to keep it in memory, which allows one to compare between different files. For now, don't check any of the boxes, and open the file.

So now the EEGLAB box should show the two datasets if you click there (see Figure 7.6). With these two datasets in memory, one has to specifically select the one you want to work on in the EEGLAB box. This can easily go wrong as soon as you have many windows open. For example, if you select dataset 1, plot the data in EEGLAB, then select dataset 2, then the earlier plot won't dynamically change to the new dataset. Indeed, if you again make a plot, and forget exactly which window is which dataset, it will be very hard to distinguish between them. So for this reason, keep naming the datasets as above, even if you don't create new file names, as it is very easy to lose track as to which plot concerns which dataset, and which dataset concerns which subject.

Figure 7.6

Resampling

As you will notice, loading, processing, and saving data can take a while, although it helps a lot if your data are saved on the local hard drive rather than an external one or a network location. Back when storage space was still an issue, people would seek to minimise the data requirements by **resampling**, meaning to change to temporal resolution of the dataset. However, another reason for doing this is that sometimes there are accidental changes in the recording setup (e.g. using different decimation factors, see p. 103). In the provided dataset, this is indeed the case even if not by accident: I have been using two different amplifiers from two different companies to illustrate similarities and differences. One allowed sample rates of 512, 1024, and 2048, while the other only of 500, 1000, and 2000. Having both in your analysis will cause problems once we average datasets (Chapter 10). That is, the exact timepoints of the datasets do not match precisely: in 1 second of data a 1000 Hz setup goes from 0, 1, 2, …, 999, 1000, while a 1024 Hz setup works as 0, 0.977, 1.95, …, 999.023, 1000, since a second is broken up either as 1000 pieces or 1024. Thus, when we average across subjects, things will become difficult. In EEGLAB, this can be easily fixed by going to Tools > Change sampling rate and setting all datasets to an equivalent number, which I did, *resampling* all datasets to 512 Hz. In the case of 1024 Hz datasets, this basically means removing every other datapoint (i.e. keeping only 0, 1.95, ... 1000), but things get a bit trickier if we do not resample a dataset by an exact integer (here: 1000/512 is *downsampling* by a factor of c. 1.953). Indeed, I would not normally do this but for demonstration purposes, but on the other hand, I would not be having this issue in a normal situation when all data are recorded with the same amplifier. As it is, my advice to you is not to downsample at all if you work with your own data, given that computers have become a lot faster over the last 20 years. If you work with my datasets, I do recommend downsampling to 512 Hz, and sooner rather than later (as I found out the hard way).

If you use my data, therefore, resample to 512 Hz and save them with an ending marking this was the second step (after importing), e.g. 'umeeg101_02_resampled'.

Editing channel locations

Once the data are available as datasets in EEGLAB, it is time to make sure that EEGLAB knows how the channels relate to physical locations of the subject's scalp. There are three basic possibilities.

If you use default locations of the 10–20 system and your recording software already showed these channels, then it's likely easiest. For example, having just loaded umeeg101_01_raw, select it again as currently active dataset. Then, go to EEGLAB's edit > Channel locations. If recognisable channel names were imported, EEGLAB will mention that there are known channel labels. Furthermore, even if the physical coordinates of the channels are missing, it will give you a good guess by calculating them based on a regular head model, which it already fills into the menu. (However, I chose [..]plugins\dipfit\standard_BESA\standard-10-5-cap385.elp, as this helped with the EOG channels.) Pressing OK should present a dialogue asking you to edit channel information. Go through the channels and see if every channel has data, as with the screenshot in Figure 7.7. Go through each channel by clicking on the arrow right (>), doing a cursory check. In particular, EOG capturing electrodes (see Chapter 6) are not always named correctly, so to place mine, I renamed them to HEOG (for horizontal eye-movements) and VEOG (for vertical ones). Click on 'Plot 3-D (xyz)' to see if the localisation seems reasonable: a three-dimensional plot should appear and dragging around it with the mouse gives you an impression of how the coordinates could fit around a head.

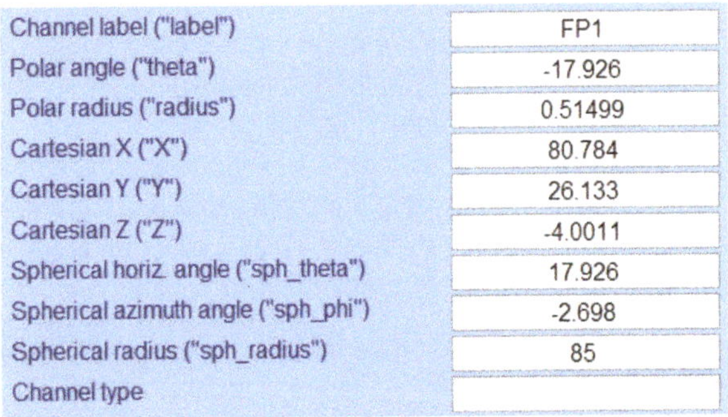

Channel label ("label")	FP1
Polar angle ("theta")	-17.926
Polar radius ("radius")	0.51499
Cartesian X ("X")	80.784
Cartesian Y ("Y")	26.133
Cartesian Z ("Z")	-4.0011
Spherical horiz. angle ("sph_theta")	17.926
Spherical azimuth angle ("sph_phi")	-2.698
Spherical radius ("sph_radius")	85
Channel type	

Figure 7.7

If you used default locations of the 10–20 or 10–10 system but forgot to name them, then a little extra work is required. For example, the other dataset (*umeeg102.bdf*), now called umeeg102_01_raw, has 64 channels but they are named A1 to A32 for one cable, and B1 to B32 for the other. However, these locations aren't random, and Biosemi uses a channel configuration file in their software. If you have such or can see the same information in the recording software, then you're good to go. Thus, if you edit in a text editor the 'biosemi channel config.cfg' I included with the raw data, you will see that the channels refer to standard names: Chan1 = Fp1, Chan2 = AF7, and so on. These are again 10–10 coordinates, which EEGLAB will recognise. So, as with the previous option, go to edit > Channel locations, and assign each channel its correct label. It's a bit of work (64 channels), but it only needs to be done once. So Channel label for channel 1 becomes 'Fp1' (mind the capitalisation!), and so on. When you're done, click on Look up locs, and let EEGLAB find the correct coordinates. Then, plot the channels to check whether it all went well, and 'Save (as .ced)' the coordinates so you don't need to manually name all 64 again, but can load the .ced file instead.

If you don't use channels based on the 10–20 system at all, for example because you used a three-dimensional surface scanner to measure locations (e.g. Polhemus), or because you have tons of channels (e.g. EGI systems), then either you will need to import subject-specific coordinates or the localisation file that comes with the EEG amplifier's configuration. Either way, you are likely best off asking the company how to read the custom locations in EEGLAB.

Once you're done and have OK-ed the channel configuration dialogue, the EEGLAB box acknowledges your work by saying 'Channel locations: Yes'.

Do not forget to save your data (e.g. 'umeeg101_03_channels_located').

Scrolling through the data

Let's have a look at the data. In EEGLAB, go to Plot > Channel data (scroll). In Figure 7.8, you see what dataset umeeg101 looks like after a few operations.

1. Move forward in time to c. 1000 seconds. Usually, the beginning of the study has movement artefacts and the ending of the data collection can be worse due to gradual deterioration of the signal to noise ratio as electrode gel dries up. So have a look at the data as they are in the best possible light, somewhere where the experiment is already running, and the participant stopped moving. We can see the experiment is running because of the presence of **markers** (triggers) that in EEGLAB are presented as colourful, vertical lines. Notice in Figure 7.8, for example, the event type S201 marks the onset of a relevant image (for a quick reminder, see p. 55, Figure 3.12), occurring at about 1003.55 in the dataset (which you can read if you hover over the plot with your mouse).

2. Turn Remove DC offset on in the Display settings. The DC offset is the baseline voltage in the channel, which is uninformative for EEG.

3. Change the vertical zoom to something like 50 – this is best achieved by selecting the number, changing it and pressing ENTER.

It's a bit of a clunky interface, as per usual with EEGLAB, but take your time to get used to it. Back in the day, it would be normal to manually go through all data and remove suspicious data with artefacts (see Chapter 8). You can still do that in EEGLAB by selecting some data, as I did with the screenshot in Figure 7.8 below and clicking REJECT, which then removes the data from the dataset. We will treat the manual removal of data next, but for now it is best to focus on getting to know your data well. At a minimum, go through the data as suggested in the essential exercises.

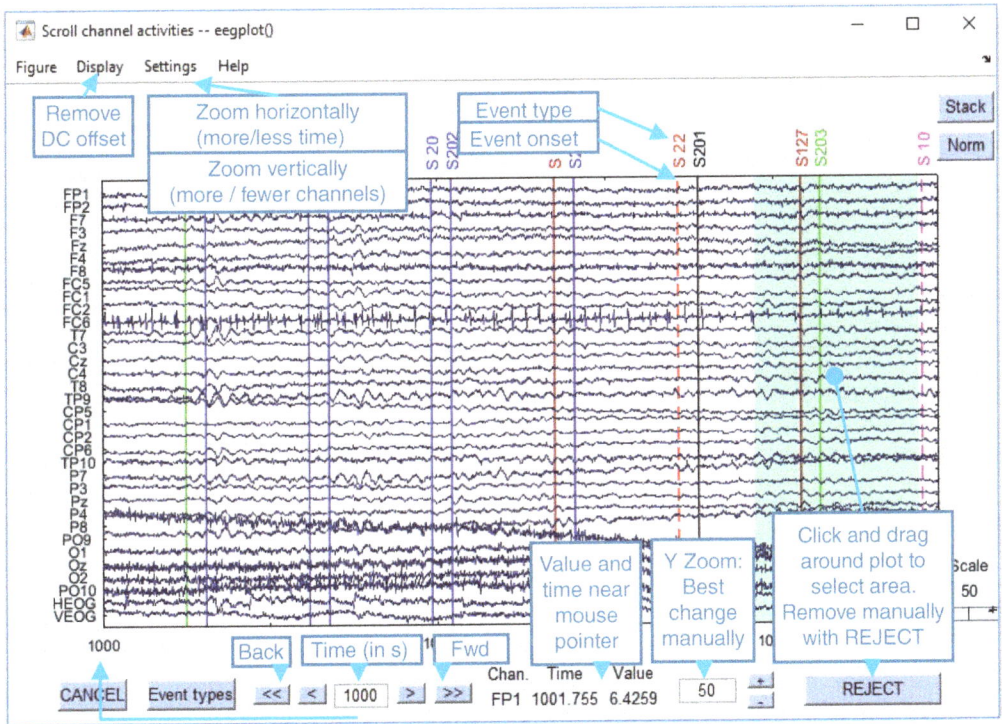

Figure 7.8

REMOVING BAD DATA

A wise person once told me that *even a cow as white as pearl will always have some spot, but that doesn't mean you can just pull its tail*. And indeed, the question of what you do with the badness of the world (which apparently includes cow spots) pertains not just to ethics, politics, and architecture, but also to EEG research. Maybe it sounds a bit theological, but I would say there's true evil (for which hell, or at least data oblivion, is reserved) as well as the sort of badness that might be corrected with a touch of kindness. Before I start raving on about the nature of evil EEG (in sheep's clothing), I would like to make the point that there is always a certain amount of completely useless data in any dataset, and it's better to get rid of it sooner rather than later.

Removing bad times: Trimming

The easiest type of bad spots of data are neatly contained in certain periods of time. Consider, for example, the two screenshots in Figure 7.9.

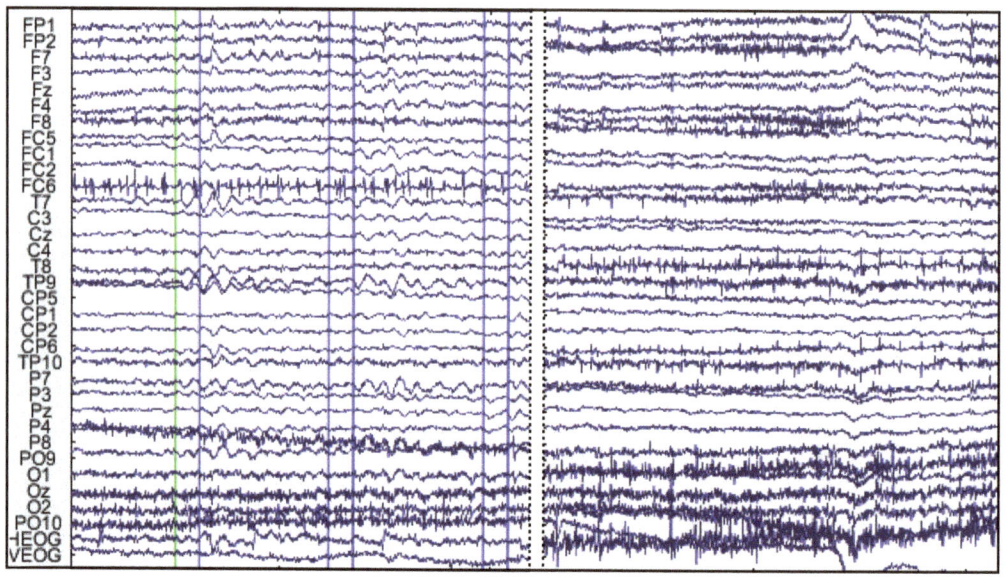

Figure 7.9

Note that the left screenshot is taken from the middle of the experiment (as can be seen from the markers) at 1000 s, and the right screenshot is taken from near the end, at 2009 seconds. Make sure you set the Y scale to the same value (here: 50), otherwise it's hard to compare. The right screen is not even *that* bad, but most subjects will start moving a lot once they find out the experiment is done, which causes artefacts. It's also possible that the elastic cap is taken off before the recording is stopped. There's no good reason to retain data beyond the duration of the experiment anyway, so I generally recommend removing all data up until 20 s before the first marker (if it occurs later than 20 s), and everything from 20 s after the last. The 20 s are there for filtering reasons, which will be covered later in this chapter.

To do this, find the first marker after the first ms (which sometimes has a 'boundary' marker), write down the second (in umeeg101_01_raw: 33) and subtract 20 (= 13). Then find the last marker and do the same but add 20 (1992 + 20 = 2012). Then go in EEGLAB to edit > select data, enter the desired range under time range (s) you wish to keep, which would be '13 2012' in umeeg101_01_raw, and then click OK (see Figure 7.10).

EEGLAB will ask you to save your data at this point, and you might as well: e.g. 'umeeg101_04_removed_timerange'). I also selected the old dataset (before the data were trimmed) to be overwritten.

Instead of selecting a time range to keep, you can also *remove* a range, which works in the same way, except that you check the box next to the time range. In this case, you would first remove the range 'o 13', save the file. We then still need to remove the final part: again, look up the last event (which now occurs at 1979!), add 20 (1999), select the data and remove the last few seconds by removing '1999 99999' (or another impossibly high number).

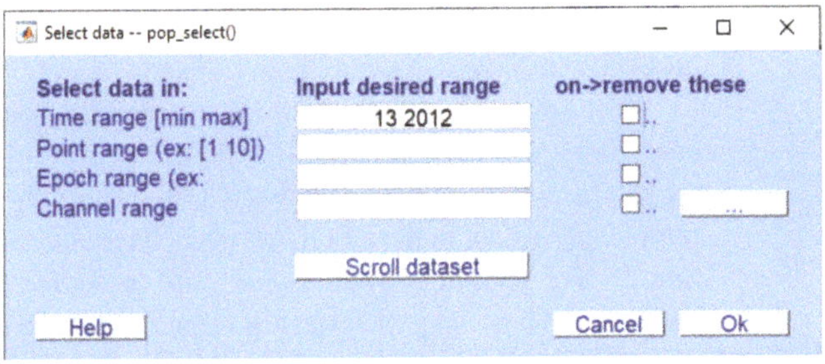

Figure 7.10

There are other periods that are best removed as soon as possible. For example, if the experiment features a break (spotted as a long period of missing markers), or the subject had a coughing fit (which looks like an earthquake). If such extend over longer periods, then they should be removed. However, if they occur in the middle of an experiment, be careful. For example, if I remove the time range from 1001 to 1002, it will suddenly appear as if the markers S7, S202 will be immediately followed by S202 again; EEGLAB will be completely amnesiac to the occurrence of the S20 (which was just before 1002). For this reason, it is best to remove very short periods of time in a different way (covered in Chapter 8). For now, if a coughing fit or such has occurred, it may be best to remove a long range before and after, for example removing the entire RSVP stream in which it happened.

Removing bad channels

As I am sure you know, there are some times that are generally bad – say, Monday morning – and some places are always awful – Antarctica – while other places are mainly horrendous at certain hours (London Bridge station at rush hour). With EEG, a place on the scalp as indicated by a channel is bad if it doesn't give a useful signal. This happens mainly if contact is lost, for example because the gel below an electrode that was already not particularly well placed has gradually dried out and lost contact altogether. Or, you happen to stand on the wire and it flies out of the elastic cap, which is not ideal, but a lot better than the alternative of the wire breaking entirely.

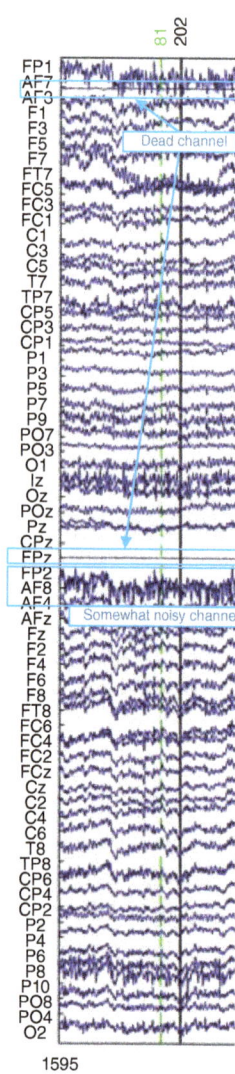

1595

Figure 7.11 Noisy and dead channels

At such a point in time, it may be useful to make a note in the lab diary (p. 130) saying electrode so and so stopped working. Some note that it happened only halfway during the experiment, but this is pointless: you are not going to use half a participant's entire dataset just to gain one channel.

It is therefore better to remove any bad channel that either shows far too much noise, or almost no data whatsover. This can be easiest to check at the end of a recording since signal-to-noise ratio is rarely going to improve over time. Of course, that comes with the important caveat mentioned previously: check the last part of the experiment where markers are still occurring, as the very ending of an experiment generally involves movement and noise.

In Figure 7.11, for example, I look at a second of data from umeeg102_02. Two channels are flatlining, although it's slightly easier to see with a larger window (if you see no signals, remember to remove DC offset in the EEGLAB plot channel data (scroll) window!). Some other channels, such as FP2 and to some extent FP1 and F7, look somewhat noisy. I would not remove these as of yet, however, as they will likely clear up after the filtering that will occur later in this chapter. However, if you can almost not see anything else because one channel is taking up much of the window, or if a channel seems to be travelling up and down, crossing many other channels, then they are better removed. Indeed, in such a case it might be hard to find out which one exactly is noisy. In Figure 7.11, for example, is FP2 the noisy one, or AF8? To get a closer look, zoom out vertically (see p. 143, Figure 7.8 'Y Zoom'). If most channels look like they are flatlining, and only one is showing something, then most likely the one is the problem – it's very unlikely that most of the brain is dead but a single area is courageously plodding on.

Once you are sure which channels should be removed, go in EEGLAB to edit > select data, and, next to 'Channel range', either directly write down their names or look them up by pressing the button. For umeeg102, for example, that would mean either writing 'AF7 Fpz' or clicking on the selection button with three periods and selecting AF7 first, then scrolling down to find FPz, then, while holding the control key (or command key on a Mac) on the keyboard, selecting the second dead channel. And don't forget to put a check in the box next to the channel range (on->remove these), otherwise you only keep these two channels rather than removing them.

As before, EEGLAB will ask you to save your data. I called this 'umeeg101_05_removed_channels'). I also selected the old dataset (before the data were trimmed) to be overwritten.

At this point, I blindly went through the pop-up asking me what I wanted to do with the new data, so it went through its default, creating a new dataset. Clicking on Datasets now gives me the following (Figure 7.12): I now have two datasets (2 and 3) concerning the same data, except 3 has fewer channels. To get rid of it, go to File > Clear dataset(s) and remove the old one (2) from memory.

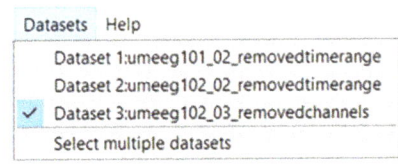

Figure 7.12

RE-REFERENCING DATA

The electric potential of any point on the scalp is always measured relative to a different point. A solid understanding of this is critical when it comes to interpreting your results. Unfortunately, there are plenty of EEG researchers who fail to appreciate this as many of them do not mention in publications where they placed a reference during recording, or to where they re-referenced. In this section, I will first provide a little bit of background to give you a basic idea of what a reference practically means, before showing how to re-reference your data.

Practical theory

Imagine you recorded a dataset of four EEG channels, the electrodes placed over the frontal, central, parietal, and occipital cortex, and used the tip of the nose as a reference. Yes, some people do use a nose reference, even though it looks remarkably silly. In any case, the original data recording could look a bit like Figure 7.13.

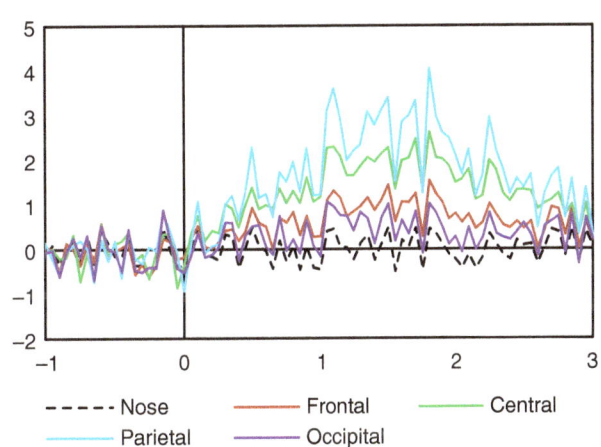

Figure 7.13

Pretty clear-cut, right? Good, strong positivity starting after the vertical bar (0), reaching a peak at about 1.6 s. Except, there cannot be data from the nose 'channel' while also using it as a reference: the electric potential recorded in EEG is always relative to a channel such as the nose, so the data recorded as relative to the nose necessarily looks like a flat line. As such, however, any noise picked up in the reference channel will show up in everything else (here shown as the similarity between the four coloured lines and the dotted black 'nose' channel). Furthermore, it becomes unclear whether there is indeed positivity: the only thing we truly know is that the parietal channel is most positive relative to our non-photogenic reference. This makes it hard for people who do not use nose-references (thankfully, most of us) to interpret the data. Accordingly, we **re-reference**, which means we re-calculate activity as relative to one or more other channels. Two common strategies are as follows.

Common channel re-referencing. Put simply, one channel is designated as the new reference, and the data are inspected as relative to this channel. In Figure 7.14, I use the same data as above, but subtract the activity of the parietal channel from each (which in itself therefore becomes 0). Now, it appears that rather than the earlier observed positivity, the more obvious explanation would be that there is a marked *negativity* in the frontal and occipital areas. Designating the parietal channel as a common reference is not very common, but more popular choices are sites that are seen as 'silent' and unaffected by brain activity, such as the bony part behind the ears, the **mastoid** part of the skull. While this view has become unfashionable (significant brain activity can be recorded from mastoid areas), many studies that used mastoid references have become classics in the literature. Thus, to replicate earlier work, it is still common to use mastoid-referencing, making it easier to see if your ERPs look like the previously observed ones.

Common average re-referencing. Rather than using a single channel (or in case of a **linked mastoid** reference, a pair), another re-referencing strategy uses

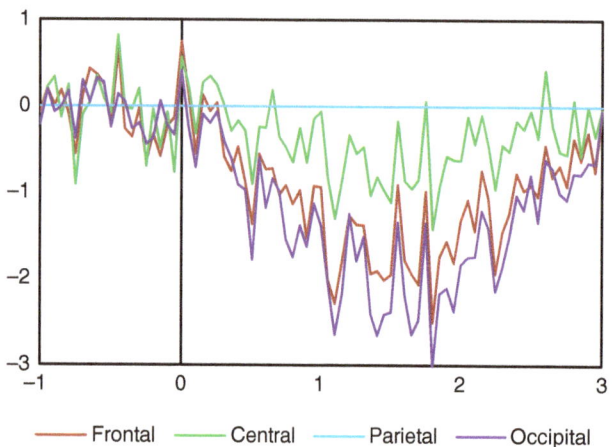

— Frontal   — Central   — Parietal   — Occipital

Figure 7.14

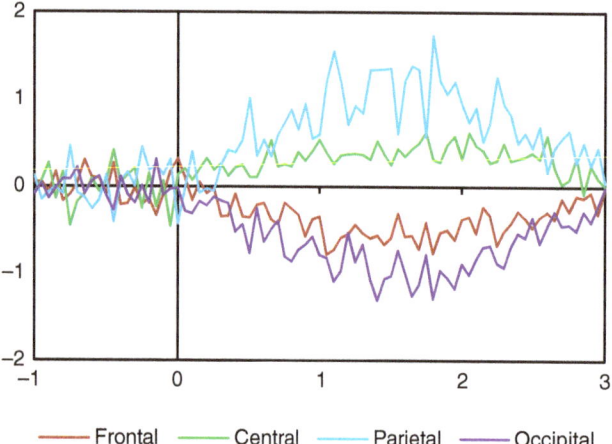

— Frontal   — Central   — Parietal   — Occipital

Figure 7.15

the average of measured activity as a reference. In the present example, it would mean we first calculate the average over the four-coloured line (leaving the nose out of the equation!), and then subtracted the result from each of the individual channels. As can be seen in Figure 7.15, the result means that all channels add up to 0: if one channel is very negative, others are necessarily positive.

There are some advantages and disadvantages to each strategy: average references are more easily compared across different studies. They also tend to show less noise. However, as will be shown in the exercises below, this can also mean that including a very noisy channel within the average computation will result in average referenced data that are noisy across the board. For this reason, it is necessary to remove bad channels from the data before applying an average reference. Finally, clearly localised EEG components are easier to pick up in contrast with individual channels. For example, it is not very clear whether the effect should be characterised as occipital negativity or parietal positivity with the average reference – they look pretty similar in size. Clinical EEG researchers thus often make use of a range of different references, trying to acquire the best possible idea of where epileptic seizures originate from.

Luckily, we are not in a hurry to diagnose people, so you can easily work with the data and try to understand how re-referencing works and how it affects the data. Let me finish with a listicle: The 6 things everyone needs to know about re-referencing (the fourth will ENHANCE your replicability!):

1. There is no such thing as a positive or negative brain potential: positivity and negativity are always relative to a (set of) reference channels.

2. Changing the reference affects the size of measured potentials (compare maximum values in Figures 7.14 and 7.15).

3. Signals look weaker the closer they are to references due to activity blurring. Two channels located next to one another measure more or less the same underlying sources so re-referencing the one to the other will reduce the measured activity.

4. Re-referencing to the common average will enhance replicability if your EEG cap has good coverage of the scalp and a large number of electrodes are placed at similar distances from one another.

5. Re-referencing to single channels remains common in the literature. What I often hear when I ask about this is that others aim to replicate known studies, which makes sense. On the other hand, something having *always been done like this* has little bearing on its correctness – there are plenty of ridiculous traditions.

6. If you want to get a good sense of the topography of a particular bump in the data, use topographical plots rather than looking at the signal data. As we will see (in Chapter 9), these plots show positivity and negativity in colours. Since they show the bigger picture, like geographical maps with mountains and valleys, they are unbiased by their choice of reference point: Mount Everest will always look higher than Scafell Pike, whether compared to sea level or the average land mass elevation.

Re-referencing in EEGLAB

Now that I have covered the basics of re-referencing, the actual process is pretty simple. In EEGLAB, go to Tools > Re-Reference the data. This will show the dialogue, which allows you to enable the common average reference, as I did in Figure 7.16. Now, importantly, this assumes that you already have removed the very bad channels as these change the average, causing the noise to contaminate all data. Similarly, dead channels should already have been removed. However, if you have channels that are outside the scope of the common 10–10 system, such as the eye electrodes marked in the figure, or bipolar electrode channels, again such as the eye electrodes marked, then enter them as excluded from the referencing. Press OK and EEGLAB does its thing again, suggesting making a new set.

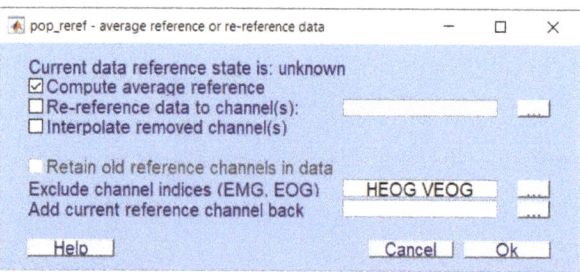

Figure 7.16

You might as well save your data again (e.g. 'umeeg101_06_rereferenced').

With some amplifiers, such as the Biosemi one, the reference is recorded with a reference that is designed to pick up most of the ground noise – the CMS-DRL mentioned in Chapter 5. That means that when you look at the data without re-referencing, it can be hard to pick out which channels are too noisy, as pretty much everything looks bad. Immediately re-referencing to the common average is not a great idea because, again, the extreme noise will contaminate every other channel. In this situation, I recommend first re-referencing to a single, known-to-be-good electrode, which can be any electrode you see looks similar to the other ones (i.e. not dead, and not moving all over the screen), but as a rule of thumb the electrodes on top of the head (Cz, Pz) work. Then, in the referencing screen in Figure 7.16, select re-reference data to channel(s) and enter the electrode, and put a checkmark under 'Retain old reference channels in data'. After re-referencing, inspect the data (plot > Channel data), but now look for channels that are problematic. Manually remove these before re-referencing to the common average. In short:

1. Check the data with its original, recording reference.

2. Remove clearly dead channels, but not the bad channels.

3. Re-reference the data to a single good electrode.

4. Remove clearly bad channels, but not the single dead channel (because that is the point you just made the reference; see p. 147)

5. Re-reference the data to the common average.

Removing channels and the average reference

One of the problems with removing channels is that the average reference is inevitably affected. For example, if all electrodes over the left side of my scalp somehow broke, then the average reference will only pertain to the right side of my head, making it hard to compare my data with data in the literature. This is a relative minor problem if only 1% of the left-sided electrodes were broken; then the average is biased a little to the right. One way to deal with this is by *interpolating* broken channels. The idea (further explained in Chapter 11, p. 247) is that if a channel is in between four other channels, then we can guess reasonably well what its activity should look like: something roughly between the four. In EEGLAB, you can do this by going to Tools > Interpolate electrodes > Select from channels, and indicating the channels you want to be 'guessed', which means the channel is deleted and added again by predicting its activity on the basis of other channels. This works best if you have more than 32 electrodes and if the interpolated channels are central rather than peripheral. However, I'm not a fan of the operation as it gives the mistaken impression that, somehow, lost data are miraculously regained. They are not. I'm therefore not recommending interpolating channels at this stage. The best way to deal with bias in the average reference is to avoid bad data to begin with. That said, if you fear a bias has crept in, you can always re-reference your data at any later point in the processing pipeline.

FILTERING

In Chapter 1, I have already discussed how EEG signals can be defined in time as well as frequency. True, brain-based EEG tends to occur between 1 and 80 Hz, depending on who you ask, while ERPs are most strongly seen between 2 and 20 Hz. In Chapter 5, I discussed how the use of a filter can affect EEG recordings: low-cut filters remove 'drift', while high-cut filters reduce the 'hairiness' of the signal. Actually, hair provides a pretty good analogy for noise and filtering for EEG, if you think of a filter as working a bit like a shower drain filter placed to avoid soap scum, hair and other such sources of *muck* from the water. Dirt tends to be bigger than water molecules, so the smaller the grid of the shower drain filter, the more dirt it captures and the cleaner the water becomes after the filter, and the less likely your plumbing is to get clogged further down the drain. So, your first instinct may be to make the holes in the filter as small as possible, passing only the very tiniest bits, but you know what happens now: you get to clean up the filter every few minutes!

Imagine the number of holes in a single filter if each were the size of a single water molecule. The *spatial frequency* of such a filter would be extremely high in order to get rid of the low spatial frequency of the dirt. Artefacts likewise tend to occur in some frequencies more than others: eye-movements and blinks produce strong low frequency (c. 1 Hz) signals, while muscle activity tends to peak in the 20–500 Hz range. From this point of view, applying a filter and removing everything from 0 Hz to 2 Hz and everything above 20 Hz should improve your data by removing the dirt and retaining the water. Indeed, in the remainder of this chapter, I will first show you how well this works. Then I will tell you it doesn't: like the fictional shower filter, the stronger the filter, the more it produces problems. As such, one should generally be careful about their application.

Different types of filter

The easiest way to see what filtering does is to look at some data and apply strong versions of each type. To filter the data in EEGLAB, go to Tools > Filter the data > Basic FIR filter. A dialogue appears (Figure 7.17), in which you are asked to define the filter. It is defined as an **inclusive filter**, or as a 'pass' filter, rather than an **exclusive filter**, or 'cut' filter, meaning that what you define here is the data you *keep* rather than remove. In my shower drain filter, I gave an example in which low frequencies (the dirt) were removed while high frequencies (the water) were retained: this is technically a **high-pass** filter. A high-pass filter removes, or 'cuts' low frequencies and is therefore also known as a **low-cut** filter. The reverse, a filter which removes the high frequency to retain the dirt – basically what is retained on the other side of the shower drain filter – is a **low-pass** or **high-cut** filter. The terminology in use can be confusing, but keep in mind that whenever I pass below a certain frequency, the 'edge' of the filter, it means exactly the same as cutting everything above this frequency.

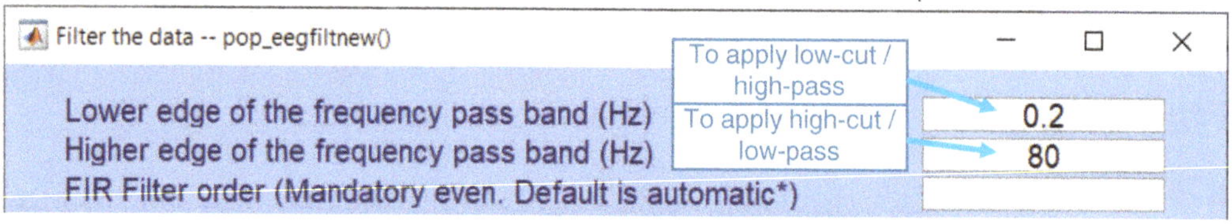

Filter the data -- pop_eegfiltnew()		— □ ✕
	To apply low-cut / high-pass	
Lower edge of the frequency pass band (Hz)	To apply high-cut / low-pass	0.2
Higher edge of the frequency pass band (Hz)		80
FIR Filter order (Mandatory even. Default is automatic*)		

Figure 7.17

Now imagine you can do both things at the same time. I'm struggling how or indeed why a plumber would implement this, but let's say you want to keep the soap, but not the water, and not the hair, from your shower. First, you could remove the water (high frequency) with a high-cut filter. You then take the filter out and scoop the contents onto a new filter, this time with a low-cut filter, which only removes the largest, lowest frequency bits. Bingo! While I promise to now cut it out with this disgusting analogy, I hope it may illustrate how a two-edged filter, known as a **bandpass** filter, works: all the data within a frequency range is passed while everything else is removed. Thus, in Figure 7.17 the filter is defined as keeping all the data between 0.2 and 80 Hz. Defining only the first, the filter becomes a high-pass filter, the second, a low-pass. Indeed, you might see all filters as adjustments of a band-pass filter: first defining a range to 'keep between 0.2 and 80 Hz', and then removing the 80, the filter becomes merely 'keep all data between 0.2 and up'. If you are still unsure, however, then click the 'Plot frequency response' at the bottom of the same dialogue, which gives you an overview of how each frequency in the original data

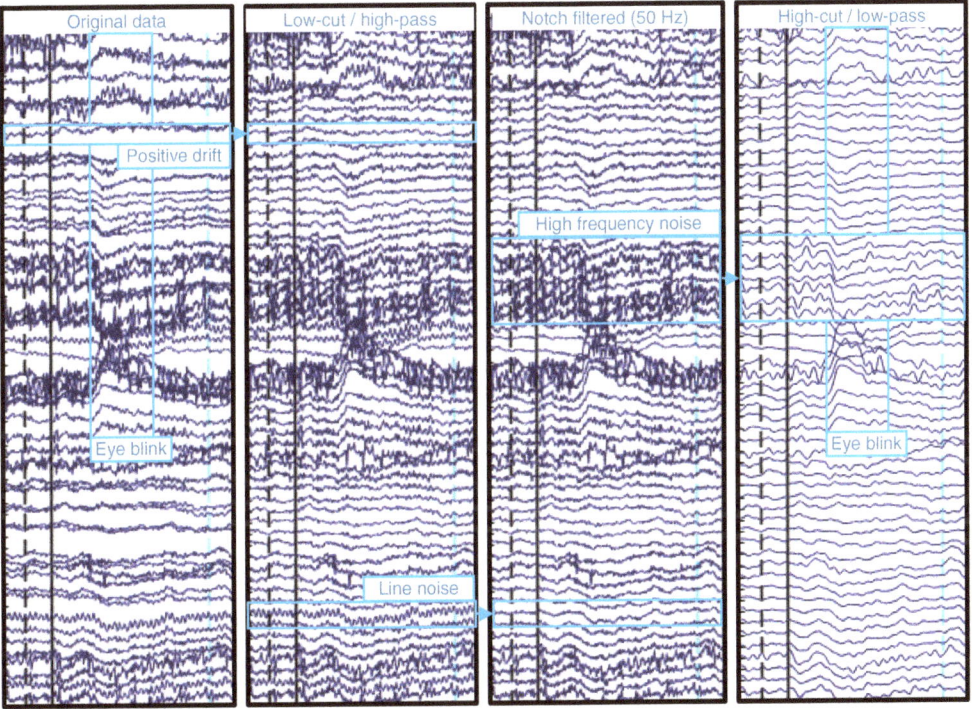

Figure 7.18 Drift, eye blinks, filtering, and line noise. In the above, from the 35th second of umeeg102, you can see what happens after I first apply a high-pass filter: it moves all lines closer together (it removes the DC offset, see p. 122, Step 4, scrolling through the data), and reduces the gradual upward or downward trend present in some channels. From the second to the third panel, I applied a notch filter at 50 Hz, and notice how little is changed apart from the line noise, which is gone in the third panel. In contrast, the fourth panel, in which a low-pass filter is applied, is much smoother. Notice that although the high frequency noise, which made the third panel look pretty rough, is now reduced, the channels with such noise still look much 'wavier' than the ones above it: applying low-pass filters can easily hide bad data rather than removing it. To emphasise the point, see how the shark fin-looking eye blink largely remains despite the strong filtering. Filtering does a lot, but it can't turn bad data into good data.

will be affected by the filter. More importantly, check the data before and after you filter. If it suddenly looks much worse than before, then chances are you entered the numbers the wrong way around.

Finally, if you want to remove the type of noise caused by power line interference, which in the US tends to be 60 Hz and in Europe 50 Hz, you will want to do exactly the opposite of a bandpass filter: the **notch** filter, which cuts a frequency range rather than passes a band. In this case, a narrow frequency band centring exactly at the line frequency should work. For example, to remove 50 Hz noise, define the band as 46 (lower edge) to 54 (higher edge). Then put a check at 'Notch filter the data instead of pass band'. In Figure 7.18, note what each type of filter does.

Using and over-using filters

Once you start to get a bit of an eye for what constitutes 'good' and 'bad' EEG data, resorting to filters becomes very tempting. As you can see in Figure 7.18, each filter to some degree 'helps', and as EEG data are so very, very sensitive to noise, without any help, you will see in the next chapter that you will likely have to ditch 50–100% of your data. Initially, it may feel like a reasonable strategy to apply as strong a filter as possible. Shower filters clog easily, however, and signal filters likewise produce new problems. I will give some examples of how filters affect EEG analysis.

Filters reduce ERP amplitudes

Time-series filters are not exactly like spatial filters in that they tend to be applied to temporal patterns, for example those that look like stereotypical sine-waves. Event-related potentials look very much like such waves: the P1/N1 portrayed show a 100 ms sinus going up at c.

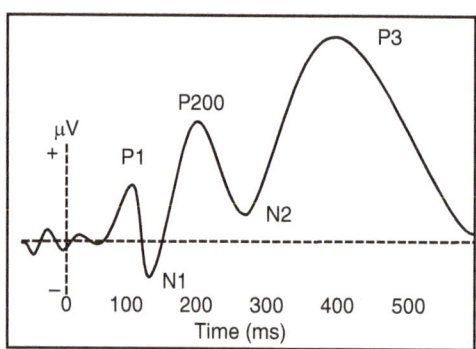

Figure 7.19

50 ms, peaking at 100 (P1), with a negative peak at (N1) at maybe 120, then reaching 0 at 150 ms. So what would happen if we apply a low-pass 9 Hz filter? The 10 hz sinus will be silenced (*de-amplified* or *attenuated*), so the amplitude of our P1/N1 ERP will be lower. The amplitude of the P3 should not be affected, as it is much slower than the earlier components. Such slower ERPs are, instead, much more affected by removing low frequencies, for example by applying a high high-pass filter. Imagine a line connecting the three positive components before going down again after the P3: the entire ERP looks a bit like a single, slow positive component with two negativities. Given that this hill has a duration of about 500 ms, a single sinus (one hill, one valley) would have a period of 1 Hz. In other words, 'silencing' the 1 Hz wave will also remove the P3 (Figure 7.19; see also Duncan-Johnson & Donchin, 1979).

Filters distort ERP components

A much more serious problem than reducing the strength of certain components is that filters can distort them in many unforeseen ways. The best known is that filters tend to work forward in time. As you may have seen during recordings, sometimes a sudden jump up or down across all channels may occur, usually because the reference is disconnected. It will take a couple of seconds after the event, but the sudden shift in the EEG will gradually be corrected, since the hardware applies a high-pass filter. However, it can only do this forward in time, so while the signal *after* the shift will be affected by the change in voltage going through the filter, the signal *before* the shift is unaffected. That makes sense, given that the hardware filter can hardly time-travel and apply a filter to correct also back in time, but it means likewise that ERPs are shifted in time as a function of filters. Digital filters, used in offline processing, have this problem to a lesser extent as they can access the entire data stream and therefore do things back in time, so to speak. Indeed, a common strategy is to first pass the data through the filter in normal, forward motion, then move them through the same filter but now in reverse motion.

More filter distortions

To further demonstrate filter distortion, I ran a 10 Hz sine wave in an otherwise static signal (Figure 7.20) through low-pass and high-pass filters. Note how the original filter (shown as dotted lines in the middle and right panels) is affected even though neither should affect the 10 Hz frequency. Instead, the low-pass filter introduced oscillations prior and subsequent to the original sine wave. The 3 Hz high-pass filter shows something similar, except the oscillation is much slower. So, the filter caused *phantom ERPs* where none existed.

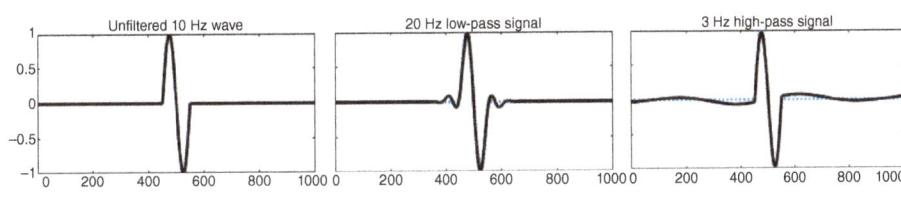

Figure 7.20

Filters do not particularly help

As we have seen in Figure 7.18, the data may look cleaner after applying filters, but are they really? After the high-pass filter, the lines moved closer together, but as we will see in Chapter 8, removing the baseline activity will do that anyway. A low-pass filter removes some of the detailing, but on the other hand, it leaves some of more egregious noise (the waviness in the high frequency block, the eye blinks) relatively in place. Why? While I mentioned eye blinks and muscle activity tend to be low and high frequency respectively, biosignals are not digital, simulated signals such as in the box above, and are not completely restricted to frequency bands, even if they have dominant and non-dominant frequency ranges. Thus, even though EMG analysis of muscle activity tends to focus on high frequency ranges, muscle activity still does affect the lower frequency spectrum (Fridlund & Cacioppo, 1986). And as you see in Figure 7.18, eye-blinks have a much larger amplitude than their surrounding EEG, even after removing their dominant frequencies.

On the other hand, noise is not necessarily as much a problem as it is sometimes thought to be. For example, if we took as a final measure of 'P3 size' simply the height of a peak, the maximum will be much affected by the degree of noise, but the average should not be. Consider the hypothetical ERP of Figure 7.21, with the 'real ERP' as the dotted line, and the actual recording as the blue one. Calculating my ERP as simply the maximum between 0 and 1 results in 1 for the real ERP and

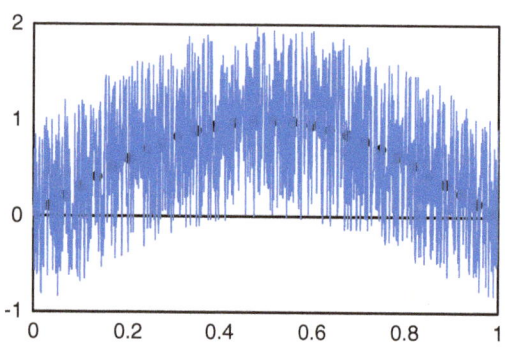

Figure 7.21

1.998 for my recording – twice as much. However, the average of each point for the signal is 0.636 and 0.637 for the noise: almost exactly the same. This makes sense: even if you don't see the black dots, you'd be able to recreate, mentally, the mental line by keeping a 'moving average' through the blue. A moving average *is* a type of high-pass filter. In other words, if the P3 is calculated as an average, then high-pass filtering doesn't add anything. On the other hand, if I wanted to know *when* the P3 occurs, then the low-pass filtering may help: for the unfiltered data above, the maximum occurred anywhere between 450 and 550 ms rather than exactly at 500 ms. Clearly, peak detection algorithms (p. 246) may suffer from random, local spikes found in unfiltered data, although, again, the remedy may be worse than the poison.

There is much more to learn: about different filters (Butterworth filters, a-causal filters, infinite impulse response filters), about the relationship between time and frequency domain, and so on. Interested readers, particularly those with an interest in physics or engineering, should definitely read up on some of the works suggested at the end of this chapter. For others, I will be satisfied if you at least get what a filter does (removing data), what it cannot do (creating noise-free data), and that one should be careful. In general, it is good practice to always try filters, for example on simulated signals, such as I will show in additional exercise 2, in order to better understand *how* filters cause distortions. However, in practice, EEG researchers seem to stick to recommended, default values. High-pass filters are recommended to keep at a maximum of 0.1 Hz (though 0.2 Hz is common). Low-pass filters are recommended to keep at a minimum of 80 Hz. Removing line-noise is always best avoided, for example by using a Faraday cage and good equipment, than filtered out. Of course, we all want super precise recordings of cooperative subjects done by experienced technicians in fancy labs, but the reality is usually less than perfect. As such, these recommendations are disregarded more often than not, but from this part of the chapter, I hope you at least do so while knowing the risks.

And after you applied the filters, do not forget to save the data ('umeeg101_07_applied_filters'). Since this was the last step, I will stop nagging about it, and hope you got the point: Save your data, and keep a document to remember what you were doing and, if possible, why you were doing it.

ESSENTIAL EXERCISES

Essential exercise 1: Preprocessing

Now it's your turn! If you are following the essential exercises only, then go through the following seven steps for every subject. Make clear notes whenever you notice something odd, or you have to do some backtracking. If, however, you are also following the additional exercises, you need only do the filtering part once, as an additional exercise aims to achieve the same result, but just by running code.

Save the result of each step with the name referencing the step (e.g. after resampling '_02_resampled').

1. Import the data.

2. If you use my data, resample them to 512 Hz.

3. Edit the channel locations and scroll through the data to get an initial impression of what the dataset looks like.

4. Remove bad times, including everything 20 s before the start of the experiment and 20 s after the end of the experiment.

5. Remove bad channels, making a note of which ones you have removed. If you are not so sure, make a note for later.

6. Re-reference the data to the common average. Inspect the data to see if you see a common type of noise: Do all the data go up or down? Is there high-frequency static noise? This may mean that you failed to remove a bad channel. Go through the previous step again before re-referencing again.

7. Filter the data, by first doing a high-pass filter at 0.2 Hz, then a low-pass at 80 Hz. Inspect the data to see if a 50 Hz notch filter is necessary. If you run into errors during filtering, try turning off 'plot frequency response' (which requires a special MATLAB toolbox).

Essential exercise 2: Filtering

So you have already filtered your data as part of step 7 in essential exercise 1, but how does the filter affect things really? Go through your data using EEGLAB plot>channel data until you find about 5 seconds of poor data that come right after some good data. Note down the exact start second of the poor data (bottom left of the screen) and the vertical zoom factor. Don't close the window, but go back to EEGLAB, and apply a high-pass filter at 1 Hz (which is too high by most standards!), creating a new dataset for the result. Again, go to channel data, inspecting the filtered data, but do this by typing in the second you noted earlier in the bottom left, and the vertical zoom factor in the bottom right, then pressing enter. Maximise both the original channel data window, and the new, filtered data so that they look exactly the same, in terms of beginning and end, zoom factor, and so on, except that the filtering is different in the new data. Do a side-by-side comparison carefully writing down how the filtering has affected both the bad signals and the good signal. Now go back to the unfiltered data and do as before but now with for an 80 Hz low-pass filter and a notch filter centred on the power line frequency (50 or 60 Hz).

Additional exercise 1: Beyond the EEGLAB GUI

Whenever you do some operation on the data, EEGLAB creates code, which you can inspect and re-use. As a result, it allows you to keep track of the history of what you have done so far, so you can be sure you have not accidentally omitted a step. Furthermore, you can easily create functioning code, with which you can create a 'batch-processing'. With a bit of coding you may avoid having to do hundreds of clicks and can just apply the same steps to each dataset. But perhaps most importantly, it gives you insight in what goes on under the hood and, as a result, you slowly gain MATLAB skills. Let's see how this works.

Step 1

Start EEGLAB and open an average referenced file of one of the datasets (as created as part of essential exercise 1).

Step 2

Filter the data with a high-pass filter at 0.2 Hz. Don't plot the frequency response, and don't save the data; simply overwrite it in memory. Now do the same but apply a 80 Hz low-pass filter. Finally, do the same but now apply a notch filter to remove the line noise. Check the resulting data to see if everything worked as it should.

Step 3

Save the data under a new filename.

Step 4

Now type in MATLAB

```
>> eegh
```

MATLAB will now output each processing step you performed as a bunch of commands, starting with:

```
[ALLEEG EEG CURRENTSET ALLCOM] = eeglab;
```

Step 5

Copy all the information to a new script (in MATLAB: top-left, new script). Remove everything but the following five lines. Change the following: (1) 'file_in' should be the name of your input file, (2) 'filepath' should be the location of the file_in on your drive, and (3) 'file_out' should be the new name you chose.

```
EEG = pop_eegfiltnew(EEG, 'locutoff',0.2);

EEG = pop_eegfiltnew(EEG, 'hicutoff',80);

EEG = pop_eegfiltnew(EEG, 'locutoff',46,'hicutoff',54,'revfilt',1);
```

Step 6

Write, in MATLAB

```
clear
```

and press enter. In the script you just made, click just in between the numbers 1–5 and each actual text, making a little red ball (**breakpoint**) appear (Figure 7.22):

Step 7

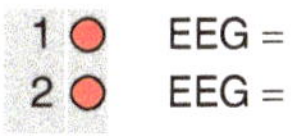

Run the script by pressing the green play button, keeping an eye on what MATLAB says in the command window. It will immediately pause at line 1 with a red circle in front of the >>, letting you know that it reached the breakpoint before line 1. Press the run button again to make it run up to line 2.

Figure 7.22

Source: Reprinted with permission of The MathWorks, Inc.

Have a look at the workspace in MATLAB. There should now be a single variable, a 'struct' called EEG. Double-click it to see what it contains in MATLAB's variable view. You can see it stores all sorts of things sorted inside different 'fields', such as the setname (the name you gave it), the data (the EEG), and the events (the recorded triggers). Click on each of these to inspect their contents. So, at this point, the data are loaded into the computer's working memory and available for MATLAB under the name EEG. Or, to put it in terms of code, the function pop_loadset returned data from the disk, which were then stored under the name EEG (any other name can be used; it simply changes the name of the variable).

Step 8

While keeping an eye on MATLAB's command view, continue the script through to the end, noting what each step does and how it relates to your original filtering and saving commands (steps 2 and 3).

To get a better idea of what each function does, write 'help [function-name]' in MATLAB. For example,

```
help pop_eegfiltnew
```

tells you exactly what the input parameters (that follow pop_eegfiltnew between parentheses) and output variables (that come before the pop_eegfiltnew in square brackets) are.

Note that you may have many more parameters between the parentheses than necessary. Depending on your EEGLAB version, you might see 'plotfreqz', with a value of 1 (on) or 0 (off) for example. A previous version of pop_eegfiltnew also worked by always including both low- and high-cut parameters. MATLAB uses an open square bracket immediately followed by a closed square bracket to denote empty space, so if you still use this version, then the filter dialogue screen where you entered 0.2 as the lower edge, but nothing as the higher one will become 0.2 and [] in MATLAB language.

But what happened with the notch filter? Note that the notch filter seems like a combination of a low-cut filter at 46 Hz combined with a high-cut filter at 54 Hz. Shouldn't that mean that only data between 46 and above 54 Hz is passed? As you can see, in the entire set of parameters – EEG, 'lowcutoff', 46', 'hicutoff', 54', 'revfilt', 1 – note the revfilt parameter and read about it in the documentation. It appears that *revfilt* inverts the filter from passing signals to cutting them. As I explained earlier in this chapter, this is indeed somewhat confusing, but for this reason it is always a good idea to inspect your data before and after filtering!

Now, let's say you have three files and want to run the same filters on them all, how do you do this?

Step 9

First, define a variable as containing all your files. Create a new line before your original script, and edit the following to make it work for you with the correct names:

```
filesin = {'myfile1.set', 'myfile2.set', 'myfile3.set'};
```

This creates a *cell* variable which holds three names. Now, to make the magic happen, you will want to create a **loop**, so that **for** each of the files, the same commands apply.

Step 10

Create a new line after filesin:

```
for cfile = 1:3
```

and a new line at the very end of the script:

```
end
```

If you have done some coding before, you will recognise this as a loop. If you haven't, the idea is that the part between the 'for' and 'end' is repeated three times, from 1 to 3 (the colon should be pronounced as 'to' in MATLAB). While doing this, a variable, which I called cfile (for 'current file'), is simultaneously updated, so that at the first run through the for…end, cfile equals 1, the second 2, and the third 3. Note that the name 'cfile' can be anything you want, and that MATLAB does not require variable declaration.

If you run the script, you will find that MATLAB will load the file and apply the filters three times, but that is obviously not very useful. You want it first to load file 1, second to load file 2, and third to load file 3. To do this, we have to adjust the function to refer to the variable 'filesin'.

Step 11

Adjust the pop_loadset function (keep in mind you would have a different filepath):

```
EEG = pop_loadset('filename',filesin{cfile},'filepath','file_path');
```

Now, MATLAB will load three different files, filesin{1} being 'myfile1.set', filesin{2}, 'myfile2. set'. Of course, it will still save each file as 'file_out.set', overwriting the same file twice.

Step 12

Use the same logic as above to fix the script so that it uses three different names to save the files.

Additional exercise 2: Signal filtering

Note the beautiful figures I made to show how filtering affected data. Now that you know a little bit of MATLAB, I can show you how to do this.

Step 1

Create a variable to contain a bit of simulated EEG (the input signal, hence 'sigin') with a couple of seconds of silence. All of this can be done in the command window or script editor, whatever your preference.

```
sigin = zeros(1,10000);
```

This creates 1 row of 10000 values, all set to zero (double-click on sigin in the variable view). We will imagine that this is our original EEG data, sampled at 1000 hz – for convenience, because it means each number equals 1 millisecond.

Step 2

Now I would like a very short 10 Hz pulse. To avoid tricky maths, I will first change 50 ms, from 4951 to 5000 to 1, and then the next 50 ms to -1. Enter the following:

```
sigin(4951:5000) = 1;
sigin(5001:5050) = -1;
```

Note that the '=' sign in programming rarely means equals like in maths, but is a command to put something in another place. So here, you can read it as requesting the value 1 to be stored in positions 4951 to 5000 of the variable sigin.

Step 3

Finally, we will filter the data and inspect the results:

```
sigout = eegfilt(sigin, 1000, 0, 20);
plot(sigout)
```

Now, there are two major parts to this exercise. First, using the help functions of EEGLAB, finds out what eegfilt did: what is the 1000, 0, 20? Second, they find out how changing the parameters affects the simulated signal: what happens if you change 0 and 20? Third, they find out what happens if you change the input signal. As you can see in the original figures, I did not actually use a square wave like 'sigin' for Figure 7.20, page 155. Try to explore different types of input signal, such as sine waves (e.g. try `sinpi(0:0.001:1)` to create a 0.5 Hz sine wave).

EXPLORE

When you downloaded EEGLAB, you were also invited to join the official EEGLAB mailing list (eeglablist: sccn.ucsd.edu/wiki/EEGLAB_mailing_lists), which is a very useful resource for any questions you might have about EEGLAB functionality or more theoretical questions on EEG analysis. For example, the topic of references (and re-referencing) is a recurring subject, with questions like why we should add a zero-filled channel before average referencing (see Makoto's preprocessing pipeline, also mentioned on p. 187). A more detailed explanation of referencing and filtering can be found in Steve Luck's *An introduction to the event-related potential technique* (Luck, 2014). Various alternative methods of re-referencing may be found in the literature (Kayser & Tenke, 2010; Lepage et al., 2014; Yao, 2001). For a practical example of why the choice of reference is important, I like Joyce and Rossion's (2005) work, which suggests two previously measured face-related ERP components (one positive, the other negative) might actually be the same single one, resulting from choice of reference. Nitschke et al. (1998), similarly explain how filter designs can practically affect ERP components, with a hearty meal of technical information for readers with a stronger physics or engineering background.

8

FURTHER PROCESSING OF EEG: ARTEFACT REJECTION AND CORRECTION

IN THIS CHAPTER, YOU WILL LEARN:

- How to segment your data into epochs
- How to spot artefacts
- How to remove artefacts
- How to recover data

In Chapter 7, I revealed my insights on the nature of badness, in terms of pearly white cows and sheep in wolves' clothing, as it pertains to EEG. Undeterred by the laws of implausibility, I will now steadfastly continue my theological march against corruption in the biosignals, how to detect it and either to cleanse it or to utterly destroy it. Yes, if you are somewhere way up in the Catholic church and have access to oodles of money, I am telling you that neurotheology is definitely a thing and that I am for hire.

If not, I will apologise and swiftly continue with the matter at hand: bad EEG vibes.

It is a truth rarely acknowledged that bad subjects do exist. Yes, we tell them there's no bad answer to psychological questionnaires, but what do you make of someone who gives a 3 on a 1–5 point Likert because they 'don't agree with the question'? Neuromarketing people sometimes claim that the *brain doesn't lie*, which is complete nonsense to begin with (if the brain doesn't lie, who does?), but brain data will suffer if the subject fails to conform, whether they do so explicitly, or involuntarily. For example, a hungover student vomiting during the EEG – true story – obviously produces movement artefacts, but so do more mundane problems, such as tapping along with a foot, losing focus, and incessant coughing. There are also static biosignal issues, such as the mentioned problem of amazing hair on a well-defined head rather than a failing comb-over on a balloon-like skull. The most problematic behaviour, however, will go unnoticed but still affect brain signals. For example, someone might misread and for some reason mentally count the number of times they saw themselves. We then don't see any visible signs of badness, what we call *artefacts*, but the data will be useless. Or someone might tap their left foot when they notice someone, causing relevance-related brain activity to be *confounded* (p. 25) with motor activity. Sadly, this means the person is best removed from the analysis, even if they were highly focused.

Apart from such catastrophes, however, there's the less insidious, *incidental* artefact. These do not ruin a dataset, but just a single spoonful of data over the course of a dataset dinner. Of course, a single bite can sour an entire dinner, and so too do epochs with strong enough incidental artefacts compromise a whole dataset. That makes it sound like these artefacts are harmless and I am some sort of culinary gonzo journalist, but imagine eating a foot-long sandwich with your favourite toppings and just 10% battery acid that you can probably eat around. So, in this chapter, I will explain how to slice around your sandwich like a mad EEG ninja, avoiding bits of battery, or possibly even go head on, *neutralising* the acid *using this amazing trick one dad told me* (and now the dentist hates me).

Before you begin: Make sure you have the data from the previous chapter. If you have been following the book, that means data that have been trimmed, re-referenced, and filtered. If not, I recommend downloading the sample data from the SAGE website, so that you can work along with me and get a good idea of what the result of each processing step should be before you start with your own data. Download this chapter's online material (https://study.sagepub.com/Spape) and extract the files in the special EEG folder. This should result in a new directory, containing three subdirectories (code, data, eeglab-gratton-master), of which data contains two separate files for each dataset: one with extension .set (with miscellaneous information like channel and event data) and .fdt (with EEG data).

EPOCHING AND BASELINES

In this chapter, we will be slicing the dataset into good and bad chunks of data. Such chunks or slices are normally referred to as **epochs** or sometimes **segments**, and the act of chopping up a continuous dataset as **epoching** or **segmenting**. As we will see in the later part of this chapter, one good reason to do this is to systematically remove data from the dataset that are **contaminated with artefacts**, without removing too much useful data. But I think it's a good idea to keep the real goal in mind, which is to reveal the effect of seeing the mental operation on the brainwaves, and epoching is critical for this.

Remember, here we use the event-related potential technique to figure out which part of the brain activity is related to the experimental condition – such as relevance detection – and *not* to anything else. For example, if the participant was thinking about their holiday, this would be an unwanted type of brain activity (unless tropical cognition is the next hot topic in psychophysiology), and thus, statistically speaking, it would be *noise*. The idea behind ERP research is that cognition relating to an event such as a stimulus happens at a stable time: detecting a face as relevant in our experiment necessarily happens after light stimulated the retina, the primary visual area was activated, and a face recognised. Noise, on the other hand, happens randomly in time. This means that if we average across stimuli, all activity related to relevance of a stimulus – occurring at a stable time relative to the stimulus – should remain, while all activity related to *holiday-thinking* – occurring at random times – should be removed. Let's see how this works in action.

Imagine a world in which there are only two mental processes (Figure 8.1): face processing (left) and thinking of holidays (right). Curiously, both processes affect brainwaves in the same way: there's a little jump of activity (positivity) followed by a slump (negativity) afterwards.

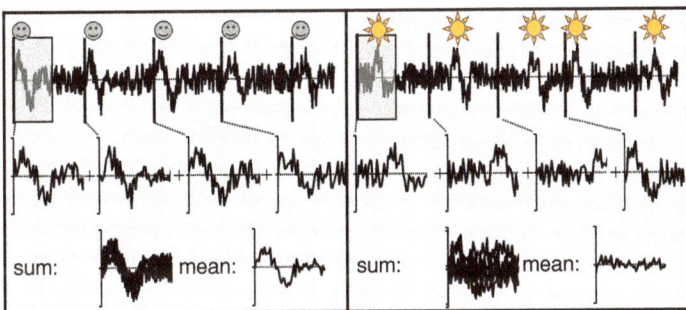

Figure 8.1

However, *face processing* is about the stimulus, so should occur just after the stimulus is shown, indicated by the vertical lines (markers) on top. Holiday thinking, however, does not relate to the stimulus, and thus occurs equally often after the stimulus, before it, or at any other time. Now, let's say we chop each of these continuous datasets relative to the onset of the stimuli. As a result, time 0 for each of these *epochs*, which is shown in the second row in Figure 8.1, is the onset of a face, which in EEG-speak is called **time-locking**. Already, you may see how, on the left side, the four bumps look like they happen at a similar time, some early positivity followed by negativity, while on the right, the four epochs have similar bumps but a more randomised time. Of course, we easily see patterns where they don't exist, but after we combine all epochs into one 'summed' pattern, it is very hard to deny that there's a clear pattern on the left where there's just a mess on the right. Notice also how averaging the epochs of all the bumps results in a smoothing effect on the left, while on the right, the original holiday-making bump is completely gone. This is because if the bumps and slumps are unrelated to the stimuli, it becomes just as likely for a bump to be combined with a slump, cancelling out one another. As a result, the bottom right is becoming a CSI-like flatline.

Baselines

OK, but is that a flat line, though, or am I just being over-eager to see a pattern where there is none? Maybe my mean on the left is just a little meaningless twirl? To answer such pesky questions, we can apply the same logic of epoching to say that *any EEG that came before the onset of the stimuli is not evoked by the current stimulus.* That means that the data before the onset should be a bit

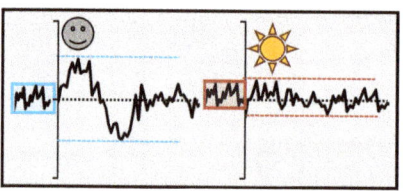

Figure 8.2

like noise, and behave like the right panel of Figure 8.1, so that averaging should cause it to go flat. As a result, ERPs are almost always plotted with a little bit of time before the onset of the stimulus, the **baseline**, which makes it easy to see what part of the signal is related to the stimulus. In Figure 8.2, I plotted the face and holiday average with a bit of baseline activity from before the stimuli. If you compare the degree to which there is waviness in the noise (boxes) with the waviness in the post stimuli areas, you are doing exactly the same as a T test or an ANOVA in SPSS: comparing the variation in the effect with the variation in noise. If the effect is about twice as large as random variation, then this customarily denotes significance.

Clearly, epoching is an incredibly powerful technique that allows you to cancel out both random noise and statistical noise. Random noise in Figure 8.2 was shown in the hairiness of the lines, which was about 50% stronger than the signal, and which smoothed out as a function of the averaging. With statistical noise, I meant processes that are unrelated to the experimental design – such as, here, holiday-thinking – and such a signal was quickly removed by averaging. As we will see, however, there is far, far more noise in EEG than is shown here, so we will need to be a bit careful before including everything in the averaging process. But before we look into that, let's first have a look at how epoching works in EEGLAB.

Epoching in EEGLAB

Thankfully, the practice of epoching, as implemented in EEGLAB, is easier than the theory behind it and we just need two pieces of information: *what is the timelocking event*, and *how long will each epoch be*? For the first question, as mentioned before, timelocking refers to the practice of creating epochs relative in time to an event – usually a stimulus – that will define the new time 0. For example, in our experiment, we will look at brain activity related to detecting a relevant face, as opposed to an irrelevant face, or a pet. Recall from Chapters 3 and 4 that at the onset of relevant faces the trigger number 201 was sent, for irrelevant ones 202, and for pets 203. So, if we look at umeeg101_07_filtered (this is how I named the last versions made at the end of the last chapter), you can see that just before second 24, a relevant face came by, while the faces at 24.5, 25.2, 26.0, and 26.7 were irrelevant (not the one the subject – me – was to keep in mind) (Figure 8.3).

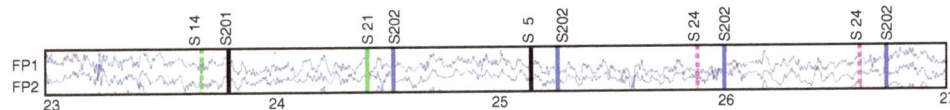

Figure 8.3

In other words, we will be timelocking to events 201, 202, and 203, each signifying the onset of a face. It might be tempting to already zoom in on one favourite condition, such as the relevant face, but it is more convenient to wait until a later stage (Chapter 9). For now, we are simply interested in what happens in the brain if someone sees a picture, never mind its exact quality.

How long should the epoch be? Ideally, an epoch should be not too much longer than the brain signals it is supposed to contain, plus a little margin to be sure. For P3-like research, all the usual potentials (see Chapter 1) tend to occur before c. 600 ms, so 800 ms would be sufficient to capture the late potentials of even slow people like yours truly. But about 600 ms after 201, note event 21 in Figure 8.3. Remember from Chapter 2, just before each face, there was a black ISI lasting 100 ms. I managed to run the experiment on a particularly slow computer (running an older version of E-Prime on Windows XP), so my 600 ms turned out to be 617 ms, and the 100 ms was at 116. So, from the point of view of event 201, at 617 there's a black stimulus onset, and at 733, the next face. We will keep that in mind in Chapter 9, but for now, let's set the end time to 800. With regard to the start time, I have already explained the use of taking in a bit of extra time before the onset of a stimulus, and 200 ms is conventional.

In EEGLAB, then, having opened the file to be epoched, umeeg102_07_filtered.set, go to Tools > Extract epochs, select the three event types to be epoched, and enter as Epoch limits -0.2 0.8 (see Figure 8.4):

Extract data epochs - pop_epoch()		
Time-locking event type(s) ([]=all)	201 202 203	...
Epoch limits [start, end] in seconds	-0.2 0.8	
Name for the new dataset	umeeg102_06_epoched	
Out-of-bounds EEG limits if any [min max]		
Help	Cancel	Ok

Figure 8.4

If you use my data, you can continue onwards from Chapter 7's final step, and save this dataset with a name ending in '_o8_epoched'.

Baseline correcting

After you press OK to the extract data epochs dialogue, you will be immediately asked whether you want to remove the baseline. The same menu appears if, in the EEGLAB box, you go to Tools > Remove baseline (Figure 8.5). Change the baseline (if necessary) and click OK.

EEGLAB already recognises the time to include negative numbers, and fills in the entire time range before the 0 as the expected baseline. That's cool. What is less clever is that immediately after extracting the epochs with a time range in seconds (-0.2 to 0.8), the baseline removal range is now in milliseconds (-200.1953 0). And what is this -200.1953? Recall that the Biosemi sampling rate was 1024 Hz, or one data point for every 0.9766 ms, and all datasets were resampled (p. 140) to 512 Hz, which means a range of exactly 200 ms is out of the question, so it takes a little more. The result won't change much this way or that because of that split millisecond, but if you are obsessive like myself, you can change it to round numbers. Just deal with it, EEGLAB!

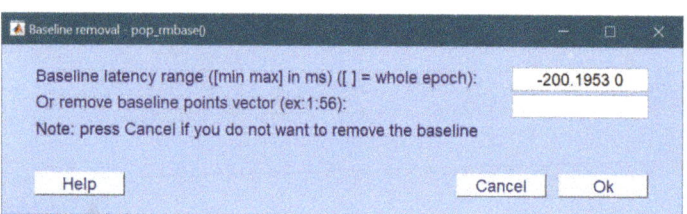

Figure 8.5

If you use my data, save the new, baseline-corrected dataset with a name ending in '_o9_ baseline_corrected'.

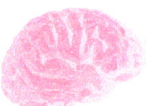

Baseline correction

Baseline correction is a pretty basic operation. What EEGLAB does is simply look at the baseline time-range (now -200 to 0), shown as the blue area in Figure 8.6, which shows the first epoch of the Pz channel of umeeg102. It then calculates the average (about -6), and subtracts this number from the entire epoch, thus moving the straight line 6 up to where the dotted line is.

After the epoching and baseline correcting, the old EEGLAB > Plot > Channel data (scroll) will give a very different view, because the data are now no longer continuous, but segmented. All the data between the epochs are lost, unless the epochs overlap, in which case the end of one epoch is the beginning of the next. We will become more familiar with this new view over the course of the next part of this chapter, which deals with artefacts.

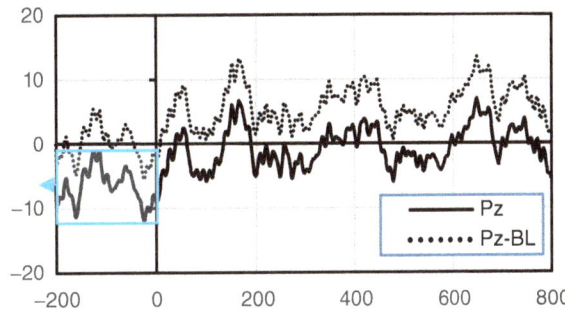

Figure 8.6

ARTEFACTS

You may think I have a bit of an obsession with artefacts, but it takes some time to really grasp just how enormous the problem really is. It is not just the notorious holiday-thinking we have to deal with, or any other sort of experimentally unrelated cognitive process for that matter, but the truth is that *untreated EEG is more noise than brain*. In many ways, doing EEG research is a bit like studying bird song with a few cheap microphones planted in a grassy field between a motorway and an outdoor dance music festival taking place despite an ongoing earthquake. The chirping you picked up may well be that of your favourite bird, but it could be the ecstatic cheering of the crowd or perhaps somebody just happened to kick over a microphone. In other words, the signal to noise is not just bad, it is beyond terrible. It would be cause for despair, but, looking at it optimistically, I find it is a bit of magic also: somehow, we manage to study the symphony of the brain *despite* the awful conditions. And yet, it is important to keep this simple truth about noise in EEG in mind whenever you read a news article that claims we will soon be able to 'record our dreams' using EEG. Such drivel is pointless because even if we could replace the cheap microphones with studio gear, scalp electric potentials give an incredibly muddy signal that combine brain activity from the entire cortex with a ton of noise. Or to put it in listicle format, *these eight types of signals combine to make your EEG*:

1. The **targeted mental state or process**. That is, the event-related potential of interest such as the relevance detection in the present study. Depending on the precise interest, that can be anything between 2 µV (e.g. medial frontal negativity), 4 µV (error related negativity), and 8 µV (P3). These are all decades old, so what most researchers are typically after is not 'finding a P3', but a change or **modulation** of the effect, such as 'how does seeing oneself affect the P3'. Such modulations are usually smaller than the original effect: seeing oneself will likely give a big P3, but seeing another relevant person should give a solid P3 anyway. Maybe, then, we could optimistically *guestimate* our effect to be 5 µV. That's the signal we're looking for.

2. Any **other brain signal** that is unrelated to our experiment. Like the holiday, of course, but don't forget that people still do things and have senses even if it is inconvenient. Motor activity can include foot-tapping, as previously mentioned, but moving the head, eyes, hands, and so on still involve the brain. And with sensory processing there is an equal amount of activity that is easily forgotten: there is a whole peripheral world of vision around the computer screen even if a participant cooperates and keeps focused on the screen, they can also hear you or the construction work, and their skin keeps on relaying

myriad messages all the time. Take a second to do this mindfulness exercise: how much can you feel if you put your mind to it? Motor and sensory information is processed by the brain with or without awareness, and while their individual influence is little (we may pick up about 5 μV of activity for a single visual stimulus), they are far more frequent than your stimulus of interest and randomly combine. Again, optimistically, let's give this source of noise a score of only 10 μV.

3.   One special case of a brain signal that usually constitutes noise is Hans Berger's famous **alpha wave**. He was able to observe this in 1924, despite poor equipment and no computer, for the simple reason that alpha waves are so large and so incredibly easy to elicit that it becomes hard *not to notice* them. Fun fact: humans get bored very easily and can lose their concentration on command: just ask them to close their eyes and think of where to go on holiday, and they will start generating alpha waves almost immediately. Sadly, they will also do this without prompting and, as you will see from my own data, pretty much as soon as a fixation signal is in view, the alpha kicks in, signifying me drifting off. Such waves are a large source of noise (say, 30 μV), but they are also horrifying for the simple fact that if a participant is not concentrating, their brain isn't either, so the experimental manipulation will likely not have as large an impact, unless it is some sort of startle stimulus. This means that even if we could simply remove alpha waves, it would not help much.

4.   **Eye-movements** are another infamous type of noise contaminating our EEG signal with a non-brain source. They are easily observed while reading, so in general you should try to avoid having long blocks of text as part of the experimental stimulus (they are fine as instructions and such). Although some people believe eye artefacts in the EEG are muscle activity, this is incorrect. While muscle activity causes the eyes to move, the consensus is that the activity we pick up in the EEG is due to the physical rolling of the eye-ball itself. We will see eye-movement activity as blocky patterns in the data, the EEG affected in a pattern that depends on the relative positioning of the electrode to the eyes: horizontal movements giving frontal positivity on one side (e.g. FT7) and negativity on the other (FT8), while vertical movements give frontal positivity up (FP1) and down (below the eye). Sooner or later, however, a powerful force of nature sets in that causes someone who looks up for a while to eventually look down again, or left to right, leading to a reversal of the pattern. This then results in 'block-like' activity patterns that are usually easy to spot (see Figure 8.10, p. 177). Eye-movements cause activity that are large to huge in size depending on the amount of movement, but typical ones are around 50 μV frontal. They are smaller, for obvious reasons, parietally, occipitally and laterally, but even there you can usually still see the activity with the naked eyes.

5.   **Eye-blink artefacts** are similar to eye-movements in that they are huge (depending on the subject and amplifier, in the 100s of μVs). Unlike eye-movements, however, they are also hard to do anything about without resorting to *A Clockwork Orange*-like measures to keep the eyelids pried open. Typically, they occur about every 5 s, although this depends on the person and their mood (Colzato et al., 2008). The standard in EEG recording is to ask participants not to blink, or to delay their need for blinking until after the trial, but I suspect that people often start to blink more if you ask them about it. It's a bit like itching, do you feel it? Are your eyes starting to feel a bit dry?

6.   Other non-brain EEG noise activity exists but is infrequent. **Blood pumping** can cause a c. 1 Hz (if the heart is going at 60 BPM, which is a bit slow, but most experiments are boring) recurring peak pattern of c. 0–5 µV, particularly near the neck if the participant wears too tight a cap. **Random muscle twitches** also sometimes occur: twitches in the eyes, head movements, and so on. **Frowning** is a particularly nasty type as it happens mainly with participants who genuinely are trying to do their best, focusing really, really hard. I try to tell them that is great, but perhaps they could keep their face relaxed, as it otherwise causes frequent, longer-lasting, high-amplitude (c. 100 µV) high-frequency noise around the fore-head and eyes. Other unintentional movements are **chewing and teeth grinding, swinging feet**, and worst of all, **coughing**.

7.   There are static sources of noise that are not due to the participant's behaviour or brain. These include **non-conductive pollutants** between the electrode and the brain, such as the meninx (the membrane in which the brain is suspended), skull, skin, skin-flakes, secretions (e.g. sweat), and air. Such sources of noise are necessarily independent sources: the random noise entering the one improperly attached electrode should not affect whatever its neighbours pick up, and for such reasons can be removed with more advanced artefact correction, such as ICA (see additional exercise 1). Without that, it is usually better to remove the bad channels entirely or to **interpolate** them – that is, remove them and deduct the activity from the surrounding electrodes. However, this does not regain bad data, and can be done at any later stage just as well, for which reason I use interpolation only very late in the analysis (Chapter 11).

8.   Finally, there is noise due to surrounding electric activity, the so-called line-noise or **mains hum**. This is due to the alternating current at the frequency of the power network, and is avoidable to some extent with fancy lab equipment (p. 98), otherwise causing steady, strong 50 Hz (in Europe and most countries) or 60 Hz (in the Americas) activity that is unrelated to your participant. There are some filtering techniques, although these have their drawbacks (Chapter 7), and while some fancier EEGLAB plugins (EEGLAB processing extension Cleanline) do a better job at it, like any source of noise they are better avoided than treated.

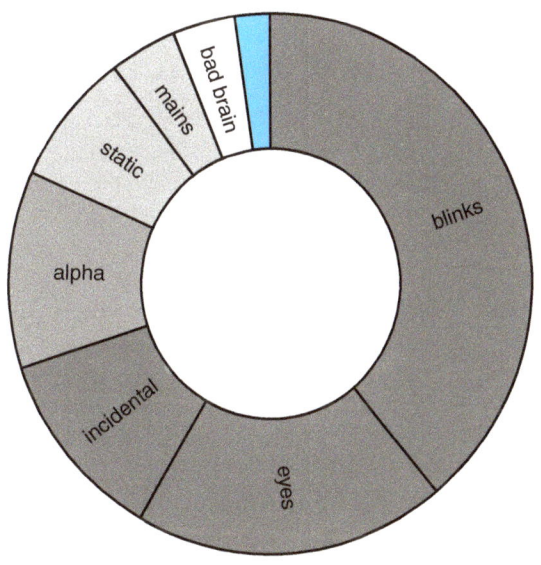

So to put this in perspective: if you see the EEG you are recording and consider all the sources of noise other than the signal you are interested in, then what we are looking at is the red slice of target activity in Figure 8.7.

Consequently, if we measure something using EEG, it is likely unrelated to the brain. As we discussed, the averaging idea behind event-related potentials already causes the noise to systematically

Figure 8.7

shrink, as long as it is unrelated to the signal. However, including an epoch with a 200 µV blink into the averaging will cause a lot more noise than a clean epoch contributes. This is why we do **artefact detection** (to find out what is grey and what is red), so as to remove epochs before they cause the ERP to be dramatically biased by contributing tons of known noise to a sensitive operation. The classic way of dealing with artefacts is by manually going over the entire EEG recording and discarding what looks like contaminated data, such as removing the darkest part of the read-out in Figure 8.7. Such **artefact rejection** is always necessary, but sometimes it may be possible to recover original data even if some artefact sources contaminated them – such as adjusting the colours of the read-out. This is **artefact correction**, which aims to reduce, as much as possible, the degree to which artefactual sources affect EEG.

In a typical processing pipeline, one first removes the worst parts of the data using artefact rejection, then applies artefact correction, and finally applies artefact rejection once again. Why do the rejection twice? One reason is that data that are beyond awful will never be great, making it pointless to even try. Indeed, if seconds of data are showing tons of noise, then there is usually some underlying reason why you should exclude the data anyway. Let's say, the data are due to a participant sneezing. Do you normally pay close attention to your environments when you sneeze? Do you even have your eyes open? Well, perhaps you are a polite sort of sneezer, but my own sneezes are famously volatile, regularly causing the entire department to start evacuation procedures. Some experimental picture won't really have registered.

The other reason is that artefact correction techniques try to detect patterns of bad data and reduce such patterns as much as possible. Some sources of noise are rare but extremely large, so optimising a form of artefact correction so as to completely remove a single outlier in 0.1% of the data is not very useful. That is a bit like organising the entire police force against a serial killer (Jack the Rip**P3**r!). Maybe it will help, but perhaps it is not the best way of balancing a limited amount of resources? So let's start cleaning out the worst of the mess first.

Artefact rejection, first pass

We started off from an epoched, baseline-corrected dataset containing all epochs, filtered and ready for artefact rejection/correction. In EEGLAB, go to plot > Channel Data (scroll) and browse through the data, noting down on paper where you see relatively good data and where you are absolutely sure there's something bad going on that is affecting most, if not all, channels. This can be pretty difficult the first time you do this, so let's start by doing it together. Starting off with umeeg101 (Figure 8.8), I zoom out vertically to a scale of 50, looking at the top of the graph, and note that the first 5 epochs look relatively OK. After a while, I seem to have got a bit tired (*seconds* into the experiment), with alpha showing at epoch 27, getting more pronounced over time, until at epoch 227, there is such a pronounced and long blink that I would bet I closed my eyes at this stage for a *micro-nap. Come on, me, wake up!* I did seem to, as indeed immediately after the blink the alpha is gone. Now, all these data are basically OK, except perhaps the nap at 227.

After a while, though, the signal degrades, and at epochs 1408–1409 I spot slight motor noise, leading into a fully bad signal in pretty much all electrodes at epoch 1410. In my logbook, I therefore note that most epochs up to 226 are OK, that 227 is a bit iffy, and that 1410–1411 is definitely bad. In the data scroll view, you could already select these epochs and click on REJECT to remove the data immediately, but I keep them for now, for purposes of illustration.

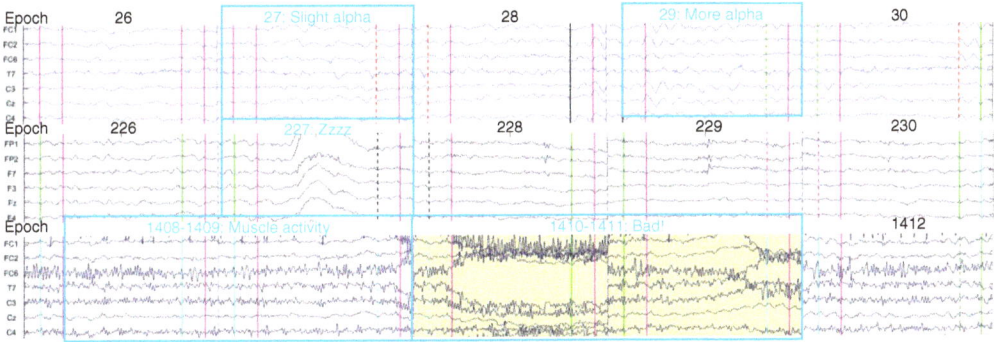

Figure 8.8   Alpha, muscle activity, bad signal

In EEGLAB, go to Tools > Reject data epochs > Reject extreme values. If you don't find this menu item, it might be hidden as a 'menu item from previous EEGLAB versions', which it is by default. To get it back, go to File > Preferences > and put a check next to 'If set, show all menu items' and restart EEGLAB. A simple dialogue appears asking for a range of electrode indices (i.e. channel numbers) to check maximum and minimum amplitude as an allowable range, and a start and end time limit. Keep the electrode indices and time range as it is, and change the minimum and maximum to something *huge* like -1500 and 1500. Press OK and EEGLAB will show the result. It probably won't show much, because the threshold was too high to catch much. Inside MATLAB, check the numeric result. For me it said:

34 channel selected

0/2170 trials marked for rejection

So that is clearly too liberal. Redo the operation but now change the criterion to -200 and 200. I found that now EEGLAB marked 15 trials out of 2170. That is better. My rule of thumb in this regard is that 1% of data will be unsalvageable. But to be sure, look at the activity scroll dialogue EEGLAB brings up and move to see if epoch 1410 and 1411 are now automatically selected (they are), if 227 is selected (it is), and if everything before 227 is left intact (all clean). Click Update marks, go to EEGLAB > Reject data epochs > Rejected marked epochs.

If you use my data, save the result with a name ending in _10_artefacts_rejected1.

Artefact correction

I mentioned there are two ways of dealing with artefacts: rejecting and correcting. Some papers I reviewed seem to think there is a third way – ignoring – but in my view that calls for a special type of *rejecting* by me – of their entire paper. With artefact correction, the aim is to regain as much data as possible by reducing the amount of noise contribution of the artefact. The problem, however, is that it is easy to see how frowning, for example, induces noise, but it is hard, from looking at a signal, to see how the signal would have been had there been no frowning. We have, as it were, a smoothie (or cocktail, but I'm writing this on a Monday morning!), containing multiple unknown ingredients, of which we know at least one, but potentially more, makes the end result taste horrible. In order to fix the smoothie, we would need to attempt **blind source separation**, figure out the ingredients so as to throw out the ones we don't like and make the result drinkable.

One of the oldest techniques that is still commonplace is by using **linear regression** and eye-electrodes. The idea is to place electrodes around the eyes in horizontal (left of the left and right of the right eye) and vertical (above and below one eye) configurations so as to maximally capture artefacts related to the eyes – principally eye-movements and blinks. Then, we use these electro-oculography (EOG) channels to predict the other channels, with the idea that all correlation with EOG channels is likely unwanted. The correlated activity is consequently removed or, as we say with regression, the residuals are retained.

While the Gratton, Coles, and Donchin (Gratton et al., 1983) method gets a little bit more complicated than this, I think it is not overly useful to go deeper into the mathematics. That is, if you are comfortable with linear regression such as you learned in your undergraduate statistics courses, then it's very similar except using continuous signals instead of survey responses or reaction times. If, on the other hand, you find statistics challenging, then talk of residuals, coefficients, and weights can just get frustrating, while it should be easy enough to understand what the algorithm does by looking at the data.

To start, find the somewhat ancient *gratton* plugin by Matthias Mittner from GitHub (https://github.com/ihrke/eeglab-gratton/ or the online material of this book). For the latter, copy the directory called eeglab-gratton-master to your EEGLAB's plugin folder, then restart EEGLAB. It requires EOG channels (such as HEOG or VEOG in umeeg101), but if you have these, go in EEGLAB to Tools > Ocular Correction (Gratton). The dialogue shown in Figure 8.9 pops up. Enter the number of the first EOG channel you want to correct against in the top line (I enter 34 to correct first against VEOG, which is the 34th channel. Note: the channel number is the same as counting from top to bottom in the plot > channel activity (scroll) view). I aim to correct channels 1 to 32 – that is all EEG channels. The other two parameters I keep to their defaults. Pressing OK gives a number of details in terms of how many blinks are detected, and how the channels each get a regression coefficient. The plugin then immediately creates a new dataset. In EEGLAB, check the new dataset to see the result of the artefact correction procedure. For example, in Figure 8.10, you can see the result with dataset umeeg101 (epochs 431–435) after first correcting against VEOG, then HEOG.

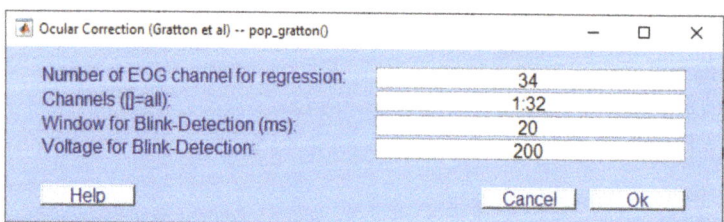

Figure 8.9

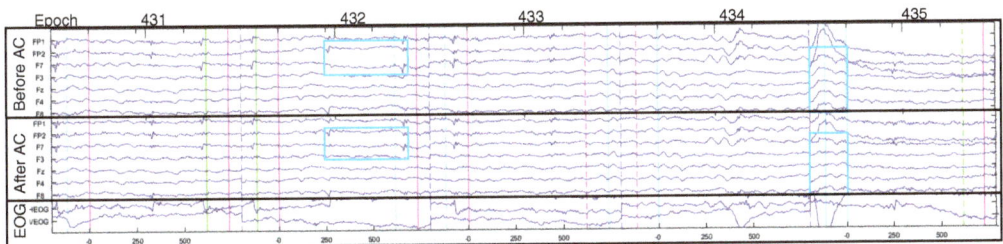

Figure 8.10 Channel activity graphs for epochs 431–435, with, in the first row, original data before artefact correction (AC), and in the second row, after correcting against VEOG and HEOG. The third row presents the original HEOG and VEOG data. In the two red boxes on the left, note the horizontal eye-movement and how AC reduced it. In the two red boxes on the right, note the eye blink. Look to see if you can distinguish more eye-movements and blinks by looking at the EOG data. Then see if you can spot the EEG artefacts and the degree to which AC worked.

Inspect the result, noting where strong ocular movements are present in the horizontal eye-movement channel (HEOG), and where blinks show up in the vertical one (VEOG). Horizontal eye movements are evident as block-like waves in the HEOG that coincide with left-sided (e.g. over F7) and right-sided (F8) activity going in opposite directions, as with the red block in the top left of Figure 8.10. Eye-blinks show up as 'vampire fangs' in the VEOG, inversely related to the very frontocentral electrodes (FP1, FP2). Note that they even show up after the correction, suggesting that the correction could have been better. However, at least the worst of the blink is gone, particularly in the electrodes that are slightly further from the source (F3, Fz, F4). It might help to tweak the parameters somewhat to get a better result, although you should avoid over-correcting the data. As to that, always make sure if the end result makes sense by eyeballing the before-and-after datasets – sometimes, a faulty artefact correction can cause the entire data to go haywire.

But what do you do if, as with umeeg102 and other Biosemi recordings, you simply do not have any HEOG or VEOG channels? One way to obtain these is to create such channels manually, for example by subtracting AF7 from AF8 or FT7 from FT8 to create a HEOG channel, simulating a bipolar electrode pair. But that is not easy using just the EEGLAB GUI, and does not help much with creating a VEOG channel, unless you used a cap that came with electrodes below the eyes as well. So for now, it may be best to leave the EEG datasets without HEOG and VEOG as is and continue with the second pass of artefact rejection. Alternatively, you could try the more advanced artefact correction techniques that are based on independent component analysis, which I will explain as part of Additional Exercise 1.

For datasets that include eye electrodes, however, save the result with a name ending in _11_arte-facts_corrected.

Artefact rejection, second pass

Whereas the first pass of artefact rejection was merely to get rid of the worst offenders, the second pass needs to leave us with a dataset we are fully satisfied with. Again, we will need to balance the need to retain many epochs to cancel out all the activity that is not related to the mental state of interest against the need to remove the largest noise contributors. Exactly what is a large noise contributor, though? This very much differs across labs' traditions, sensitivity of equipment, and, weirdly, skull thickness. That is, given that the EEG signals travel through the not very conductive skull, its thickness is inversely related to EEG signal. For such reasons, it is impossible to present an absolute threshold for saying, 'Ah, it's unlikely real data could ever have such a high number'. On the other hand, if everybody just does their own thing, it will quickly become impossible to compare research between labs. To cope, I present a relatively simple procedure that has a mixture of relative and absolute criteria, with the aim of maximising the number of datapoints to keep and minimising the number of suspect epochs. I start with the following three *golden rules* based on my experience:

---

1. If you are removing less than 5% of your epochs, your conclusions will be more related to noise (and holidays) than about your mental state of interest.
2. People blink about once every 4–20 seconds. People also frown, touch their head, move their head, produce alpha with boring experiments, and so on. The former can be corrected with artefact correction, the latter not so much.
3. If data recording went smoothly, and the experiment did not require movement, then an expected data loss of 20% is expected.

In other words, a good subject/dataset will have 5–20% data loss due to artefact rejection, and a bad subject will have more. These 'rules' are hardly set in stone, of course. They are mainly to help you cope with having to remove lots of the precious, hard-won data.

•   Start with the artefact corrected file we just made and in EEGLAB, go to Tools > Reject data epochs > Reject data (all methods).

•   In the large dialogue, look at the upper part, which is a copy of the former find extreme values from the section 'Artefact rejection, first pass'. Leave the default values as they are and press Calc/Plot. Wow, super colourful! (Figure 8.11).

As indicated in Figure 8.11, all epochs in green – likely all of them – are rejected for failing to stick to the default thresholds of + or – 25. Note that the six channels in red are failing this test with regards to the first epoch (shown). However, note that these channels are all quite far to the side of our elastic cap, which means (1) these tend to be noisier areas that might need a less conservative threshold, and (2) these tend not to be our areas of interest to begin with. In EEGLAB, go to Edit

> Channel Locations (or in MATLAB, type EEG.chanlocs.labels in the command window) and note all the channels that are less than 7 to the side, and not in the very front or back of the hat (i.e. not AF, FP, O, and I). For umeeg101, these include F3, Fz, F4, FC5, FC1, FC2, FC6, C3, Cz, C4, CP5, CP1, CP2, CP6, P3, Pz, P4. Then find out which *channel numbers* these have – that would be 4, 5, 6, 8, 9, 10, 11, 13, 14, 15, 18, 19, 20, 21, 24, 25, 26 for me.

- Go back to the large Reject data (all methods) dialogue (Figure 8.12), but now enter under electrodes the numbers you obtained above, leaving out the commas. Far fewer epochs are now green. Click on UPDATE MARKS to get the number – I got 1285.

That is still a bit much. And indeed, if you go through the Calc/Plot screen, you will notice a number of occurrences at which EEGLAB says there is an artefact, but I wouldn't be so sure. To get a better idea of the balance between removing too much signal and retaining too much noise, I always recommend making a little plot. You already know that with a threshold of + or – 25, 1285 out of 2170 are removed, i.e. 1285/2170 = 59.2%.

- Do the same as before, but move the threshold to 100 (and -100), note the number of epochs removed, and write down the percentage. Do the same for thresholds of 10, 20, 30, 40, 50, 60, 70, 80, and 90 and create a graph of the result.

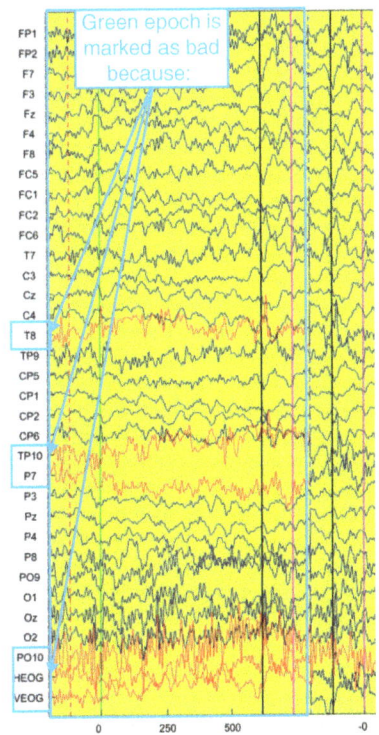

Figure 8.11

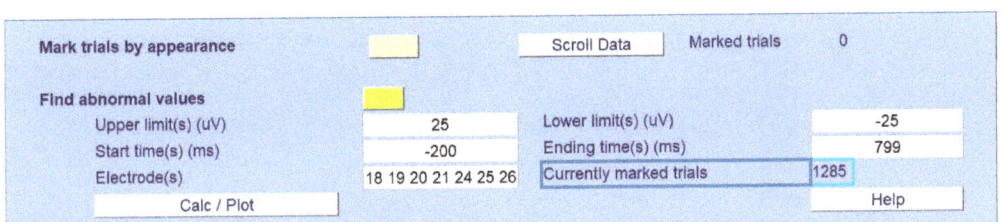

Figure 8.12

You can see my result in Figure 8.13. Note the three points:

1. Below 5%, which I had suggested earlier as sounding unlikely given the many times EEG is disrupted by random, unsalvageable artefacts, which corresponds to a threshold of about 70.

2. 20%, which I suggested feels *reasonable*.

3.   The *knee*, corresponding to the point of the graph at which the direction of movement starts
     to go the other way. That is to say, up to a threshold of c. 30, each slightly higher threshold
     gave us many new clean epochs (40% are removed at a threshold of 30, but only about 20%
     at a threshold of 40 – half as many), but at some point the added value of increasing the thresh-
     old a bit declines. If you balance the added value of more epochs against the cost of likely
     including more noise, this can be visualised as the point at which the graph seems to bend,
     hence the knee. Yes, if you have ever done principal component analysis in SPSS, the Scree
     plot is exactly where I got the inspiration for this from.

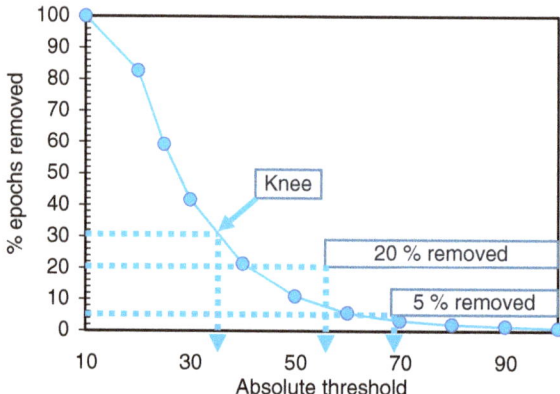

Figure 8.13

Choose, based on the considerations above, some threshold. Both 2 and 3 are pretty defensible, or you might split the difference and calculate the average between the knee and the 20%
criterion, at about 45.

Does this feel very arbitrary to you? Well, it probably is, but the further you go from your normal, SPSS-type exercises, the stranger the field of statistics gets and the fewer clear-cut answers
there are. For what it's worth, I tend to stick with a few EEGLAB defaults for further removal of
the artefacts:

Enter the threshold in the Reject (all methods) dialogue, hit Calc/Plot and mark epochs.

Then go through the abnormal trends, improbable data, and abnormal distributions, making
sure you select all electrodes, except the eye-electrodes – so for umeeg101, that is 1:32.

As a result, for me, 346 epochs are removed for having abnormal values, 82 for abnormal
trends, 188 for being improbable, and 198 for having abnormal distributions, which would mean
814 in total. But, luckily, artefactual epochs are like cows in that they rarely have just the one spot,
and so, too, do the various criteria overlap.

Finally, click on Reject marked epochs. EEGLAB will ask you whether you are sure. Make
sure you note in MATLAB's command window how many epochs it suggests are removed:
598/2170 for me, or about 27%, which is a fairly large number, but not unreasonably so. So click
Yes and save the new dataset.

After cleaning and correcting the EEG data, a process sometimes called 'pruning', you will
end up with a dataset that is fully pre-processed: trimmed, filtered, re-referenced, epoched, artefact corrected, and artefact rejected. This will be the starting point of the real EEG analysis, which
I will explain in Chapter 9.

Make sure you save these datasets with a name ending in something like _12_pruned.

ESSENTIAL EXERCISES

Essential exercise 1: Further processing

If you followed through from Chapter 7, you should have a number of datasets that do not have severely distorted data, and are re-referenced and filtered. If you haven't, feel free to download the sample datasets from the SAGE website (https://study.sagepub.com/Spape).

   Now continue processing each dataset, saving the dataset anytime something has been changed. Make sure you inspect the data from time to time and make notes.

1. *Process*: Epoch the data, *time-locked* to the critical events.

2. *Process*: Baseline-correct all epochs using a reasonable, neutral baseline period.

3. *Process*: Do a first-pass artefact rejection, removing the seriously bad epochs from further analysis.

4. *Inspect*: Note whether datasets have EOG channels and write down the channel number.

5. *Inspect*: Find epochs that look like they are **contaminated** with (1) eye-movement artefacts, (2) blinks, (3) muscle activity, (4) other artefacts. If possible, discuss with others and decide based on group consensus.

6. *Inspect*: Based on the above considerations, identify epochs that you feel are representative of being good, bad, and *iffy*, writing down their number so you can later retrieve them.

7. *Artefact correction*: If there are EOG channels, attempt correcting artefacts using linear regression and the *gratton* plugin. Inspect the data before and after, in particular with regards to the good, bad, and iffy epochs identified in step 6. If, however, you aim to do the additional exercises, you need only try the gratton plugin once, and then compare to the more advanced artefact rejection procedure that is explained in additional exercise 1. In this case, complete additional exercise 1 before continuing with essential exercise 2.

Essential exercise 2: Artefact rejection

Using the mixture of relative and absolute criteria I presented above, go through each dataset as described in the section 'Artefact rejection, second pass'. I would recommend drawing, either like me in Excel, or just on paper, a graph that shows the number of epochs to be rejected on the Y axis and the absolute criterion on the X axis. Don't forget to always use both positive and negative values in the EEGLAB dialogue (e.g. -25 and 25). Keep in mind that you will likely have to remove about 20–30% of data, unless you have just completed additional exercise 1, in which case you could expect about half as many.

Additional exercise 1: ICA based artefact correction

One of the most important recent developments in EEG methodology was the application of independent component analysis (ICA) to EEG. ICA is a technique for blind source separation of a multivariate signal into additive, maximally independent components. Imagine, for example, you are talking on your mobile phone, there is a source signal (you talking), but usually also plenty of background noise, which the microphone will pick up just as easily. The traditional way to deal with this was to apply a filter to your microphone, removing all frequencies that are not in the typical speaking range, but for similar reasons as argued earlier, this necessarily led to a degraded signal, making you sound 'tinny'. For such reasons, many phones these days also contain microphones placed at different locations on the phone so that they will receive more background noise and less of your voice, even though they will likely still have a trace amount of it. In other words, if we consider each microphone as providing a signal, each will have a contribution of both background and voice (Figure 8.14).

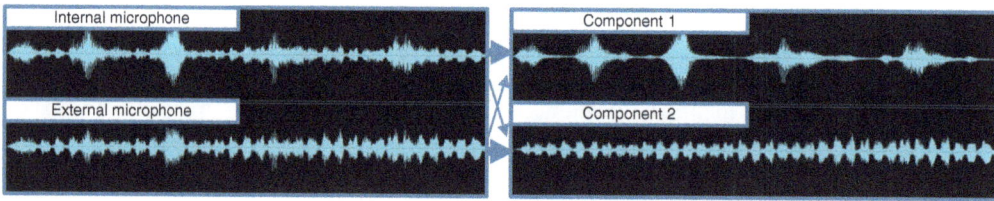

Figure 8.14

The job of ICA, then, is to separate from the two signals in the left panel of Figure 8.14 so that the two underlying components, in the right panel, are revealed, which should result in a separate audio signal that can then be transmitted to the person you're talking with. Note that the two actual signals on the left are *decomposed* (i.e. the inverse of mixed) into the two components as determined by the weights (illustrated with the thickness of the arrows). Together, the two components on the right contain the same information as the two signals on the left, except that the signals have an amount of variance shared whereas the two components are uncorrelated. This is what ICA does: maximise the degree to which components are independent from one another.

If you remember that EEG is a little bit like recording something using many microphones that pick up very similar information if they happen to be close to one another, you might concede that my analogy is pretty apt (for once). It also explains the two principal reasons for doing ICA: source analysis and artefact correction. As explained earlier, multiple EEG signals capture part of the same neuronal source or *dipole*, for example with frontal positivity and parietal negativity relating to two polar opposites of the same dipole. In other words, there is an amount of redundancy in our explanation involved if we keep talking about the positivity and negativity as if they are completely separated entities. At the same time, by estimating the location of a single dipole explaining both poles within the brain, we might be able to say where in the brain the magic happened. This rather tempting proposition led to many EEG groups using ICA to reach a deeper understanding of where the EEG is coming from, working at the *source* rather than *electrode level*. Personally, I find this to be an admirable, if challenging, endeavour as it requires making different ICA decompositions comparable between participants (how do my ICA components relate to yours?), and studies or labs (how do our ICA distributions compare to classic P3 research?).

In this extended exercise, I will explain how to apply the much less grandiose, but clearly useful, application of ICA to artefact correction. Like the two microphones picking up both the speaker and the background noise, our EEG electrodes contain both real brain-waves and various types of artefact. Luckily, artefacts tend to be independent from real EEG in time and location: a single EEG source (of, say, a P3 component) will likely be picked up by several electrodes at the same time, while eye blinking normally does not occur at the same time as a P3. Let's see how this works in EEGLAB.

Step 1

Open the re-referenced, epoched, baseline-corrected, and first-pass artefact rejected file created earlier. ICA doesn't particularly care about the epochs, it basically jams them all one after the other, but by using the epoched data, we can be sure it is optimised towards the EEG generated as part of the experiment. However, it is best practice to first remove the early baseline: again, correct the baseline, but now 'use the whole epoch', which removes the average of the entire epoch rather than the -200 to 0 range. Then go to Tools > Decompose data using ICA (older EEGLAB: Run ICA). There are multiple different ICA algorithms implemented in ICA, but the default 'runica' works fine, so stick with the default options and click OK.

Now sit back and enjoy. Get some coffee, go for a walk. It will take a while – about an hour for me. I am told that *binica*, the compiled version of runica, is much faster than *runica*, but alas, it just works for operating systems that aren't Windows. So, if you are using OSX, Debian or Ubuntu, try binica. I am not too bothered, though: it's good to take a break every now and then, so I'm getting that coffee!

You may have seen the step-counter in MATLAB saying 'step 94 – lrate 0.000143, wchange 0.04207507, angledelta 130.1 deg': it goes on like that until step 500 or learning rate drops to less than one in a million. Once it's done, there is little fanfare in EEGLAB other than that the main box now says ICA weights: Yes. Let's see the result of the decomposition in step 2.

Step 2

A recent addition to EEGLAB has been the automatic classification of ICA components, which uses fancy machine-learning trained on many other datasets to predict whether your components are related to the brain, muscles, eyes, heart, line-noise, channel noise, or other sources. In EEGLAB, go to Tools > Classify components using ICLabel > Label components, use the Default version of ICLabel, and use the default options in the next dialogue as well to conjure up an overview of all the components (as many as there are independent channels). Go through the components, writing down which ones clearly do *not* relate to the brain.

This takes an amount of personal training, but ICLabel helps a lot by providing a little indicator of how likely it estimates a component to be brain-related. It is good not to totally depend on this, particularly while you are learning, so click on any of the numbers in the big component property window shown at the end of step 2. For each component, you see, clockwise, (1) a topography of the weights, showing how each electrode combines into the component; (2) a breakdown of ICLabel probabilities; (3) a time-based view of the component's activity; (4) the power spectrum of the component; and (5) the epoch view with the activity shown in colour. As you go through the large amount of information, keep in mind some ground rules:

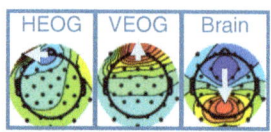

Figure 8.15

- Rule 1 Components are sorted by default in terms of their variance accounted for. This means generally that the first ones tend to relate to the eyes, as they cause a huge amount of independent activity. It also means that removing the last few components will have relatively little impact.

- Rule 2 The topography can often tell you if a component is related to the eyes, resulting in frontal colour, suggesting 'HEOG' or 'VEOG' dipole sources as shown in Figure 8.15, while brain topographies often show more smooth landscapes over the top of the head.

- Rule 3 The independent's component activity may show the typical shapes that characterise artefacts, such as the blocks (HEOG) and vampire fangs (blinks) discussed previously.

- Rule 4 Real EEG tends to have a power spectrum that generally goes down apart from a small bump around 10 Hz (alpha activity). Artefacts related to eyes tend to look similar but without the bump, and a more pronounced beginning. Muscle activity often has more high-frequency activity, although the differences are not that clear-cut, as shown in Figure 8.16.

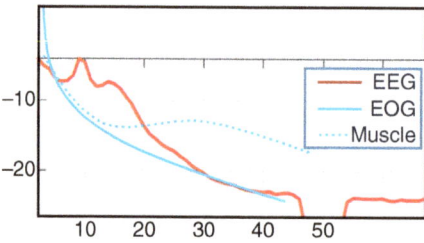

Figure 8.16

- Rule 5 Finally, artefacts are rarely event-related. Thus, for example, in Figure 8.17, you can spot that pretty much every single row starts with blue, going to yellow, to blue and to yellow again, showing up as a nice pattern of colourful bars in the epoch activity that average into the clear ERP to the bottom of the figure. Note that this criterion is neither necessary nor sufficient: people often manage to blink when something exciting happens, though rarely this consistently.

In any case, write down all the components you are pretty sure are bad, and those you are not too sure about but are certain contain *no* event-related or brain (e.g. alpha) activity. Make a note if you're not too sure about some of them. For example, for me and umeeg101_07_ranICA.set, I wrote: 1 2 3 9 10 11 12 14 15 16 22 25 30 33, the underlined ones I was unsure about.

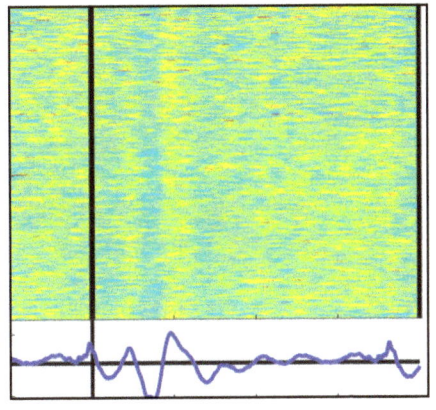

Figure 8.17

Step 3

In EEGLAB, go to Tools > Remove components from data, enter the numbers you wrote down in the List of component(s) to remove from data, and press OK. Don't worry, EEGLAB does not immediately remove them. Click Plot single trials to see how removing the components will affect the data. Figure 8.18 is what it shows for the previously noted epoch 227. In the central channels, the signal remains pretty similar, but the VEOG and HEOG activity is completely removed. In addition, some noise is gone from the lowest channels.

What about those uncertain components 14 and 22? What I usually do is not to complete the removal immediately, but put the different selections side by side, so first including all, then including all but 22, then including all but 14 and 22. Creating full-screen windows of each result, I hover back and forth between the various options, looking to see whether a component coincides with eye-movements and such known activities or whether removing a component appreciably affects things. Based on this, I make a final selection.

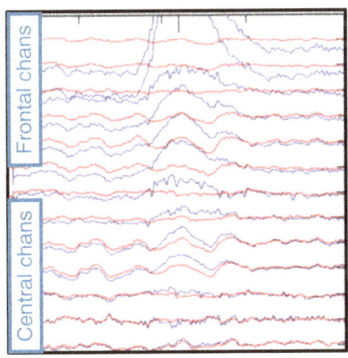

Step 4

Go back to the window asking you to 'plot single trials' and so on, and hit accept. Don't forget to subtract the baseline (−200 to 0) from each epoch again before saving the result. Make sure the filename indicates it is now artefact corrected (or 'pruned with ICA' as is the default in EEGLAB). Go through steps 1–4 for the other datasets as well.

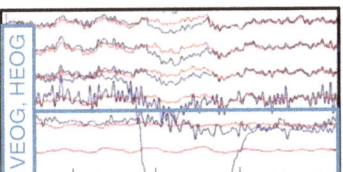

Figure 8.18

Additional exercise 2: EEGLAB batch scripting

In Chapter 7, you started making simple MATLAB scripts to make your job a little easier. Making things easier for yourself is, of course, very important, but arguably the more critical part of creating code is to make things transparent for yourself and others. I previously noted that things may feel arbitrary: why are these channels bad and those components removeable? A reviewer may well feel the same and accuse you of cherry picking, while you yourself are starting to doubt as well: maybe your results are just a fluke? For such reasons, it may be good to be able to show your entire pipeline, from start to finish, which can just recreate the same results with ease, allowing you to say that your observed effects (or lack thereof) are not simply due to a weird choice in the parameters.

So, create a batch script – either by building on top of the previous one or by making a new one that takes the previous dataset files and continues work. The batch script should do the following:

1. Load the dataset.

2. Epoch and baseline correct data.

3. Remove very extreme epochs.

4. Run ICA.

5. Save the dataset.

6. Load the dataset again.

7. Remove individual-specified artefact components.

8. Reject artefact contaminated epochs.

9. Save the dataset.

This is a very hard exercise, so I will give three hints. First, the easiest way of getting a head start is by doing the preprocessing from the beginning of this chapter to the end, typing eegh in MATLAB and then copying the result to a script (see Chapter 7 additional exercises). Second, I would suggest splitting up steps 1–9 into 1–5 and 6–9, then manually opening the saved dataset of 5, inspecting the ICA decomposition, writing down the numbers for each subject in a format similar to the filename *cell* described in the previous chapter:

```
badcomponents = {[1 2 3 9 10 11 12 14 15 16 22 25 30 33], [1 2 5 7...]};
```

This cell type variable has now the bad components (for me) of subject 1 in badcomponents{1}, and for subject 2 in badcomponents{2}, so you can retrieve them later on.

Then, you will need a for-loop that does steps 6 and 7. You can work from what you wrote in Chapter 7 to include:

```
for cfile = 1:4
    EEG = pop_loadset('filename',filesin{cfile},'filepath','file_path');
    EEG = pop_subcomp(EEG, badcomponents{cfile})
end
```

Third, if you do the artefact rejection using the big rejection dialogue and inspect how this plays out in the EEGLAB history (eegh), you will note that it starts off in a way that is relatively easy to understand with pop_eegthresh (threshold-based rejection), pop_rejtrend (trend rejection), pop_jointprob (improbable data), and pop_rejkurt (abnormal distribution). Following, EEGLAB combines the result of all these using eeg_rejsuperpose and removes them with pop_rejepoch. However, it hard-codes the specific epoch numbers, like

```
EEG = pop_rejepoch( EEG, [9 24 26 28 29 30 32 33 35 39 42 43 49 51 58 60
```

…and so on, but that will of course not be correct for every dataset in a batch-job. So, instead, you need to edit the code first to look up the epochs that should be rejected, and then use this variable within the pop_rejepoch command. To find the bad epochs you can use:

```
Badepochs = find(EEG.reject.rejglobal == 1);
```

Sadly, the cool relative thresholding procedure discussed in 'Artefact rejection, second pass', is now also harder to achieve. For the final exercise, see if you can implement it somehow, anyway!

EXPLORE: ADVANCED PREPROCESSING PIPELINES

The pipeline of preprocessing, as explained over the last two chapters, is necessarily oversimplified for ease of explanation. Once you get into it and you want to expand your knowledge, I would urge you not to take my suggestions as some sort of golden standard. Rather, it seems to me a reasonable minimum or even an average amount of work from the impression I get when reviewing work by others. However, if you truly want to get much better than most of your lecturers (or more likely, by then, colleagues), I would heavily recommend looking up *Makoto's preprocessing pipeline*, currently at sccn.ucsd.edu/wiki/Makoto%27s_preprocessing_pipeline (but easier Googled), a 15–23 step programme with many thoughtful explanations of why certain parameters improve the quality of your analysis.

Part of Makoto's pipeline uses (as I am writing this) Cleanline, which is a much better way of removing 50/60 Hz noise than just applying a notch filter to the signal. The plugin is available using EEGLAB's extension manager.

Gratton, Coles, & Donchin (1983) describe their linear regression-based artefact correction algorithm in the article listed in the References.

There are many good sources on ICA, although I am going straight ahead and recommend Wikipedia for the mathematics and basic idea. Applications of ICA are described by giants in the field Aapo Hyvärinen and Erkki Oja (2000), and when it comes to artefacts in EEG by Jung et al. (2000).

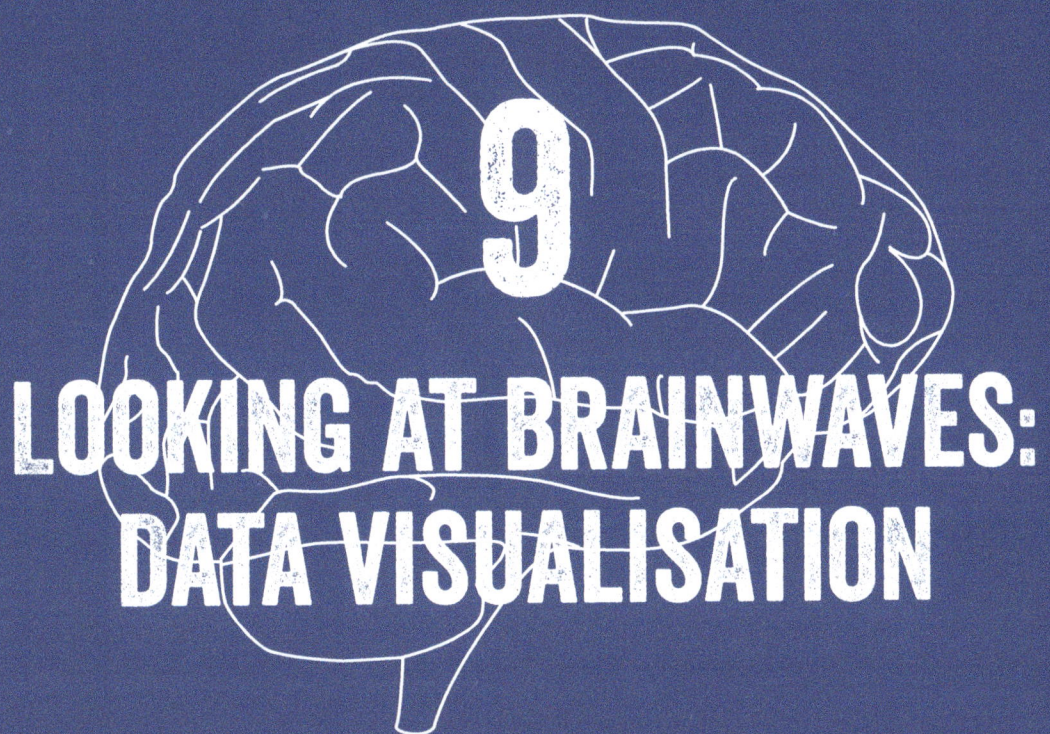

9

LOOKING AT BRAINWAVES: DATA VISUALISATION

IN THIS CHAPTER, YOU WILL LEARN ABOUT:

- How to sort EEG into conditions
- How to create single-subject ERP graphs
- How to identify ERP components

I am a pretty small fellow. No, let's go back, that should read: I am a tiny guy. I know this not only because I rarely get to see any live concerts, as inevitably some sort of barrel of a brute places himself straight in front of me, but also because of PE. Are there any PE teachers reading this? I hope not, but if there are, I congratulate you for changing your subject to one where you win by creating new knowledge rather than by making the most of whatever fortunate genes you were bestowed with at birth.

Indeed, the celebration of such happenstance was in my childhood made into a graphic display by drafting the children to create a single line ordered by height before requesting two popular pupils to choose teams. I do not know if this is a specifically Dutch practice, perhaps the evolutionary catalyst for making my compatriots generally gigantic, or whether it serves an established pedagogical purpose to give children the quick basics of bullying. Having been the bookmark ending both the line and any team's choice as a football, basketball, or baseball player, I can tell you the strategy certainly serves its purpose.

To make this not yet another lengthy moan of a childhood drama story, I will quickly close the anecdote by saying that my partner and I agree: never trust anyone picked first in PE. And so, we were both picked last, and remain incredibly clumsy, but that's fine if you've got each other.

My story hopefully tells you something about the power of visualisation. Getting children to make a line sorted by height is not merely a practical exercise that stops them from screaming their lungs out, it also tells a story and creates a hierarchy. The short ones stand out, as do the tall ones, even though previously it might not have been that obvious. Indeed, consider the famous Jane Elliott experiments that separated children into blue-eyed and brown-eyed groups causing them to become little monsters. Making differences obvious by visualisation causes us to change our thinking about the group itself.

It is a good thing, then, that while EEG corresponds to brain activity, it does not itself care how we think of it. In this chapter we will be doing a lot of sorting and visualisation with EEG. We have already done the dirty job of sorting epochs into good ones and fixing up the recoverable ones, and removing the worst ones entirely from the dataset – something I wished my PE teacher had considered to be an option. Now we will separate our data into useful teams of data to contrast and compare by using visualisations of the data at the single-subject level. Their game is to finally start answering the question: What happens if you see the person you had in mind?

Before you begin: Make sure you have the data from the previous chapter. If you have been following the book, that means average re-referenced, filtered, epoched, and pruned data. If not, I recommend downloading the sample data from the SAGE website (https://study.sagepub.com/Spape), so that you can work along with me and get a good idea of what the result of each processing step should be before you start with your own data. Download this chapter's material (https://study.sagepub.com/Spape) and extract the files in the special EEG folder. This should result in a new 'Chapter 9' directory containing in turn two folders: code and data. The latter contains seven datasets with 'Pruned_with_ICA' in the name, with two separate files for each dataset, one .set and one .fdt file.

SORTING OUT THE EPOCHS

Up until now, we have worked with just one file per subject. Well, you made copious amounts of backups, ending up with many files numbered in accordance with the processing step, but there was always one final file. Indeed, that is the convenience of using a numbering scheme as opposed to calling something 'final_finalest_v2_nowreallydone', as most of my early essays were called by the time I failed to make the second deadline extension. But from this point on, we will be going from serial to parallel processing in order to compare conditions. That is, the one participant's *master* file will be split up into separate parts, each containing different **analysis cells**. Let me explain.

EEG, like most neuroimaging methods, rarely looks at brain activity as such, as this inevitably contains a mixture of all sorts of cognitive, affective, motoric, and other processes. Even in fMRI, you might have a nice blob in the prefrontal cortex (related to memory, non-speech sounds, calculation, pain, and others), accompanied by insular activity (reward, heat, planning, emotion, pain again, and humour). Does that mean you found a hot new rewarding recognition task for planning the number of humorously occurring non-speech sounds of pained students doing multiplication? Probably not. We therefore do contrasts, first with a baseline, which in fMRI could be a resting state reading, and in EEG is generally the couple of milliseconds before stimulus onset, and then between conditions. The most common method of analysis is a factorial design in which we aim to determine the independent contributions of independent variables (**factors** in ANOVA) in explaining differences (**variance**) in the dependent variables. Each factor can have multiple conditions, and by multiplying (**crossing**) the number of conditions between factors we get analysis cells.

For example, in our current experiment, we have three potential factors: *relevance* (2 conditions based on whether a face was to be kept in mind, i.e. relevant vs irrelevant), *stimulus type* (2 conditions: human face vs pet), and *ownership* (2 conditions: own face vs other face). To make a three-way ANOVA, crossing the factors would therefore yield $2 \times 2 \times 2 = 8$ analysis cells. Unfortunately, unless you are much better at herding cats than I am, this design is impossible: the analysis cell of seeing a relevant face that is a pet and also your own face is missing at least from my recording. So we need to make some choices in creating designs, and because the build-in EEGLAB analysis tool requires files to contain only epochs that match the definition of the analysis cell, this needs to be done now.

There are some possibilities:

- The simplest design is simply to compare relevant with irrelevant faces in a one-way 2-condition analysis.

- But given that in the present experiment pets are never relevant, and are decidedly odd compared to the other stimulus material (it is, after all, an oddball study), we might extend that to a one-way 3-condition analysis, with faces that are relevant, irrelevant, or odd (pets).

- However, *ownership*, as defined above, *can* be crossed with *relevance*, resulting in a two-way ANOVA with *relevance* (relevant vs irrelevant) and *ownership* (self vs other) as factors.

In the rest of the book, I will concentrate on the second option as it is slightly more exciting than the first, while giving you the chance to explore by moving into more difficult terrain by following the third design. So, how do we start?

Approaching epoch selection

To put it into a more practical perspective of what needs to be done, consider Figure 9.1.

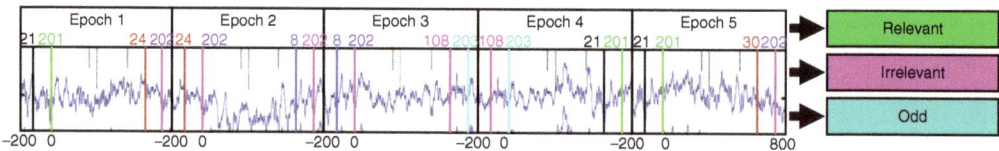

Figure 9.1   Five epochs. Every 1000 ms (including 200 ms baseline activity) epoch should be sorted into three files that each match an analysis cell. I did my best to make it slightly easier on the eye, but EEGLAB's seriously colour-challenged designers made my job nearly impossible. The problem is that EEGLAB fuses each epoch one to the next, and assigns random new colours to each new marker (vertical lines), while time and marker boundaries are presented on the same screen. I would recommend zooming in on a single epoch to get a grasp of the layout before you want to get an overview of the marker information.

Recall from the last chapter what the various markers mean: 201 indicates onset of a relevant picture, 202 an irrelevant picture, and 203 a pet. The preceding number is the actual picture. So the first epoch is relevant and concerns picture 21. Now, you might think that using the last chapter's *extract epochs* could work as we do need to 'extract', from the above, the 201s, 202s and 203s, which would mean 3 relevant, 4 irrelevant, and 2 pets out of our 5 epochs… Hang on, that can't be right!

The problem is that each epoch is 800 ms, while pictures are coming at 700 ms, so at about 700 ms in epoch 1, there's the marker 202, which in turn forms the basis of epoch 2. So the activity at the end of epoch 1 is the same as the beginning of epoch 2.

What we need to do, then, to get the relevant image data is to pick out the 201s *if and only if* they occur at 0 ms, which is the case in epochs 1 and 5, but not in epoch 4.

How to select epochs in EEGLAB

Somewhat confusingly, EEGLAB has four menu options that hint they might select epochs. I forget it myself all the time, which makes me look like a newbie in front of the classroom, so let's try to remember:

1. *Extract epochs* time-locks continuous data and segments them into epochs (p. 169). Because you normally time-lock data based on an event, you *could* theoretically use this instead of the better approach (approach 4), although it will produce some warnings.

2. *Select data* edits the channels or time-points under consideration, for example removing bad channels (p. 146), or keeping the part of the experiment that was still good (p. 144). It does allow selecting epochs, but requires you to enter the specific numbers ('1 5' would give the two relevant epochs). So if you have a complete list of the many hundreds of epochs that are irrelevant for each participant, then this would also work, but it's not very convenient.

3. *Select data using events* removes a time-range around specific events that you input. This one typically catches the eye as it sounds about right (we do want to remove data, and we do use events for that, right?), gives a much simpler dialogue box, and comes before approach 4. But it is much more similar to approaches 1 and 2.

4. *Select epochs or events* is my go-to menu option for keeping (or removing) epochs based on a set of conditions, which I will now explain.

Figure 9.2 is the dialogue summoned by going to edit > select epochs or events. The idea is that we select epochs of a certain kind (here defined as type 201) based on their occurrence. By only entering type 201 without the latency, all epochs are selected that have somewhere a 201 in them (e.g. 3 in the 5 epoch figure), whether or not the 201 is the 'current picture' or the next one. So we include the 201s *only* if they happen at time 0 (or latency 0). Slightly confusingly, however, this must be indicated as a range rather than an exact point in time, so ask it to search between −1 ms and 1 ms. The number does not need to be precise as long as (1) the subsequent markers are not included (so -700 to 700 would not be OK), and (2) it's not between 0 and 0 as nothing can occur without time actually passing. To be fair, EEGLAB will still work as expected, but it simply feels wrong!

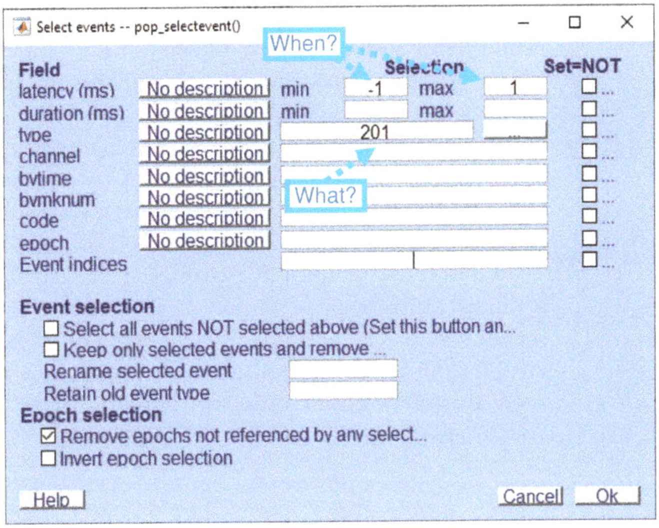

Figure 9.2

You can also make more interesting selections based on the other properties of the markers, but these depend largely on your EEG file-type, and in my experience are rarely used within this dialogue. There is also the possibility to get rid of events themselves (event selection), and to invert the epoch selection (given the above selection, that would remove the epochs time-locked to 201).

VISUALISING SINGLE-SUBJECT DATA

Now we finally get to the meat of the matter, or if you are vegetarian like myself, the soy chunks. That is, we will be looking at averaged EEG data evoked by precisely timed events: evoked potentials. There is sometimes a bit of confusion as to whether we should be talking about evoked potentials, event-related potentials, or a specific modality that is involved, such as with visual evoked potentials or lateralised response potential. Indeed, many of us use them almost interchangeably, but there are slight differences in usage depending on the background of researchers and their interests.

Any EEG that is time-locked to a measurable event can, through averaging out the noise, be demonstrated as elicited (i.e. evoked) by the event and therefore necessarily both be *evoked by* and *related to* the event. But the association has slightly different connotations, much like you're presumably related to your mother, but her involvement in the process was likely more than most other relationships, even with equal genetic contributions. In the literature, however, I rarely spot evoked potential anymore without explicit mentioning of the involved modality. Still, if the external cause is so clear, you might as well spell it out. Thus, as the offspring evoker is called a mother, we're talking about visual evoked potentials, auditory evoked potentials, motor evoked potentials, and so on.

So, if we know the external event, why would we talk about their being merely event-related then? The answer is that we are psychologists and not physicists or physiologists, even in our studies of psychophysics and psychophysiology. Both of these disciplines work with neuroscience methodologies, but while we are interested in the experience of vision, physicists tend to be more interested in the 'external reality' (e.g. the spectrum of light), and physiologists by the biomedical reality (the visual organ). The implication of this difference in nuance is that our focus on mental functions, such as attention or memory, reduces the relative importance of the original signal. Thus, the EEGs in the present study are evoked by visual stimuli, but they are related to so much more: their relevance, ownership, and of course whether the visual stimulus is a pet.

I find it thus helpful to talk about visual evoked potential if we do not make an attempt to figure out exactly what part of it is due to mental aspects, which generally means the 'condition' in our designs. Then, when we're talking about how relevance affects this, I switch to the event-related potential, with the P3, for example, being a component of the event-related potential, which we measure as the difference between relevant and irrelevant conditions.

The visual evoked potential: General ERP components

So, let's have a look at the visual evoked potential, which we defined as the EEG evoked by any picture, regardless of the picture's relevance, and so on. However, in experimental design, there are two main ways of defining 'regardless':

1. The mixture of all conditions: Here, we make no distinction at all between any sort of condition and just start off with the entire set of epochs time-locked to visual events, such as you ended up with by the end of the last chapter (or from downloading them, see https://study.sagepub.com/Spape).

2. The minimal control condition: Here, we will start with an 'empty' baseline and do all our follow-up contrasts against this. This works best with a set of conditions with a clear hierarchy of complexity, for example if we first were to look at 'irrelevant faces', then 'relevant faces', and finally 'relevant faces that represent ourselves'. Often, this sounds quite ideal to begin with, and I try it often, but usually it emerges that this is a naïve way of looking at things (see boxed text opposite).

So what do you do? I urge you to try both (in essential exercise 1), and see how the conclusions depend on the approach. As for me, I will use approach 1 with file umeeg101_10.

Silent baselines are rarely quiet

The problem with this way of thinking is that it rests on the idea of activity, and its opposite, inactivity. It is for us as researchers a particularly enticing idea that the brain is basically waiting in silent anticipation for our experimental stimuli to come in, and then spring to action, instead of doing all sorts of things that have nothing to do with our designs. Thus, for example, it took a very long time for fMRI researchers to notice that sometimes areas are *de-activated* in comparison with 'empty' baselines, and such *blobs* were therefore regularly discounted before the idea of a **resting state network** – activities related to not doing something – took hold. In EEG, we have had an easier time coping with the idea of 'negative activity', often seemingly preferring it even to positivity when plotting negativity up and positivity down in graphs. Still, the promise of silent baselines remains tempting, as demonstrated for example by the linked mastoid reference discussed in Chapter 7 (p. 148), which was supposed to be the secret to compare our electric potentials against.

Load the dataset into EEGLAB, making sure you know which number it is. For example, if I already had three datasets loaded and open the new one, it will appear under EEGLAB > Datasets as 4. Then go to Plot > Sum/Compare ERPs,[1] and a dialogue appears (see Figure 9.7, p. 200). Enter that number (4 for me) next to Datasets to Average. Furthermore, remove the minus sign from the Plottopo options (i.e. to 'ydir', 1), so that positive will be plotted up. This reflects an ancient yet ongoing division in the EEG community whether negative should be plotted up (some engineers and clinicians prefer) or down (which makes looking at graphs easier). By default, EEGLAB shows the former with 'ydir', −1, but this confuses some of us (certainly me), so I recommend switching it from −1 to 1. Leave everything else as it is. Hitting OK should show an overview of the visual evoked potential across all recorded electrodes, as shown in Figure 9.3.

At first, all those little blue scribbles may make understanding it a daunting enterprise, especially if you have a higher channel recording. But the layout helps as it can be read while imagining seeing the subject from straight above with top front, bottom back, and left nearing the left ear and right nearing

---

[1]If you are not seeing this menu item, you likely don't have EEGLAB showing all menu items from previous versions. Enable it by going to File > Preferences.

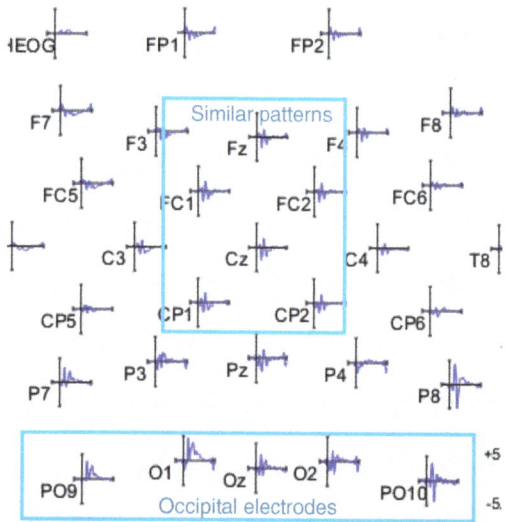

the right ear. Compare some channels with their neighbours: Cz looks quite like Fz, FC1, FC2, CP1, and CP2 with two spikes up and a little negativity between them. We can also use some prior knowledge: early processing of visual activity takes place in the occipital cortex, so we might see early activity there. And behold, nearly every O channel seems to have a negative spike right next to the onset of stimuli (see Figure 9.4).

To really get the hang of things, however, I recommend taking out a notebook and clicking on an electrode, which causes EEGLAB to pop out a larger graph of the selected channel.[2] In the panels in Figure 9.4, for example, I clicked on the O channels to have a closer look. Once one is opened, if you go to Tools > Data Tips, you can click anywhere in the figure and a precise indicator of the value within the figure appears as an X (time) Y (voltage) coordinate.

Figure 9.3

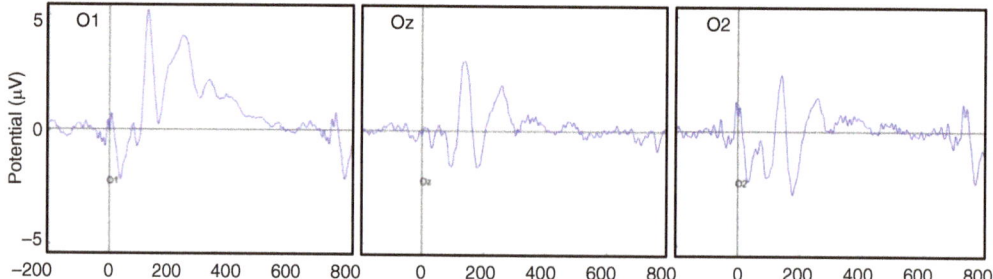

Figure 9.4

Let's give that some practice by looking at the channel figures and investigating their similarities and differences. How do you know if something is a real potential and not a random fluctuation? As scientists, we like to pretend this is very difficult and requires a trained eye, and to some extent this is true, but there are some good rules of thumb.

1. A potential is generally more than two times bigger than the activity observed in the baseline (p. 168). Indeed, if you think of the baseline as 'stuff that occurs by chance', and the range before baseline for O1 in Figure 9.4 is at its most extreme −0.72 (read this with the data cursor),

---

[2]Sometimes MATLAB pops them right in the corner of the screen and you can't move them. In that case, resizing them even slightly moves them back to a more manageable spot.

the first positive peak for O1 at 8 ms is likely not a potential (Y = 0.69), while the first negativity at 37 ms (Y = −2.17) likely is, being almost three times the size of the baseline. With O1, we then observe some wiggling that looks like the baseline between c. 63 and 104 ms, followed by a massive quick peak at 126 ms (Y = 5.23), that goes down until a relative negativity at 161 ms, before a longer positivity, which climbs until about 242 ms, thereafter drifting off into nothingness. Solely on the basis of O1, then, we have two negativities and two positivities, but it would be terribly premature to call these N1, P1, N2, and P2.

2. A potential is generally not observed in only one electrode, because the scalp diffuses the potential (p. 8). So if we follow the same logic as with O1 but with Oz, then we see negativity at 94 ms, positivity at 134 ms, negativity at 178, and finally positivity at 258. Comparing these with the O1 potentials, I would say the 37 (O1) and 94 (Oz) ms do not coincide, but 126 vs 134, 161 vs 178, and 242 and 258 are very similar and likely reflect the same components.

3. Finally, because neural activity produces an electric dipole, polarity inversions can often be detected as negativity on one side of the head and positivity on the other. Thus, with Figure 9.5, we can see for Fz (+ at 38, −128, +181, −253) while for Cz (+91, −130, +186, and −260). The last three latencies are quite similar between the two electrodes, but note how they mirror Oz's (+134, −178, +258). Indeed, Fz's +38 neatly mirrors O1 (−37) and Cz's +91 Oz's −94. They thus look to be just two sides of the same coins.

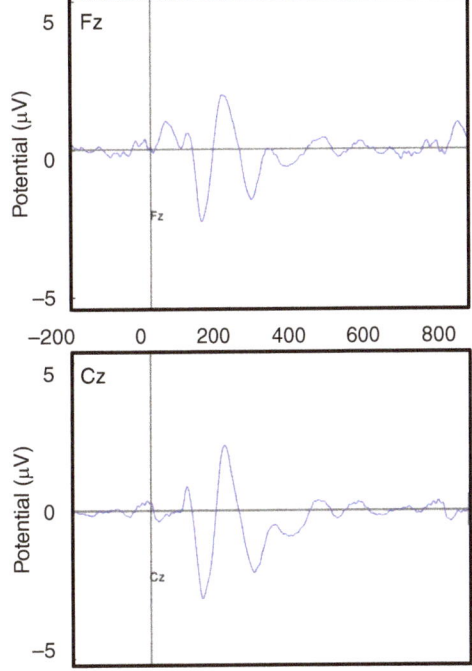

Figure 9.5

My figure doesn't look anything like these at all. Now what?

As a rule of thumb, a good averaged evoked potential has less noise in the baseline (activity before onset) than in the temporal range after the onset of the stimulus. If in your graph the baseline looks identical to the active period, then

(Continued)

something probably went wrong: (1) Were there enough epochs in the dataset? If the number of epochs is very small (more than 60 is ideal), then averaging will not reduce the noise enough to make a difference (p. 167). (2) Are there too many noisy trials remaining in the dataset? If there's one trial included that has 100s of µV of artefact activity, then it will have a disproportionally big impact on an effect that is less than 5 µV. 3) Otherwise, there might be something technically going wrong. For example, did the time-locking work? If a trigger was produced without corresponding to an external event, then inferring evoked activity becomes impossible. This is one reason EEG people are such a punctual bunch!

Looking at topoplots

Now let's see what the identified components of the visual evoked potential look like mapped over the entire scalp. Topographical maps of electrophysiological activity, or topoplots in EEGLAB slang, were invented by EEG hero Grey Walter (see p. 4), who wanted to see in one quick overview how EEG activity was distributed over the scalp. Decades of developments in how this can best be visualised have now crystallised into a basic two-dimensional fold-out of a scalp or a three-dimensional head-model in which the channel activity for specific latencies is projected into colours that blend between the channels positioned on the scalp.

To see how that works in EEGLAB, make sure you have the correct dataset activated, then go to Plot > ERP map series, and choose In 2-D. 3-D, if you're interested, works similarly but requires awkward rotating to see everything, so I'll leave you to figure that out yourself. The only necessary parameter is the first: 'Plotting ERP scalp maps at these latencies'. Let's enter some of the latencies we have just found: '37 38 127 128 258 260'. The result I get is shown in Figure 9.6.

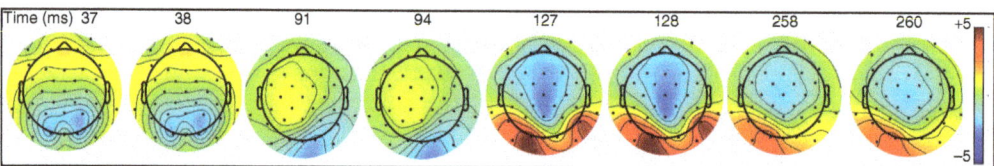

Figure 9.6   Topoplots at eight latencies for one recording session

The first two latencies show negativity over the parietal-occipital cortex, slightly lateralised to both sides, against prefrontal positivity. The second pair could be characterised by a further and more medial occipital negativity and a left/central positivity. The third is very different, with central/midline negativity against strong bilateral occipitoparietal (PO9/PO10) positivity.

Finally, the last pair of latencies has more dispersed central negativity, mainly against left occip-itoparietal positivity. Clearly, there is marked resemblance between each of the four pairs and it might make sense to think of these as four components of the evoked potential rather than eight. If we look at the literature, the last three *could* be the C1 or P1, N1, and P2 or N2, but we're talking about one recording session of one subject, so again that is premature at this point. As such, I imagine this extremely early N37 (to give it at least some name) is likely to be a fluke and won't stand comparison across subjects.

Isn't it kind of obvious that you get the same pattern of activity if you plot nearly the same moments?

Of course it is, but there is more to it than that. In fact, if you look at the evoked potential across a range of times, you will notice that there is strong resemblance for tens or even hundreds of milliseconds in topography before suddenly the pattern completely changes into a new one. This observation has inspired the idea that EEG researchers should be interested in **microstates**, quasi-stable patterns of activity pattern in the EEG that are typically analysed by using strong band-pass filtering and automatic grouping of activity patterns using statistical analysis techniques such as k-means clustering. To get a sense of this, try plotting instead of specific latencies a wider range such as by entering in the 2-D plot dialogue 10:10:400, which is MATLAB for plotting the entire range FROM:STEPSIZE:UNTIL, so from 10 to 400 in steps of 10.

The event-related potential: Comparing two conditions

Unlike the previous effect of seeing *an image* on the evoked potential, many evoked potentials that are related to cognition are revealed only by comparing across conditions. The P3, for exam-ple, is quite often almost entirely absent with stimuli that are frequent and irrelevant (marker 202), and these are by far the most common in the present study. So, if we would take the present selec-tion of epochs based on a mixture of 201, 202, and 203 images (relevant, irrelevant, and odd), the number of them that are potentially (sorry) P3 evoking gets swamped by the number without sP3, and the resulting positivity versus baseline would be very minimal. Indeed, if you look at the evoked potential overview (p. 196, Figure 9.3), there is not much going on after c. 260 ms, with the

Cz channel ERP (Figure 9.5) merely giving a negativity (c. 400 ms) before flatlining for the remaining time. However, that might be because the large number of irrelevant images do not just have *no activity*, but show a relative negativity at that point in time, in contrast to the relative positivity for relevant and/or odd images. What we need to look at, then, is not the absolute voltage of the visual evoked potential, but the difference caused by the relevance of the event-related potential. We can do that with one of the more confusing dialogues in EEGLAB (Figure 9.7).

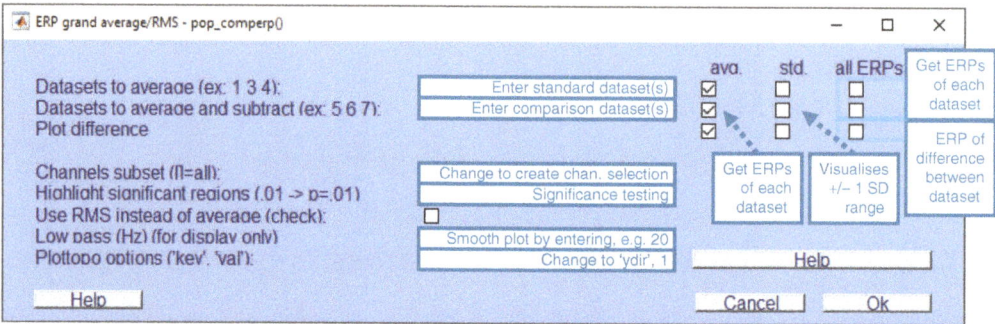

Figure 9.7　Compare ERPs dialogue

Earlier, you just entered a single dataset to the top, and that works fine. But the logic that underlies the comparison visualisation is pretty hard to follow and the results can be unpredictable. The main idea is that you enter a dataset (or multiple datasets) as the condition of interest to be compared against 'comparisons'. For example, you enter the one you intend as the experimental condition (e.g. relevant), in the top field, then the comparison or control condition (e.g. irrelevant) in the comparison field. Then, putting a check under 'Plot difference' will make a comparison between the standard and the comparison. But you can enter more than one dataset in each field. Keep in mind that avg., std., and all ERPs all refer to the same row, so entering dataset 1 3 4 in the standard field, and selecting avg. will plot the average over datasets 1 3 and 4, leaving you with a single line. Enabling also all ERPs will add the individual three ERPs of datasets 1 3 and 4 as well. I would recommend exploring the interface a bit and see if you can get something that looks reasonable out of it. But to make things a bit easier, here are two examples you can start with, given that I have three datasets loaded: dataset 1 (relevant), 2 (irrelevant), and 3 (pets).

1. Standard: '1', ✓ avg. Comparison: '2', ✓ avg. Plot difference: ✓ avg.: This creates a nice figure with relevant vs irrelevant, and shows the subtraction ERP of relevant – irrelevant.

2. Standard: '1 3', ✓ all ERPs. Comparison: '2 2', ✓ avg. Plot difference: ✓ all ERPs. This creates a rather complex figure with ERPs for relevant, odd; then subtracts irrelevant from relevant, and also from odd (hence adding it twice), showing these individually. Then finally shows the average of irrelevant and irrelevant (to avoid the duplicate).

So, you can make very complex comparisons with this dialogue, but I cannot really recommend it, especially since it is hard to compare and contrast four ERPs within the same figure (see pp. 259–260 for figure guidelines). As such, I would recommend either entering multiple entries in Datasets to average, and hitting all ERPs, *or* entering a single dataset to average, dataset to subtract, and plotting the difference, as with the first example.

As to the other options, they seem largely self-explanatory. Entering channel numbers in channel subsets allows you to create a less daunting overview. Highlighting significant regions seems to be old EEGLAB functionality requiring data from multiple subjects to be loaded that is now implemented using the STUDY structure (see Chapter 10). Using RMS visualises variance instead of average for no clear reason. Low pass filtering allows you to smooth your ERP figures by entering a value to filter below – I suggest 20 as a possibility – which may help presentation without the filters reducing the quality of the actual data (p. 154). Finally, there are the additional functions enabled within the underlying EEGLAB MATLAB code, for which you can get an overview if you type 'plottopo' in MATLAB's command view.

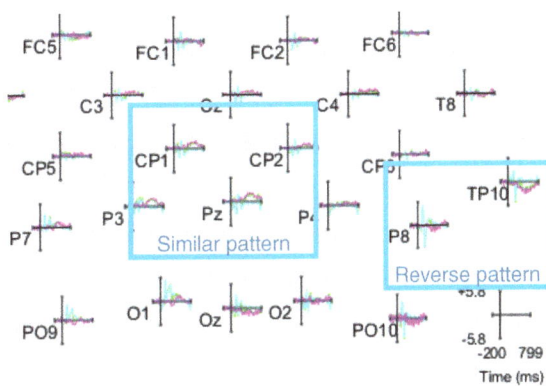

Figure 9.8

With that in mind, let's have a look at our main contrast. I'm using the suggested first example above and I drew ERPs for relevant and irrelevant images (Figure 9.8). The **difference wave**, or the ERP that is the subtraction of two other ERPs, in this case relevant – irrelevant, is shown in purple. While the differences are pretty slight compared to the evoked potentials themselves, in turquoise and green, we can identify a similar pattern, a late positivity over parietal and central-parietal electrodes. We can also spot the suggested mirrored, or inverse activity over the right-lateral temporoparietal lobes, which is near the mastoid area. Let's have a closer look at these patterns by clicking on Cz, Pz, and TP10 (Figure 9.9).

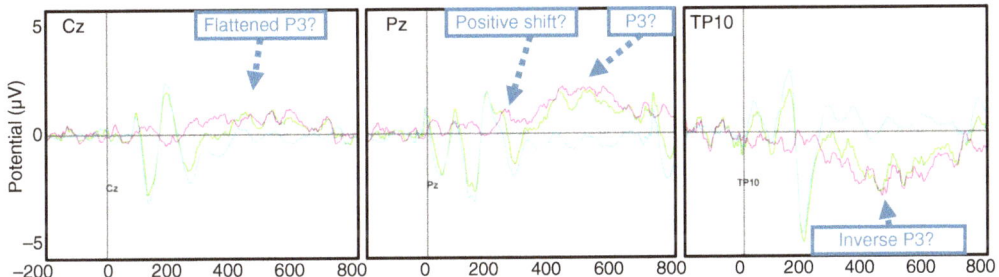

Figure 9.9

In Cz, there is hardly any difference between the relevant and irrelevant images, as the fluctuation of the difference wave before the baseline is almost as strong as after the baseline, although after c. 280 ms it tends to be more consistent in the positive side of things, culminating with peaks of 0.97 μV at 371 ms, 0.94 at 452, 1.08 at 519, and 1.16 at 576 ms. With Pz, a slight **positive shift** is shown, a sideways displacement of the ERP with a peak at 261 μV, then from around 300 ms we see a gradually developing positivity peaking with 2.17 μV at 433 ms, remaining more or less at that level until it starts to drop from 564 ms onwards. Meanwhile, the TP10 shows very little until, from about 240 ms, it starts showing marked negativity, peaking at 448 ms with −2.98 μV, then gradually returning to base level.

Of course, despite these very precise-looking numbers, this is hardly a very scientific approach. I imagine that you might be looking at the same lines and feeling it's just a big mess and this is all just confirmation bias here. That said, we do know from decades of research that the P3 is a positive response in the ERP that occurs c. 300 ms after noticing infrequent, relevant events, such as here demonstrated with the green line. Classic studies, however, tended to use a mastoid reference for obtaining EEG, which is located very nearby TP10. Imagine what the same ERPs would look like if you applied a mastoid reference. Mathematically, that is the same as subtracting the purplish/pink line of TP10 from that in the Pz: at 448 ms you'd get positivity of 1.95 (current Pz ERP with average reference) minus *minus* 2.98 (current TP10 ERP with average reference) = 4.93 μV with mastoid reference. That is a pretty big difference, about 20 times larger than the variation before the baseline!

So maybe this is evidence that my brain back then responded to seeing a face I had to keep in mind, good material for publishing in the *International Journal of Studies of Me* – or indeed, in this book – but, given the evidence, how sure are we that brains generally respond to seeing faces that are kept in mind? In the next chapter, I will turn to this very question and test hypotheses at the population level. Before that, however, it's important first to get a good grasp of what counts as an ERP component rather than random EEG activity and an understanding of just how much variability there is between epochs, conditions, and people. The exercises are designed with this purpose in mind.

ESSENTIAL EXERCISES

Essential exercise 1: Single subject analysis

Part of doing EEG research is to get an eye for what your results should look like. This exercise is for you to get the hang of inspecting visual evoked potentials, while at the same time making sure your data are good enough to proceed.

1. For each subject, open the cleaned dataset you ended up with in the last chapter. Or if you do not have your own data, feel free to use mine from the online material (https://study.sagepub.com/Spape). Use the EEGLAB Sum/Compare ERPs dialogue to visualise each dataset.

Do they each show regular, post-baseline activity that looks like a stereotypical ERP pattern? Which channels look best, which ones worse? Which datasets look best to you?

2. If there are any datasets that look as if the activity after 0 ms is not different at all from the activity before 0 ms (i.e. very bad), go back to the analysis steps from the previous chapters. There are three common problems: (1) you may have removed too few artefacts, so a few trials with a lot of noisy activity may have spoiled the averaging procedure; (2) you may have removed too many epochs, so the averaging was not effective in reducing noise of single-trial activity; and (3) a bad recording or bad participant. For the former, if there was too much noise across the board, the data may be unsalvageable: garbage in, garbage out. As for the latter, sometimes participants are so tired (maybe it was a Friday afternoon or the morning after a student party?) that the sheer amount of alpha they produce makes reading anything else impossible (see also boxed text below). In any case, the only solution to this may be removing a lot more epochs than normally, or removing the dataset entirely. Removing c. 20% of your recordings for such reasons is not abnormal in the EEG literature!

3. Now that you're satisfied with the datasets, pick three and inspect the visual evoked potential of each dataset for the mixture of all conditions (see p. 194: 1). What are the first few clear positive and negative potentials that peak before 300 ms? To do this, first look at the large overview and note where you see early deflections. Select the electrode and note down exactly when positive and negative peaks occur. Write these down in a table like the one below. Select more electrodes that you feel will give similar results and add these to new rows:

Channel	Peak (ms, pol, µV)	Peak (pol, ms, µV)	Peak (pol, ms, µV)	Peak (pol, ms, µV)
Oz	94 (−1.56)	138 (+3.09)	178 (−1.62)	258 (+2.04)
O2	33 (−2.17)	90 (−2.02)	141 (2.49)	176 (−2.76)

4. Identify how these components can be grouped, for example by shading them as I have done above, and by creating topoplots, as explained in the chapter. Group them together within a new table that also gives a description and a tentative name you could give to the observed potential (negativity peaking between 91 and 94, maybe 'N93'?). Then look up references that describe something similar. Here, I suggested the observed N93 could be similar to the C1 first described by Spehlmann.

Potential	Latencies and sites positive	Latencies and sites negative	Description	Literature reference
N93	/	91 (Oz), 94 (O2)	Early occipital negativity	C1 (Spehlmann, 1965)

5. After you have identified a few such components of the visual evoked potential, do something similar for the two other datasets. What similarities and differences do you observe?

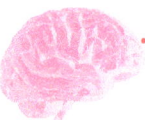

Alpha spotting

If you see participants with about 8 clear positive, and 8 clear negative potentials in the 800 ms after the stimulus onset, you are probably not finding evidence for a whole new range of components. That is, 8 sinuses in 800 ms translates to 10 in 1 s, or 10 Hz, which is in the middle of the traditional *alpha frequency band* (pp. 3, 172). In other words, these potentials were probably not evoked by the stimulus, but rather the brain was taking a little micro nap – or sometimes a not-so-micro nap.

Essential exercise 2: Single subject ERP analysis

Now that you have an idea of the basic visual evoked potential, we will take a step towards the more psychologically interesting question of seeing a stimulus you are currently keeping in mind. As part of the analysis, you will also be saving the condition-specific data, which you will be further using in Chapter 10. Note: this part of the exercise will partially be repeated using only MATLAB script in additional exercise 1.

1. Again, for each subject, open the cleaned dataset with the conditions mixed. Using the technique explained at the beginning of this chapter (p. 193), create a new dataset with epochs only including irrelevant stimuli. Then, go to Edit > Dataset info, and change the subject code to the subject's number, and the task condition to 'irrelevant', before saving the dataset to disk.

2. Go back to the mixed conditions dataset, and do the same as 1, but for the relevant and odd conditions.

3. Compare and contrast relevant with irrelevant conditions. Can you see a P3 potential? Are there any other potentials related to relevance detection? To organise your observation with some structure, you could use the same set of tables I introduced in essential exercise 1, but over the *difference wave* rather than the evoked potential. However, due to the longer latency of the P3 and other late potentials, it can be a bit hard to get a clear reading, so you might use smoothing in the dialogue (or doing it in your head). Alternatively, use the former US president's unique method of meteorological modelling to work out whereabouts the P3 should be, by printing it out and using a sharpie! No, of course that is not a good way of doing things, but the main thing here is to get a good idea of the ERPs: We'll leave the actual statistics for the next two chapters. Whichever way you do it, make sure you stick to your own set of rules as you move from dataset to dataset, and determine if the P3 is similar, different, or completely non-existent in each dataset.

4. Finally, do the same but for the oddball condition. Compare these datasets with the irrelevant condition. Are pets similar to irrelevant faces, or perhaps to relevant ones (as they are also infrequent)? Are there any very early potentials that look different from the oddball trials? Do you think that is due to a perceptual difference (e.g. something visually different in pet faces compared to human faces), a cognitive difference (e.g. its categorical infrequence), or some affective difference (e.g. the fondness one feels for pets)?

Additional exercise 1: EEGLAB batch scripting 2

All this clicking and mucking about with the various EEGLAB dialogue boxes can be tiring and the consistent routine makes me feel like I'm actual doing a computer's work: a lengthy series of small steps in a processing pipeline. What's worse is that I make many errors in such procedures, perhaps because I'm a pretty disorganised person, or lack the procedural skills PE never taught me, but even so, very few of us are robots (I think). So let's make the computer do the mindless work for us, the better to concentrate on the EEG researcher's favourite hobby, eyeballing squiggly lines.

Step 1

To start, open one file and go through the first part of essential exercise 2, selecting only relevant epochs. Go to Edit > Dataset info, change the subject to the subject's number (for me: 101) and the condition to rel. Then save the file, with a sound ordering system (as explained in Chapter 7, p. 139), and the condition name (rel). Finally, type in the MATLAB command window:

```
>> eegh
```

…and have a look at the function's output. There's likely quite a bit of stuff, but only the following four lines are of interest:

```
EEG = pop_loadset('filename', 'myfilein.set', 'filepath', 'mypath');
EEG = pop_selectevent(EEG, 'latency', '-1<=1', 'type', 201');
EEG = pop_editset(EEG, 'subject', '101', 'condition', 'rel');
pop_saveset(EEG, 'filename', 'myfileout.set', 'filepath', 'mypath')
```

Note: Most lines will be a bit longer than this. I removed all non-essential parameters from functions. myfilein.set, myfileout.set, and mypath should describe your input file, output file and their storage location that hold for you. Lengthier documentation appears with the online material.

Step 2

You could easily adjust these four lines to work through your entire sample, first by copy–pasting. But try to use the same logic introduced in Chapter 7, additional exercise 1, and Chapter 8, additional exercise 2, to batch script this with a loop. More elegant, however, would be to use two loops simultaneously: one to loop through the pool of subjects, and one to loop through the datasets. Here's a handhold to get you started:

```
filesin = {'filein1.set', 'filein2.set', 'filein3.set'};
conds = {'201', '202', '203'};
for cfile = 1:3
    for ccond = 1:length(conds)
        EEG = pop_loadset('filename',filesin{cfile},'filepath','mypath');
        EEG = pop_selectevent(EEG,'latency','-1<=1','type',conds{ccond});
        EEG = pop_editset(EEG,'subject',filesin{cfile},'condition', conds{ccond});
        pop_saveset(EEG,'filename','myfileout.set','filepath','mypath');
    end %for ccond
end %for cfile
```

What we now have is a **nested loop structure**, a loop inside a loop. Thus, the code will run over 3 subjects, and within each subject run 3 times (9 times in total). To add clarity to this, I put a little comment (%) after each end, so the reader sees what is being ended here (in this case, the ccond for loop is closed first, then the cfile for loop). Notice that I exchanged the 'for ccond' to run from 1 to *length(conds)*, instead of 3. While both would work just as well, using a function to obtain the length of cell variable *conds* is recommended as it makes the code more flexible. *length(conds)* returns, as you might guess, the length of the cell variable conds (which indeed has 3 entries, therefore returning 3).

Technically, length is not the same as size, but rather refers to the largest dimension of a n-dimensional matrix, which is an incredibly MATLAB-geek way of describing that if I have a paper of size 21 × 30 cm, its length is 30.

The above code works, but there are certain issues. Most immediately, all files should load nicely (add strategic *breakpoints*, p. 159: 7) to determine whether this is true, but they all save to the same filename. Adding an additional variable, like filesout, will only help somewhat, because we need to end up with 3 files × 3 conditions = 9 files, not 3. To fix this, you could use string concatenation, which is code slang for mashing text together. For example, try in MATLAB's command window:

```
>> mashingtexttogether = ['mashing' 'text' 'together']
>> all3subjects = [filesin{1} filesin{2} filesin{3}]
```

(without final semi-colons)

Concatenation in MATLAB is slightly unlike other coding languages, as it formalises text, like almost everything else, as matrices. Text, such as 'mashing', is a 1-by-7 matrix, or 1 row and 7 columns (e.g. row 1, column 4 contains 'h'). Using square brackets and a space between entries is a MATLAB abbreviation of using *horzcat* – horizontally concatenating entries pushing them one next to the other. *horzcat*, in turn, is a specific part of more general *cat* (mentioned because I like cats), during which we concatenate matrices (e.g. 'mashing' and 'text') over the second dimension (the columns, not the rows). Sometimes it is pretty mind-blowing how mathematics can make things that are ostensibly simple so terribly difficult!

Step 3

In any case, use this knowledge on concatenation to fix the nested for-loop so it will create 9 unique filenames (or if you have 7 subjects, 21 filenames). Add additional variables so the concatenation in pop_saveset makes filenames look like '101_11_rel.set' (or subject name_file version_condition.set). Finally, clear up the code so a dataset is loaded only once rather than for each condition. Of course, this creates a certain problem: if you first select from a file the relevant epochs, and then select from *that* dataset the irrelevant ones, you will not find any. To do this, you will therefore first have to load the EEG file in variable EEG, and then make a new variable (e.g. cEEG) to avoid overwriting the previous selection.

Additional exercise 2: Seeing yourself in the Brain Mirror

Remember that one of the pictures in the experiment was the participant's own face? As I wrote previously (E-Prime p. 59, OpenSesame p. 85) it is most convenient to have that picture number always be the same, so you will now be able to analyse whether the brain responds in a particular way if someone spots their own face.

Step 1

Using the select epoch dialogue (p. 193: 4), create datasets for any stimuli that are relevant and irrelevant, as before. Then divide these stimuli into images that showed the participant's own face, and those that did not. Remember that triggers describing the images were sent just before the images themselves (and thus not between -1 and 1 ms). Save the images in a format describing a 2 × 2 design with four datasets for the participant: rel_self, rel_other, irr_self, and irr_other. Note that the rel_self condition is pretty rare given that 1 out 6 images is relevant, and only a few of them are of the participant. Should it happen that a subject does not have enough rel_self epochs, then skip this exercise for that participant.

Step 2

Run eegh in the MATLAB command to show the used EEGLAB commands, then adjust the loop from additional exercise 1 so that it creates datasets for each participant. This will quickly become problematic, because odd images are never of the participant. For this, see the following hint.

If you want to make an exception to for-loops, for example because you want to remove some extra epochs from one subject but not the others, or as here, because you want to further divide only relevant and irrelevant epochs but not odd ones, you can use the **IF-THEN-ELSE** statement. As with any other programming language, **conditionals** check a rule (the condition), and do something only if the rule is checked, and (optionally) something else if it is not. Thus, for example, I had my conditional EEG already loaded in cEEG – see the last part of additional exercise 1, and then make a sub-selection of the previous selection, but only for condition 1:

```
if ccond == 1
    ccEEG = pop_selectevent(cEEG,..);
else %if ccond
    disp ('not first condition!')
end %if ccond
```

Again, I made two comments to indicate the ending of the specific IF statement to add clarity. I also replaced the rest of the second line to avoid spoilers for this exercise. The last two lines (displaying 'not first condition' if the ccond fails to be 1) are not necessary to make things run, but are for purposes of illustrating the complete IF-THEN-ELSE statement.

Skipping bits of code, and the difference between *is equal* (==) and *is* (=)

In MATLAB, == describes an equality sign, while = means that the part before the = sign needs to become the part after. Isn't that the same? Well, no: asking MATLAB "ccond == 1" is a yes (true) or no (false) question, while telling MATLAB "ccond = 1" commands it to load 1 *into* ccond. It is quite capable of fulfilling this command whether or not ccond *used to be* 1, 2, my cat, or an EEG dataset. In programming languages, signs such as smaller than (<), greater than (>), equal to (==), smaller or equal to (<=), unequal to (!=), and so on, are called **boolean operators**, functions that return true or false. MATLAB is a bit funky in that it does not do true and false, just 1 and 0, which results in odd-looking code. For example, let's say you already did a bit of code and don't want to run it every single time. We could make a variable, 'skipthis' and put it in a conditional:

```
skipthis = 1; %load 1 (true) into skipthis

if skipthis == 1 %already processed last night!

    % the part you want to skip

end %if
```

But we know skipthis is 1 (just having assigned the value), so we might as well move that down:

```
if 1 == 1 %already processed last night!
```

However, that looks ridiculous, of course 1 equals 1! But, as said, a boolean operator returns true or false, which in MATLAB is in itself coded as 1 or 0, and indeed, running 1 == 1 in MATLAB's command window will give '1' (true) as a result (tautologies being tautologies). So the code becomes:

```
if 1 %already processed last night!

    % the part you want to skip

end %if
```

Which looks odd to me, but it's a very nice way to switch bits of code in my MATLAB script on or off using just 7 characters.

Step 3

Now that you have relevant, irrelevant, odd, rel_self, rel_other, irr_self, and _irr_other datasets, do the same analysis as with essential exercise 2, but try with one subject to explore whether seeing your own face provokes a P3 response, *even if it is irrelevant*. Alternatively, the difference wave between relevant and irrelevant might be different for seeing your own or seeing some other face. Or perhaps the brain couldn't care less about the ownership of the face? Give an informed answer to this set of hypotheses and once you have defined it, see if another dataset corroborates the evidence.

EXPLORE: EASY EPOCH SORTING AND FANCY MEASURES WITH ERPLAB

EEGLAB works well enough for me to do the simple ERP analyses I require, but a vastly more sophisticated (also free/open-source) toolbox is called ERPLAB (https://erpinfo.org/erplab). ERPLAB (Lopez-Calderon & Luck, 2014) is somewhat like an extension of EEGLAB and comes with the pedigree afforded by ERP master Steven Luck (see Explore in Chapter 1) whose lab supervises the project. The most interesting advantage of using ERPLAB is in its more advanced way of sorting out epochs. That is, if in EEGLAB we want to compare two conditions, we will need to save them in multiple datasets, which you will now be familiar with. The disadvantage, apart from having heaps and piles of files you swiftly lose track of, is that if somewhere along the line you removed some conditions, you won't easily go back without rerunning the preprocessing steps. To avoid such problems, ERPLAB introduces specific tools for keeping track of how each epoch relates to experimental info and conditions (ERP bins), such that later on the data can be dynamically recombined. It makes for a thoroughly useful toolbox, especially if your analysis is a bit more complex than ours, and if you are not keen on doing a lot of custom MATLAB scripting (I am!). It also has best-in-class tools for detecting and exporting ERP measures (p. 246), such as peak latencies, amplitudes, area amplitudes, and so on.

10

GREAT BRAINS THINK ALIKE: GRAND AVERAGE ANALYSIS

IN THIS CHAPTER, YOU WILL LEARN:

- How to average ERPs into grand average ERPs
- How to compare conditions
- How to identify differences in ERP components

Throughout this book, I have introduced each chapter with a little, personal anecdote that had some marginal bearing on the topic of the chapter, or the book, or at least brains in general. But I confess I am struggling with this one for two reasons.

One is that we are currently in what I can only hope will later be called the year of the coronavirus, rather than the year marking the beginning of the Viral Age, the successor of the Iron, Industrial, and Modern Age. This means that I am writing from home – although as you will read below, I would have been anyway – being bombarded by news from other writers who can't get far enough to find any other topic to write about than Covid-19. I vaguely recall there used to be topics of interest other than virus-related ones, and the thing is that EEG has very little connection to the only topic of importance. Sure, there is 'Black Lives Matter', but even if I could, theoretically, run a study to show that brains are neither black nor white, or more usefully, to show how prejudice affects perception, it is hardly practically feasible or ethically reasonable to bring anyone to our lab for testing. Indeed, the only ones who managed to successfully shoehorn a topical concern into EEG research are the group spearheading a document outlining how to do EEG in the Viral Age (Google: 'erpinfo.org protocol for reducing Covid-19'). Keywords: rigorous disinfection, minimal contact, good facemasks, goggles, and face shields for both experimenter and participant. In summary, best don't.

The second is that as I was waiting things out, I had a little family addition. Good timing, you might say, especially with that insanely long Scandinavian-style parental leave I get around here, and indeed it would have been – were it not for Covid-19 making it almost impossible for fathers to be present at the hospital. Almost but not quite, thankfully, and indeed I did get the chance to see some clinical EEG in action due to an odd, though ultimately harmless, complication. Fathers, I am told, are often horrible when it comes to keeping their fingers off the techno-gadgets in the hospital, and doubly so when they reckon they know something on the topic. Did you know I am writing a book on the topic?! Cue a long sigh from the mother. Unfairly so, given that she was well informed of my preoccupations since we started dating (p. 110, Figure 6.1). But all's well that ends well, and the healthy baby is sleeping quietly next to me as I am writing this in the middle of my parental leave, which suddenly looks a lot like anyone's normal workday in the Viral Age.

Early on, I had one concern. When the baby was born, I still had a beard, which steadily progressed from ruggedly manly (said no one) to professionally academic (ditto) to homeless drunk (unspoken though implied). Would the little one recognise me even if I shaved it off? If only I had an EEG amp at home, I could easily have tested this with a bunch of faces, modified with and without beards, and a sample of 4-week-olds, in an only slightly modified version of the experiment describe in this book! What we want to know, then, is whether infant brains in general recognise a face without a beard as the same one with a beard, so I could safely shave. In other words, we would be averaging across different participants, getting an idea about how brains *in general* behave. For such a question, we turn towards the grand average, which is the topic of this chapter.

Before you begin: Make sure you have data from the previous chapter. If you have been following so far, your data are average re-referenced, filtered data that are free of artefacts, epoched and sorted into conditions. The number of conditions depends on whether you are following the essential track (conditions *relevant*, *irrelevant*, and *odd*) or the additional track (*self*, *other*, and combination conditions: *rel_self*, *rel_other*, *irr_self*, *irr_other*) of the book or somewhere in between. If not, you can download the sample data from the SAGE website. Indeed, I rec-ommend doing this even if you *do* have your own data, as it allows you to work along with me and get a good idea of what the result of each processing step should be before you start with your own data. Download this chapter's online material (https://study.sagepub.com/Spape) and extract the files in the special EEG folder. This should result in a new directory, containing two subdirectories (code, data), of which data contains datasets for each subject and condition com-bination. See the readme.txt file for more information.

VISUALISING MULTI-SUBJECT DATA

In the last chapter, we looked at what happens in the brain if an individual participant sees an image – the visual evoked potential – and how this evoked activity is affected (or modulated) by attention, here operationalised as whether it was kept in mind or not. Task-relevance is known to affect the event-related potential, particularly the P3, and thus we noted that a part of the evoked potential could be related to this process. In EEG jargon, the P3 component of the event-related potential seemed modulated by task-relevance.

I say seemed, here, because this would be an incredibly hasty and unscientific conclusion. What we know is that in one person (me), at about 450 ms, there was a strong positivity after images that matched the one kept in mind vs images that did not. However, if you did the essential exercises, you will have noticed that there were substantial differences between subjects: some had earlier effects, some later, and some likely did not show anything much at all. There are many reasons why there is such an incredible amount of variation between subjects (p. 24, Comparing brains and EEG norms), but by and large we want to be able to say something about human brains in general, rather than particular (or in my case: peculiar) brains in general, for which reason we want to average across participants. Remember, however, that ERPs, as covered in Chapter 9, are already averages of epoched EEG, so calling them 'average ERPs' seems tautological, for which reason we call such 'averages of averages' *grand* averages instead.

EEGLAB uses the STUDY structure to calculate and plot grand averages. Interestingly, it can do so for more than just ERPs – which is what we have been mainly covering in this book – but also time/frequency analysis, which I introduced in Chapter 1 (p. 24, Comparing brains), and will explain here from an analysis point of view. The STUDY structure also allows for basic, statistical analysis of sim-ple one-way and two-way factorial designs, which I will go through in the second part of this chapter.

Creating the STUDY design in EEGLAB

To look at grand averages, we first need to create a new STUDY in EEGLAB. The simplest way is to first load all the datasets you want to analyse or average. Let's begin with an example in which I would merely want to look at the general grand average, regardless of stimulus content. In the previous chapter (see p. 194), I explained different contrasts, and various forms of what we could call a 'control' condition in the context of neuroimaging research. For now, let's have a look at what I then called a 'minimal control condition', which we might define as a basic *irrelevant* stimulus, so one that is common (not odd) and one that does not match the one the participant is currently keeping in mind. Accordingly, I make sure my EEGLAB is currently empty (or go to File > Clear study / clear all), then load all irrelevant 'irr.set' datasets in EEGLAB. Selecting File > Create Study > Using all loaded datsets will cause EEGLAB to attempt to create a study with all the subject, group, and condition information from the loaded datasets. Pressing OK to ignore the now usually outdated warning about memory requirements, and we are greeted with an important dialogue box (Figure 10.1).

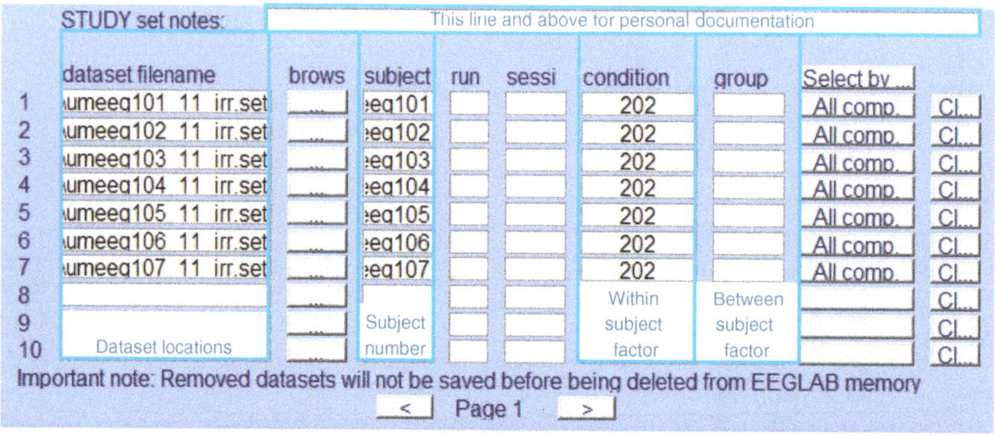

Figure 10.1 EEGLAB study building dialogue

If you did everything *by the book*, it should all be ready to work. As you can see, I have seven subjects, and each dataset has a specific subject number. All conditions are also set to target '202'. I find that a bit less transparent, so I would recommend changing it all to 'irr', or 'irrelevant'. However, whatever you do, for these fields it is critical to be consistent: no 'irr', 'Irr', and 'irrelevant' to denote the same condition. The reason is that EEGLAB will look up how many different names there are in that column, and it will call every new name it finds a new separate condition, so if you use *irr* and *Irr* interchangeably, it will tell you, after pressing OK, that it finds '2 conditions (some missing)'. We will come back to what this 'some missing' refers to in terms of experimental designs, but for now we have a single condition (irr), and each dataset should reference a different subject number. As you can see, I used 'umeeg101' to 'umeeg107', but anything will work, as long as they are all different across datasets. Run and session are variables that are otherwise ignored by EEGLAB, while condition and group I will further explain in 'Contrasting conditions'. Finally, there is a special column in which you can select the components to be included in the analysis, which is mainly of interest if you want to do your analysis at the level of source components following independent component analysis (see Chapter 8, additional exercise 1), a technique I would only recommend once you have a good grip of basic ERP analysis.

If you had already used Edit > Dataset info for each dataset and defined the condition and subject, everything should be immediately filled out correctly. Otherwise, edit subject to reference subject number and change all condition fields to 'irr', then press OK to continue. The EEGLAB box should now tell us that we have a STUDY set, with 7 subjects, 1 condition, 1 session, and 1 group, and that it is now using quite a bit of memory (for me 1655 MB, but it will depend on how much you removed in previous chapters). We can also see there is a pretty large variability in terms of channels per frame, for me: 32, 33, 34, 57, 60, 61, and 62. This has, of course, everything to do with the fact that I was using a 32+2 (EOG) channel amplifier *and* a 64 (no EOG) channel one, which is not the normal lab situation – or at least, I would not recommend it (few reviewers will appreciate the surprising novelty). Now, to fix this, and do various other final computations, go to Study > Precompute channel measures (Figure 10.2).

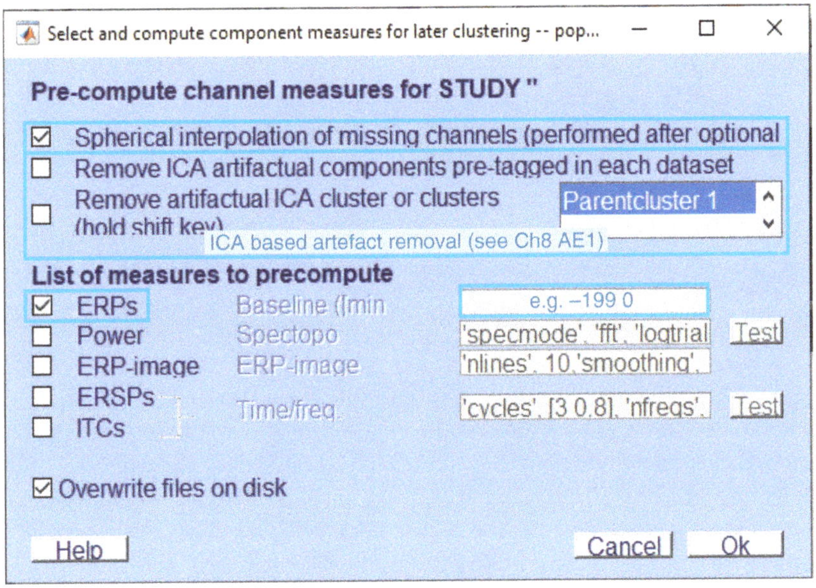

Figure 10.2

The dialogue enables spherical interpolation of the missing channels, which roughly means that missing channels are estimated based on surrounding activity given the assumption that the human head looks a bit like a balloon (i.e. spherical interpolation, see Chapter 11, p. 247). Right below these, there are several options for removing ICA components related to artefacts, which you may already have done in Chapter 8's additional exercise 1, or, if you have not, then this is not the best moment to start, so best to leave these unchecked in either case. The list of measures to precompute is quite extensive, allowing power and several time/frequency decompositions as well as this book's main focus, ERPs. I will explain Time/Frequency analysis later on in this chapter, but for ERPs we can specify a baseline and that's it. In fact, specifying a baseline is pointless if you already did this (p. 170).

After pressing OK, EEGLAB starts to interpolate a huge number of channels, and about a minute later, its precomputation is finished and we can have a look at the grand average evoked potential.

Looking at grand average ERPs

Following the precomputation of channel measures, you can now plot grand averages. Go to Study > Plot channel measures, and notice the slightly curious dialogue box (Figure 10.3).

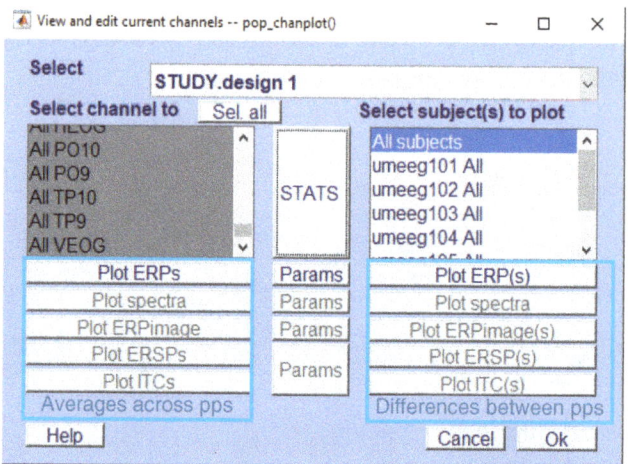

Figure 10.3

Curious, that is, because the left side is mirroring the right. For convenience, I summarise the distinction between left and right above as asking EEGLAB to either plot one graph summarising the average as a single line (or single line per condition), or to plot individual graphs for individual people. It may sound as confusing as it looks, so try both and notice the difference. For now, we can only Plot ERPs (as nothing else was precomputed in the last section), so press that, leaving everything equal.

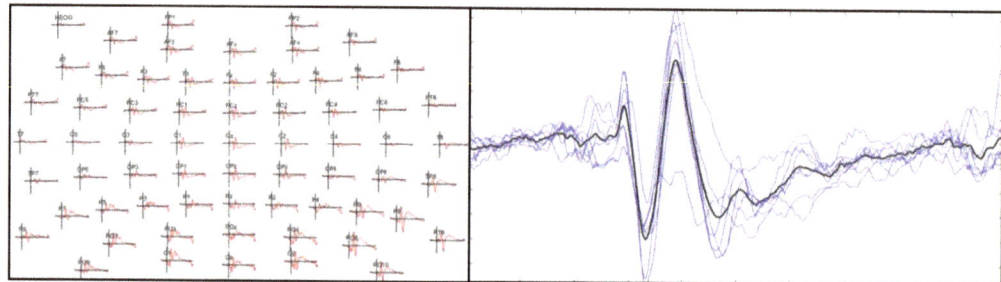

Figure 10.4   A left Plot ERPs and a right Plot ERP(s) sample output. Keeping all channels and all subjects (see Figure 10.3), the *left* Plot ERPs gives an overview that is similar to the last chapter's Sum/Compare ERPs (see p. 200), except as an average across participants. Selecting only 1 channel in the 'select channel' menu above, Cz, and clicking the right 'Plot ERP(s)' across 'All subjects' gives the graph on the right.

The black line in the centre of the right panel in Figure 10.4 is, in other words, the same as the red line in the middle of the panel on the left (Cz). Of course, I could have selected all channels, and the right Plot ERP(s) command would give a similar figure to the panel on the left, except with 7 additional lines, each summarising an individual participant, but clearly that is an overkill of information, so I opted for these instead. As you can see, there are some similarities between participants, in particular in the first negativity shown and the second positive peak (note: unlike the individual subject method of plotting, the STUDY plotting uses positive up by default).

Individual differences in the EEG: Causes and what to watch out for

There are also differences: one participant seems to have hardly any P2, and for another it lasts quite a bit longer. Such differences are quite natural and should not be cause for worry. However, if you spot one outlier who has huge spikes compared to all others, they will necessarily influence the average a lot more than others. This is very similar to what you will remember from statistics in general: outliers affect averages. However, in EEG, the situation is much worse as the black line above is not one average, but *one series* of averages (given the sample-rate: c. 512 averages). Furthermore, these averages are not independent: knowing that participant 6, say, has +20 μV at 412 ms over Fz will make it pretty easy for us to guess their activity at 414 ms over FCz. Why? The data are not independent observations but concern **time series** of the same participant that show similarities in time and space. In time, because EEG in general is more defined by lower (c. 2–10 Hz) than higher (20+ Hz) frequency, meaning that activity within c. 50 (1000 ms / 20 Hz) ms will be quite similar. Moreover, our low-cut filter applied in Chapter 7 (of 80 Hz, see p. 105) should mean that both waves of signal *and* noise shorter than 12.5 ms should be gone. In space, because spatial blurring or smearing causes the EEG in neighbouring channels to be quite alike. That causes two problems: one is that basic statistics that assume independence fail to function correctly, a problem I will discuss later (see p. 270). The other is that outliers will affect most of the rest of the analysis, never mind the specific timepoint or electrode. Thus, if you spot true outliers – classically defined as data that are more than 2 standard deviations removed from the average, but, as a rule of thumb, waves in Figure 10.4's right panel that look twice as big as those from other datasets – then *this* is the best moment to take a step back in your analysis and find out if the issue could have been solved, for example by using stronger artefact rejection or correction procedures (see Chapter 8).

We could now proceed by identifying peaks using the strategy outlined in the previous chapter (pp. 196–197), visually identifying positivities and negativities, except across participants rather than on the basis of merely a single subject. Remember, many ERP components only show up as differences between conditions, but as a kind of sanity check, it makes sense to identify also the positive and negative peaks of the visual evoked potential. That is, try to identify the P1, N1, P2 and N2 using the discussed techniques, keeping in mind that these should be not at all (P1/N1) or minimally (P2, N2) affected by stimulus relevance as they are more related to lower-level features (certainly the 1s) than more cognitive components (see Chapter 1, p. 13).

TIME/FREQUENCY ANALYSIS

Previously, I discussed how EEG is not merely a bunch of brain regions firing at the same time causing peaks in an averaged waveform but reflects the sum of oscillations synchronising and desynchronising. I like to see it as a kind of music, an orchestra of many tiny guitars playing inside the auditorium of the brain. Striking a note on a guitar will cause the string to vibrate (oscillate) at a certain frequency (the lower A string oscillating at 110 Hz). Likewise, you might imagine that seeing a face will start a melody of brain regions: from the basal ganglia and visual cortex of early visual processing, via temporoparietal areas related to attention, to frontal areas related to consciousness and decision-making, each area contributing new sounds. Obviously, I'm grossly oversimplifying, not just by the functional relationships of the areas, but also because unless you are Frankenstein, you will find it hard to find brain areas that are completely silent. That is, each string was already oscillating even before you struck a note, which **desynchronised** the preceding tone before the new one could be heard. So, what is more important: the new sound or the stopping of the old one? A continuing neuroscience debate focuses exactly on this question, with proponents of the latter theory holding that the disruption of the common alpha rhythm – often seen as a sign of inactivity – or **event-related desynchronisation**, signifies an area becoming more active (Pfurtscheller & Da Silva, 1999).

Music of the brain

You might have noticed that I am very fond of daft analogies, but the comparison between EEG and music holds reasonably well and seems a popular starting point for artists. There is at least one patent out for translating EEG into music to induce and control mental states (Knispel & Wright, 1989), plenty of human–computer interaction studies on EEG *sonification* (Leslie & Mullen, 2011; Miranda, 2006), and tons of artworks making use of such techniques (I am unsure how artists index their work's impact, but I have been present at more than enough installations of this kind!). The idea is that (1) the brain's EEG is indeed composed of reoccurring patterns of oscillations, just like music, (2) these do show a clear, even if not clear-cut, relationship with mental functions. Putting (1) and (2) together in the big blender of bonkers, it makes 'sense' that listening to brain oscillations will influence the brain like some sort of echo of mentality, a *dreamachine*, and perhaps synchronises the audience to the same tune! That's, like, groovy, man! Of course, I only got this sceptic *after* initially being quite excited about the idea and trying it out in MATLAB, which is very easy. Assuming you have loaded an epoched dataset in EEGLAB as dataset 1, have a *listen to the music of the brain* by typing the following in MATLAB's command window:

```
sound(ALLEEG(1).data([5 18],:,1))

%for additional exercise learners, some hints:

%sound plays the .data of ALLEEG(1) – dataset 1 –channels 5 and 18, the entire

%length (:), epoch 1.

%note that MATLAB will play at default audio sampling rate, rather than the

%recorded EEG sampling rate, causing it to have an increased pitch than what might be
%heard if EEG was actually sound.
```

...It is not exactly Andrew Lloyd Webber.

Leaving the neuroscience to the more mathematically and theoretically minded, psychologists and cognitive neuroscientists focus on the question of the relationship between the mind and the brain. So, how can we figure out whether a cognitive function, like attention or memory, will be reflected by a specific type of oscillation in the EEG? Since Grey Walter (Chapter 1, p. 5), the classic way to approach this question is by using **Fourier analysis**, which decomposes any time series into a frequency spectrum, informing us for each frequency how much it is present, in power (the square of its amplitude). The downside of this technique is that it loses temporal information, so there is little way of knowing whether an event such as seeing a face evokes a change in spectral power. If that is necessary, we use a form of analysis that keeps some temporal precision while still decomposing the time series into its spectral properties: **time/frequency analysis**.

Wavelet analysis is the most common form of time/frequency analysis. It is a relatively recent mathematical invention, and – certainly to me – quite a complex one. But at least the essentials are easy to grasp (see Figure 10.5). At its core, the technique uses an ideal shape of an oscillation, transformed to optimally capture the data at each frequency. These oscillations are then run through the data, matching at each timepoint the model oscillation with the data. The outcome is a time series of power for each of the analysed frequencies.

Even if this feels like it's quite a lot to take in at the moment, rest assured that EEGLAB makes it easy to work with. So let's have a look at how it works in practice, starting with *good ol'* spectral analysis.

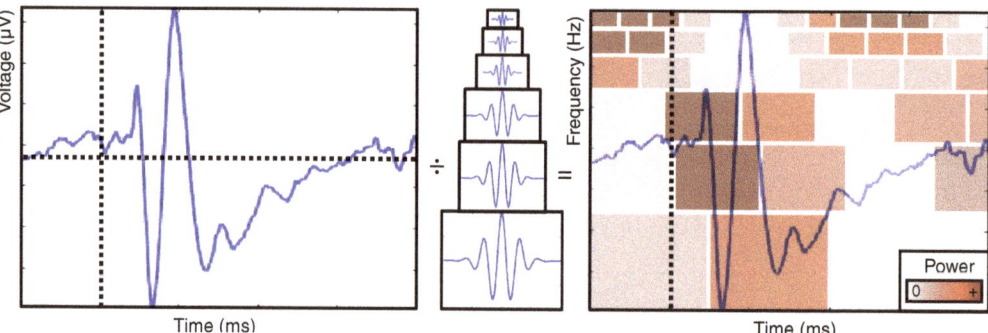

Figure 10.5 The idea behind wavelet analysis. Imagine we have a normal time series signal such as the ERP to the left. Wavelet analysis uses model forms of waves, such as the *Morlet* wavelet in the middle. To demonstrate, I visually matched each point in time of the actual ERP wave with the Morlet waveforms: if I figured that, at one of the 'bricks' X location, the ERP resembled the wavelet, I coloured the brick a dark shade of red; if I was slightly unsure, a little less red, and so on. The wavelet analysis returns a similar 'wall' of many bricks, making it easy to see that right after the onset of the stimulus (the vertical, dotted line), a lower frequency oscillation begins. These oscillations end soon after and are followed by higher frequency oscillations. Note that the bricks are larger at the lower frequency, which is because lower frequencies necessarily result in longer wavelets, much the same as a simple sine wave of 10 Hz having a period of 1000 ms / 10 Hz = 100 ms.

Spectral analysis

To make EEGLAB perform spectral analysis over our entire sample, go back to EEGLAB's main window, Study, Precompute channel measures. Selecting the Power spectrum, keeping the default parameters as they are, press Test to see what a spectrum analysis over the first 20 trials of the first dataset would look like (Figure 10.6). You get one line for every channel, and it is not a whole lot of data, so chances are that it will be messy. Likely enough, however, you will be able to note the typical EEG spectrum, as well as the results of the various filters you applied up to now. The spectrum of EEG normally shows higher power at lower frequencies than higher frequencies, apart from a small peak at c. 10 Hz (alpha). However, we see a noticeable *fang* in the graph at 50 Hz, which is where we applied the 50 Hz *notch* filter (see Chapter 7, pp. 153–154) to filter out the power line noise. Furthermore, after 80 Hz, the power takes a steep drop across all channels, and this is clearly the result of the 80 Hz high-cut (low-pass) filter that was previously applied (Chapter 7, p. 157). So no surprise there.

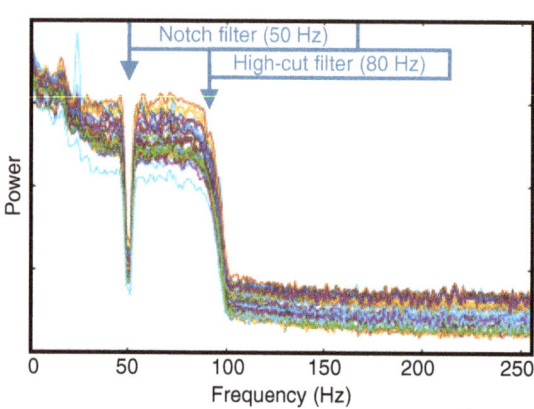

Figure 10.6

You may notice some channels have a bit more power in general and others less (the light-blue channel lower than most), but that is quite normal. The light-blue channel also has a strange spike in power, but remember that we're dealing with only 20 (out of *thousands*) of epochs of only one participant, so don't worry.

The limits of spectrum analysis

What *can* you learn from the Test analysis then? The critical feature to look for is whether the resolution of the figure is adequate to the research question you have posed. We can see, for example, by using MATLAB's data tool, that there are data for 6.02 and 7.03 Hz, but not for 6.5. So if the research question centres on mysterious 6.5 Hz oscillations, then clearly you are out of luck. Furthermore, there are also no data below 1 Hz, which makes sense as the analysis is performed over 1 second of data, so the frequency analysis cannot know whether a frequency below 1 Hz is present: a full 1 Hz sinus takes exactly 1 second. Similarly, we have no data for extremely high frequencies, specifically for those above 256 Hz. The idea is that one needs at least some digital datapoints to estimate a single, analogue wave, and so, as a rule, a Fourier transform cannot provide useful data above the **Nyquest frequency**, which is half the sample rate (i.e. the temporal resolution: 512).

After precomputing the spectral analysis, you can plot them for the entire study. Again, go to Study > Plot channel measures, and notice that the Plot spectra, underneath Plot ERPs is now available. First, however, click on Params and set a useful frequency band, for example the entire frequency range in which EEG dominates (c. 0 to 80 Hz). Clicking on Plot Spectra will, after a little loading, show the results of the analysis, with many plots that should all look like Figure 10.7. Now the EEG spectrum is much more obvious: a one-sided *dune*, gradually diminishing in power with a clear bump within the alpha band range. I looked at it a bit more precisely with the data tool and found the bump's edges were indeed between 7 and 12 Hz, so very close to Hans Berger's original observations (Chapter 1, p. 3).

Having looked that up, we can also find out where the alpha brain rhythm dominates, by going back to Params, selecting under Multiple channels selection the option 'Plot averaged topography', and changing

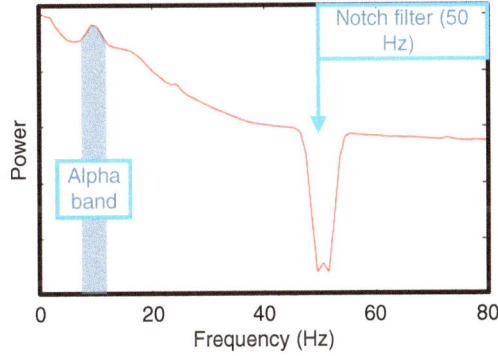

Figure 10.7

Figure 10.8

the frequency band to be plotted to a range between 7 and 12. This produces a pretty figure (Figure 10.8), which shows increased power over the occipital and parietal-occipital areas. Ha, I thought, that is right in line with what we know about alpha! But then I looked at a few other frequency bands (delta: 1–4 Hz, beta: 13–30 Hz), and they did not look markedly different. This brings me to two important points: (1) it's almost always better to look at differences between conditions than a single 'where is it?' blob, and (2) it's very easy to see what you want to see in your results. I'll spend the second half of the chapter talking about differences between conditions and will save the more insidious methodological issues for Chapter 11.

Time/frequency analysis

The pipeline for the combined time/frequency analysis is very similar, although it requires a bit more tweaking of parameters. I have explained the essentials as wavelet analysis as a process that matches slices, or 'bricks' in my analogy, of the waveform with idealised oscillations, the wavelets, to create a 'wall' that shows how the waveform compares over time with each frequency. Now, given that a lower frequency wavelet will have a **cycle** – the period or 'rate' of a wavelet – of a longer duration than one at a higher frequency, our **temporal resolution** at the lower frequency end of the spectrum is very low, as exemplified by there only being three bricks in the lowest row in Figure 10.5 (note, however, that in T/F analysis the wavelets are usually run in overlapping windows). This means we cannot easily capture fast fluctuations at low frequencies, and possibly we lose information as to the change versus baseline. For example, if I ask EEGLAB to use a single cycle for the wavelet analysis, it will start with the lowest frequency of 3 Hz, creating wavelets of 373 ms, so, given that my epochs each start from 200 ms before the stimulus onset, EEGLAB has the first wavelet runs from −199 to +174 ms, so we cannot know whether the result is due to the evoked activity of the stimulus. On the other hand, having very short-range wavelets can result in some spurious activity due to very occasional spiking in the EEG. An oscillation, as opposed to a random change in the EEG, will usually last at least a little while, much like any note on a guitar.

As you might notice, things are getting rapidly vaguer, as this is where EEG research experts start to give you few solid answers and a lot of 'well, that depends, but give it a try and see if it works'. That is not always useful, so keep in mind that (1) we are looking to optimise the temporal resolution so that we have enough information in both the frequency space and the temporal space, and (2) we do that mainly by changing the number of cycles. Let's see how that works (Figure 10.9).

Based on such criteria, I thought the [1 0.5] seemed like a reasonable fit for our current research design. However, if you want to do analysis based on wavelets, it makes sense to keep that in mind during all your design decisions, as it would be useful to have a much longer baseline and epoch length. Finally, running this wavelet analysis was already taking an hour for the first subject, so I stopped it (CTRL+C in MATLAB), then tweaked the parameters some more to change them to:

```
'cycles', [1 0.5], 'nfreqs', 50, 'maxfreq', 80, 'ntimesout', 50
```

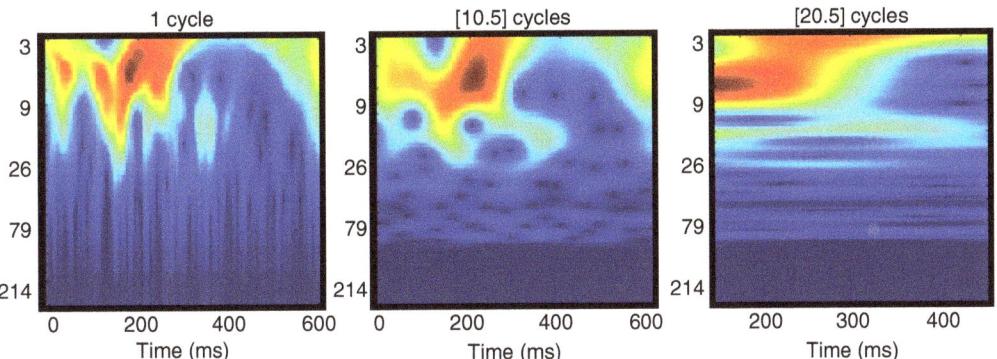

Figure 10.9 Testing out different numbers of cycles. Here we have three wavelet analyses, with the first using a constant 1 cycle. The second uses a variable increase in the cycle, meaning that it will start at 1 cycle for the lowest frequency (3 Hz) but gradually going up to use multiple cycles at higher frequencies (up to 42.67 at 256 Hz). This reduces the vertical spiking at the higher frequencies seen with the first panel, smearing them out across time. However, if we start already at 2 cycles, as with the third panel, we start to lose a significant amount of temporal information: note the much smaller X axis.

This reduced the number of frequencies involved in our analysis to those that could be of interest to us (i.e. below the low-pass filter of 80 Hz). I also reduced the number of times to a mere 50 as running a wavelet of 373 ms in hundreds of steps across less than 700 ms means a whole lot of redundancy. The reality is a little bit more complicated than Figure 10.5 suggests as EEGLAB 'slides' the wavelets in overlapping steps across the actual waveform, but in the end, the amount of additional information is very marginal. So we do not lose anything critical by reducing the number of times and the maximum frequency to 80, other than the prettiness of the picture.

Roughly an hour later, then, my computer had finally decomposed all data into time/frequency. Similar to the spectral analysis, we can now use the Study > Plot channel measures, then clicking ERSP (left or right) to conjure them up for the screen. Doing so will show a flood of images like the one in Figure 10.10.

Indeed, they all look quite similar, an early increase in power at the lower frequency (4–10 Hz), followed by decreased power over the higher alpha (11–12 Hz) and lower beta (13–18 Hz) end. I found this pattern to be clearest at the right-lateral parieto-occipital area (hence PO8 is shown here), which is pretty neatly in the region sometimes identified as the EEG's face area (e.g. where the N170 often dominates). Coincidence? Probably. Again, it is very easy to see what you want to see in your results.

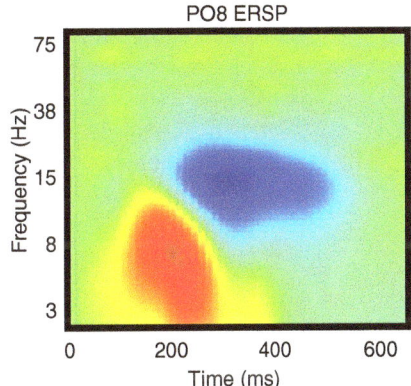

Figure 10.10

CONTRASTING CONDITIONS

Looking at grand average ERPs or time/frequency EEG analysis without regards of the exact condition often *sounds* good in a press-release sort of way: 'These 10 Jaw-Dropping Brain Areas Lit up When A Mum Saw Her Baby's Face: The Seventh Will Make You Cry Along'. But in science we generally try to discern exactly what causes something to happen and, as long as a person isn't dead, their brain areas will light up – or be active anyway – all over the place. Indeed, a satirical journal article presents evidence that a dead salmon's brain showed statistically significant brain activity (Bennett et al., 2009) either due to social perspective taking in the zombie fish, or due to endemic statistical problems in the field, take your pick. So even if we were interested in pursuing a career in *Neuroclickbaitology*, we would need at least two groups – mothers and non-mothers – and two conditions – faces and non-faces. And even then, lots of things are not faces, and lots of people are not mothers. And then things get suddenly very complex. Do we compare faces with black shapes, other objects (e.g. cars), or objects that might just look a bit like faces (the front of cars, perhaps)? Likewise, in the between-participant variable, do we compare mothers with fathers, children, childless women? While I mentioned this point before (see Chapter 2), it is important to remember that in science we try to maximise the precision of our explanation.

To do so, we first need to stop thinking about brain areas magically lighting up, but rather think of these as a contrast against something. In cognitive neuroscience – although I would argue in Psychology in general – a within-participant factor is often necessary. A common, though pretty poor practice is to compare against a resting state, but luckily in our design we had a more useful *control condition*. If we want to know what happens in the brain if you have a person in mind and you see that relevant person, it makes a lot of sense to compare against seeing persons you don't have in mind, i.e. the irrelevant ones. This EEG pattern marking relevance then becomes a new, much cleaner measure, which we can further analyse. For example, we might then ask if mothers are better at relevance detection than other, lesser beings (see chapter introduction: I am biased). In other words, what comes out of our comparison of conditions in the one-way analysis (relevant vs irrelevant) can be further investigated by adding new variables to the analysis in the multifactorial design. If that turns out to be the case, we speak in EEG slang of a 'modulation': the EEG related to relevance is *modulated* by the additional variable.

The secondary variable is not necessarily a between-participants factor (group) such as motherhood, although, as we will see, EEGLAB's otherwise very convenient study design takes that as a given. For now, this is slightly inconvenient as we simply do not have that many participants: in the data made available for the book I provide seven datasets, and four them are recorded from the same person (me). But for illustration purposes, let us imagine a researcher would be interested in knowing if the author of this book is sane, and uses the relevance detection effect as a proxy of sanity. The methodology is shoddy: 4 times one subject, 3 persons as a control group? The prediction is odd: why would sane people have a bigger P3 (well, see Bramon et al., 2005)? Or perhaps I'm just giving myself a safe get-out clause, in case it turns out my P3 is found lacking.

Comparing conditions in a one-way design

Remember how you made a STUDY design at the beginning of the chapter. We will now do the same, but load both the irrelevant and relevant datasets. Look back to Figure 10.1 on page 214 and simply change the condition names under condition to something readable (like 'irrelevant' and 'relevant') for the irrelevant and relevant datasets respectively. Note that each dialogue only supports 10 datasets, so if you have 7 participants and 2 conditions, you will need to click on the right arrow next to Page to enter all the data. Once you have entered the data for the subject and condition columns, and after clicking OK, EEGLAB should indicate that there are now a number of subjects (7 for me), conditions (2 per subject for me), as well as some other information that summarises the entire selection of datasets rather than just the one currently selected.

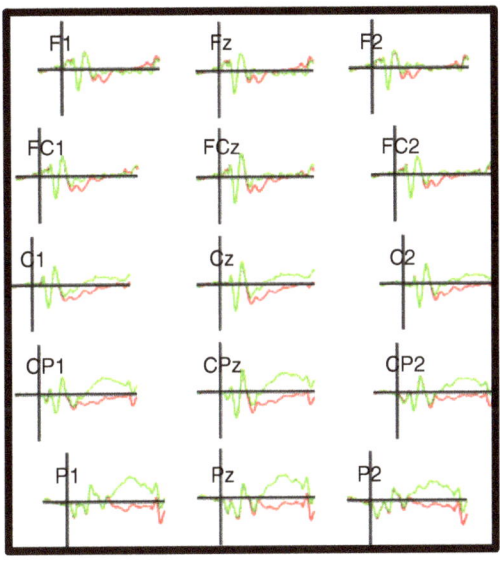

Figure 10.11

Now going to Study > Plot channel measures; we can see how the two conditions compare against one another. Clicking on the Plot boxes on the right side (below the individual subjects) starts to create an insane number of ERPs on the screen, so I would stick to the left buttons. Immediately clicking on Plot, however, will show the ERPs in two separate figures, which makes things very hard to compare (try it once, though). Instead, go to Params next to the ERPs, and check the option for plotting the first variable on the same panel. Doing this will show the ERP of each condition as an individual line within the larger array of figures, as shown in Figure 10.11.

Look at Figure 10.11, there's actually an effect! It sounds obvious that there's an effect, right? Of course there would be, you wrote a whole book about it! I mean, any normal statistics book goes about getting the data just so that the effect comes out significant or not significant and then you start writing advice. That is a sound approach, but also quite boring. I went about it the other way. First, I bluffed my way with the publishers, told them that it will obviously work, *I am a scientist, let me show you!* Then I wrote the first nine chapters just guessing it will surely, probably, maybe-please-any-higher-being-help-me, work out, and then *bang* the result is there, *exactly as it should be*! It is amazing, and I get to keep my underpaid, but fun, job.

Because it is there, right? I mean, that green wiggly line sits higher than the red, somewhere after the second negative dip? The difference is largest over the parietal areas? OK, so relevant images should amplify the P3 response, so a positivity in the ERP after about 300 ms is usually prevalent over the parietal areas. We can do a whole lot of statistics, and we will, but a solid effect in the main contrast doesn't generally disappear – the way that vague, non-replicable effects do sometimes appear by applying questionable statistics.

So let's do some stats! Hang on, wasn't I supposed to be the fun sort of teacher who rarely brings out SPSS, and even if then, at least brings in pictures of cats against the encroaching anxiety? No, we'll do the real stats in the next chapter, but let me at least show you how to do Bad Stats, without any SPSSy requirements.

Go back to the Study > Plot channel measures dialogue and hit the massive STATS button. In the menu that pops up, put a check next to 'Compute 1st independent variable statistics if any', which will cause EEGLAB to compute statistics for the condition variable. Below, we get to enter the specific type of statistics to be used, which is by default 'Use parametric statistics' – the normal branch of statistics[1] – and the Statistical threshold, which is the *alpha* level under which the null-hypothesis of the two conditions having equal ERPs is being rejected. But remember, we don't have just one test, like your standard statistics class does, but a number of channels (here: about 61), and timepoints (here: 512), so EEGLAB will calculate many, many tests (31,232). That means there is a lot of chance that the null-hypothesis will be rejected despite being true. I will discuss the problem of balancing type I and type II errors later (p. 270), but for now let's at least agree that the Psychology default value, of 5%, is probably *far* too liberal, so let's change it from 'exact' and set it to .001 for now.

Note: Keeping it 'exact' will output ERP like plots with the exact *p* values – like the SPSS 'sig' standard – on the Y axis, which is nice otherwise, but tends to be information overload.

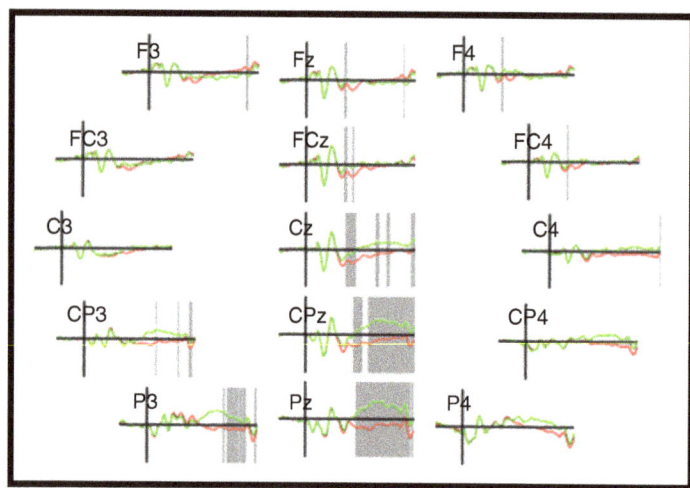

Figure 10.12

After also changing the Params to have a check next to 'Plot first variable on the same panel', I also selected only a few channels in the Select channel to plot panel: F7, F3, Fz, F4, F8, FT7, FC3, FCz, FC4, FT8, T7, C3, Cz, C4, T8, TP7, CP3, CPz, CP4, TP8, P7, P3, Pz, P4, P8 (see p. 102 for electrode position names). This gave a reasonable overview of where the main differences between conditions were found (the middle three columns shown in Figure 10.12). In grey are all the times that show a significant difference between the two conditions, and indeed the clearest part of

---

[1]Yes, fellow stats nerds, I know that is an imprecise description. Parametric statistics use population models with a predetermined set of parameters. For example, the standard-normal distribution has a shape defined by the standard deviation, a single parameter.

it seems to be in the later part of the ERP around CPz, Pz, and P3.[2] In other words, seeing a person on the screen elicits a healthy, P3-like response.

Comparing conditions in a multifactorial design

We can do the same for more advanced, 2 × 2 designs with the EEGLAB study method. As mentioned at the beginning of 'Contrasting conditions', this 2 × 2 design in EEGLAB automatically means one between-participants and one within-participants variable, which means a mixed design (between and within). One might study whether there is a difference in seeing a relevant versus irrelevant face in men and women, children and adults, politically left-leaning and right-leaning individuals, but it will generally require more datasets than we've got available. To show how it can work, however, I can think of only two ways of creating *groups*: separating either those scanned with a BrainProducts 32 channel amplifier from those with a Biosemi ActiView II, and separating people who are me, from people who are not me. Neither is methodologically sound, but being the self-centred person I am, I naturally went for the latter option (Figure 10.13).

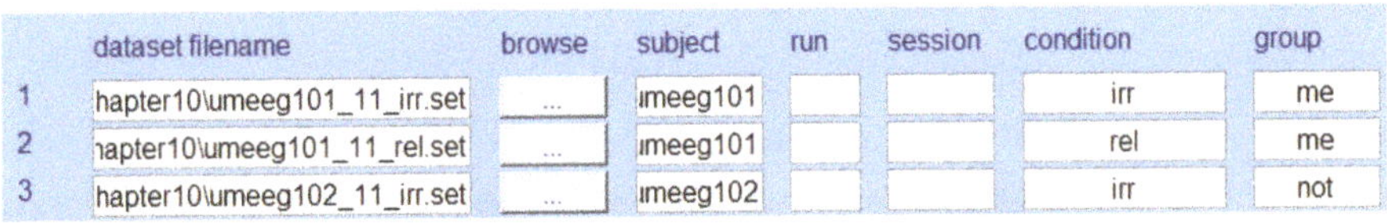

	dataset filename	browse	subject	run	session	condition	group
1	hapter10\umeeg101_11_irr.set	...	umeeg101			irr	me
2	hapter10\umeeg101_11_rel.set	...	umeeg101			rel	me
3	hapter10\umeeg102_11_irr.set	...	umeeg102			irr	not

Figure 10.13 Mixed design in EEGLAB Study. The information on who is who can be found at the beginning of Chapter 7. Make sure to enter subject numbers, condition names, and group identity.

After precomputing channel measures – again – I can finally find out whether my brain is special. I will leave it to you to fiddle around with the plotting menu, but if you, like me, cannot find a good way to plot a 2 × 2-way figure in EEGLAB, don't worry: it is pretty hopeless and the Plot variables on same panel option seems a bit broken. To be fair, what would it look like? Four different lines, two in one colour and dotted, two in another, straight? Perhaps a difference between relevant and irrelevant, plotted for each group? There are so many possibilities for making different figures, I find it much easier to first sketch a figure on paper, and then use various other software (or MATLAB without EEGLAB) to create the figure. We will do this in Chapter 11. In the meantime, however, I managed use EEGLAB's STUDY plot to create a somewhat reasonable graph (Figure 10.14) by concentrating on only the most obvious channel, Pz. It also showed some p-values for the within-subject difference between irrelevant (left) and relevant (right), and for the difference between groups me (upper) and not-me (lower). However, the more interesting question, whether the difference between my relevant and irrelevant ERPs was unlike that of others seems to be answerable with *apparently not*. I am, after all, not special, or at least my brain says so. I disagree, but underline that more research is not needed (see p. 286, final point).

---

[2]Despite the name, the electrode P3 has little to do with the P3 component of the ERP.

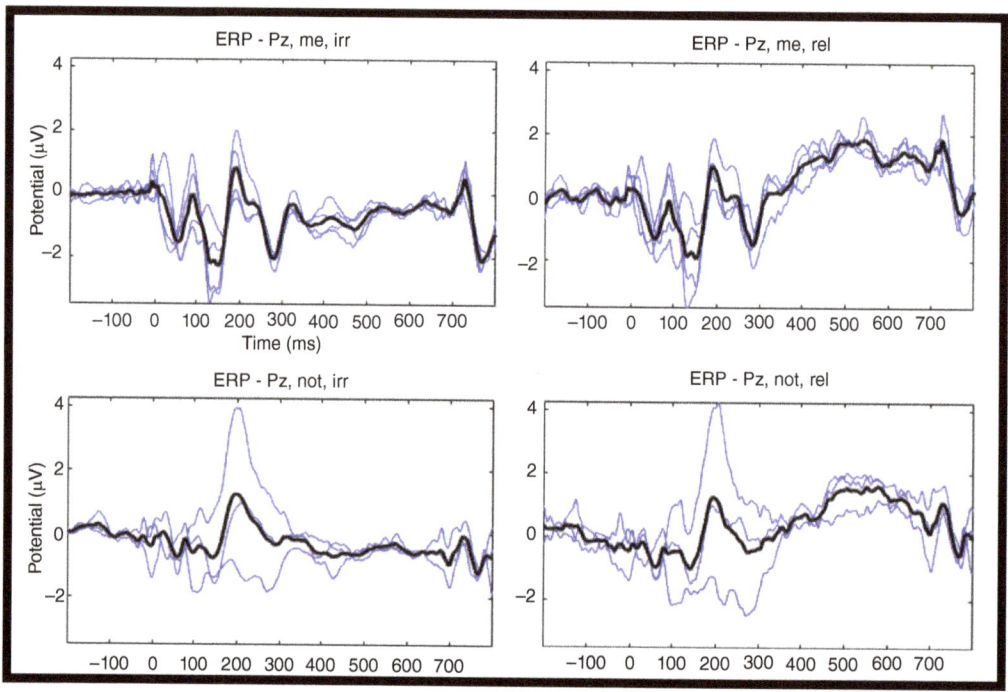

Figure 10.14

I used the right panel in Figure 10.14 for plotting ERPs, so each blue line is one dataset. The one thing I found surprising was that my own ERPs (top panels) looked so very similar, even though they were all from different days and using different machinery.

Interpreting differences in the ERP

To make a meaningful answer to our research questions as posed in Chapter 2, we cannot simply show a nice ERP figure (such as Figure 10.12). Why not, you might reasonably say, a good graph says more than a thousand stats, yet tutors and reviewers usually prefer 'hard' numbers anyway. Our various exploratory hypotheses (see p. 33) therefore require us to extract data from EEGLAB to pour into SPSS or other statistical package to see if any secondary factor, such as seeing yourself or someone else, modulates the primary factor (relevance). For this reason, it would be useful to define **regions of interest** (ROI) that become single metrics we can test statistically in the next chapter. As such, we can follow an approach similar to how we defined the ERPs in Chapter 9 by contrasting the relevant and irrelevant conditions, but now by using the multi-subject analysis we just conjured up. The object is to reduce the thousands of combinations of electrodes and time-points to a handful of numbers.

There are many ifs and buts here, and I will discuss them at length with regards to type I and type II errors in the next chapter, but as a rule of thumb, if an area in the ERP is significant (i.e. marked in grey in the results figure) for a continuous period of time in multiple locations, it is less likely to be a chance finding than single points of scattered significance (i.e. grey bars). Thus, we can ignore the few significant spots in Figure 10.12 on page 226, such as in CP3, F3, and F4, concentrating on the electrodes that show longer lasting differences.

Note: If there are too many grey areas, or too few, you may want to adjust the statistical threshold.

Using the MATLAB data cursor, I thus went through the 60+ plot channel measures figure and jotted down the following notes, with a few ground rules: (1) I ignore a single grey bar in the midst of non-significant areas as a likely false positive; (2) I ignore a single white bar in the midst of a large area of grey as a likely false negative; (3) I ignore all areas in the most lateral parts (e.g. F8). The result was the following:

F2	280–291, 328–353, 367–377
FCz	277–300, 341–347
FC2	273–304, 320–345
Cz	291–298, 308–329, 337–365
CP1	376–690, 705–714, 724–800
CPz	351–412, 458–628, 636–742, 754–800
P1	380–800
Pz	371–763, 779–787
P3 and POz	similar picture to P1 and Pz, if slightly fuzzier

Now, keep in mind that these numbers look deceptively precise, being in milliseconds. They are not, however, as you will have noticed how much they change with slight changes in threshold, or whether one starts to interpret my rules a little bit more liberally or conservatively. On the whole, though, we can see two basic clusters:

ROI 1: Frontocentral electrodes (F2, FCz, FC2, Cz) show an earlier (c. 270–350) difference between relevant and irrelevant, more or less in the N2 area (coming just after a noticeable positivity, the P2)

ROI 2: Centroparietal electrodes (CP1, CPz, P1, Pz) show a later (c. 370–800) difference, quite nicely matching what one might expect from a P3 component.

These, then, would be ROIs one can work with, but the literature tends to use certain conventions that group the areas of interest together. For example, midline areas (Fz, Cz, Pz) can be contrasted with left (F3, C3, P3) and right (F4, C4, P4) areas, or frontal (F3, Fz, F4) with parietal (P3, Pz, P4). Another way to do analysis, which we will discuss in the next chapter, is to make a crossed design of electrodes, separating both laterality (left, centre, right) and frontality (frontal, centre, parietal) into a 3 × 3 design. Finally, some EEG researchers take areas that have previously been associated with certain processes, such as manual movements (the left and right motor areas overlying C4 and C3 respectively) and configural face processing (T6, P8). Our own areas are slightly frontal right and parietal left, so one possibility would be to include more, including F1 and FC1 in the first ROI and CP2, P2 in the second. Alternatively, the less-is-more strategy would be to include less and stick with FCz, Cz for ROI 1 and CP2 and P2 for ROI 2.

Complicated and somewhat arbitrary as that may be, this will be only the first of many more such decisions to make: What kind of metric do we use (e.g. peak maximum or mean amplitude), and how do we enter areas, windows, and experimental factors in our statistical analysis? These questions I will answer in the next chapter, in which I will start by getting the data out of EEGLAB and into a more conventional statistical programme for further analysis. But to conclude this chapter, I will give a few suggestions of what to look for when viewing grand averages.

Baseline differences or ERP differences?

Do you see any significant difference much earlier than you expect it? If so, chances are that there is something wrong with your baseline. If, for example, your baseline is short (e.g. 50 ms), and for some reason the ERP was more positive there than expected, this difference in the baseline may have distorted the rest of the ERP. For example, in Figure 10.15, the red ERP in the baseline is more positive than the green – except for the last 50 ms or so. This causes it to give early differences we would not have expected: a lower N1, P200, and N2. The problem immediately goes away with a slightly longer baseline. Just try to imagine what the figure would look like if the red line were corrected by the amount it is currently too far 'up'.

However, such easy solutions are not always available, usually due to issues with the experimental design. Imagine, for example, we have an oddball study with 10 relevant vs 50 irrelevant stimuli in repeating lists of cycles of 6 stimuli each *without repetition* (as opposed to less restrictive randomisation as suggested for E-Prime on p. 49 and for OpenSesame on p. 75), such that a relevant stimulus can never occur after another relevant stimulus. This would mean that any relevant stimulus can only occur after an irrelevant one, therefore after a red line such as in Figure 10.15. In contrast, any irrelevant stimulus can be preceded *either* by a relevant (in 1 out of 5 or 6 depending on programming) or by an irrelevant stimulus. That means that a strong (prior relevance-related) positivity *can* occur in the baseline of the irrelevant but not the relevant ERPs. Consequently, the irrelevant ERPs may look distorted, with unexpected effects in the early components.

A somewhat related issue you might have observed in the ERP figures is at the very end of the ERPs. Did you notice the very strong bumps in most ERPs between

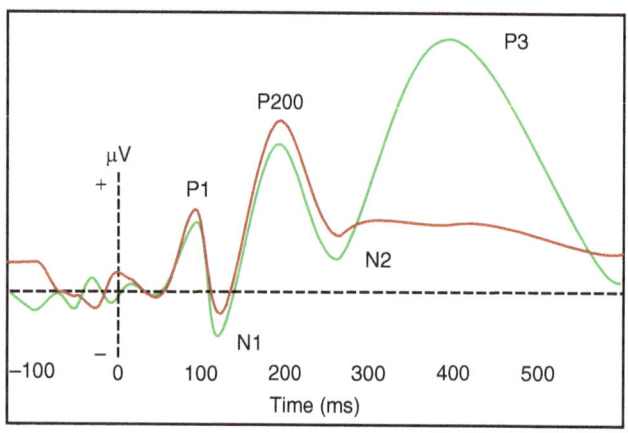

Figure 10.15

c. 700 and 800 ms? As I discussed elsewhere (p. 169), this is because the next image should have appeared at 700 ms, preceded by a black screen of 100 ms. The black screen, at 600 ms, is a visual change and thus evokes P1s, N1s, and so on, in its own right. In other words, if the normal P1 occurs at c. 60 ms, we can expect another P1 at 660 ms (plus timing errors).

While we have a strange bump at the very end of the ERPs, these are unsurprising, but with some designs they may appear only in one condition and not another. That is, in our current experiment, people passively observe images of a similar length with a new image always following, but in choice reaction paradigm, you might show a stimulus and request a response (e.g. yes or no, left or right). This can lead to problems, especially if the stimulus is terminated by the response, perhaps showing a blank screen as soon as the keyboard is pressed. If there is a reaction time difference, then we can expect a distribution of P1s (and many other components) to affect the late part of the ERP with faster reaction times causing earlier P1s. While that is a rookie mistake, it can be hard to completely avoid. For example, one might use a constant duration for the stimulus and not have button presses terminate it. However, a response, whether it has visual consequences or not, will still show correlated brain activity due to motor preparation and execution. Likewise, P3-latency effects that correlate with the reaction times may still affect the ERP and, as I will discuss, these can even result in illusory differences in preceding components.

The take-home message, then, is *always remember to consider ERPs in their physical context, both sensory and behavioural.* We tend to focus too much on our variables of interest and target measures, but an evoked potential comes with both a history and a future.

Which ERP component is affected?

It can be challenging to decide exactly which ERP component is affected and oversimplistic measurements may hide that fact. For example, looking at the ERPs in Figure 10.16, we might calculate the N2 as the average amplitude between 250 and 350 ms, and the P3 as that between 350 and 450. However, these are merely calculations we have decided upon by applying a set of rules for declaring a ROI, as explained in this chapter. Importantly, ERP components are *latent* processes with a neural origin that give rise to the observed, *manifest* changes in the evoked potential readings. As the latent processes may have overlapping effects, it is not always easy or even possible to know which specific configuration of latent components caused their manifest effects as they sum together in the measured ERP.

For example, imagine an experiment with two conditions (red and green) shows ERPs like the left panel in Figure 10.16. A careless conclusion might say that the green condition shows an increased (more negative) N2, followed by a P3 that has a lower amplitude. But this is far from obvious. In the middle panel, I show how you can look at ERPs as a sum of multiple observed, latent components that partially overlap. Note how the N2 latent component slightly overlaps with the latent P3 component, even if that is no longer observable after I summed them together in the bottom (Σ) ERP. As the P3 is therefore like a mountain that starts already within the valley that is the N2, it could be that the P3 is merely enhanced in the red condition, rather than that both the N2 valley *and* the P3 mountain are lower in the green. Indeed, if we plot the difference wave as the subtraction red minus green (dotted line in the graph), it seems a more *parsimonious* explanation to say the difference is simply due to an amplified P3 than a diminished N2 *and* amplified P3.

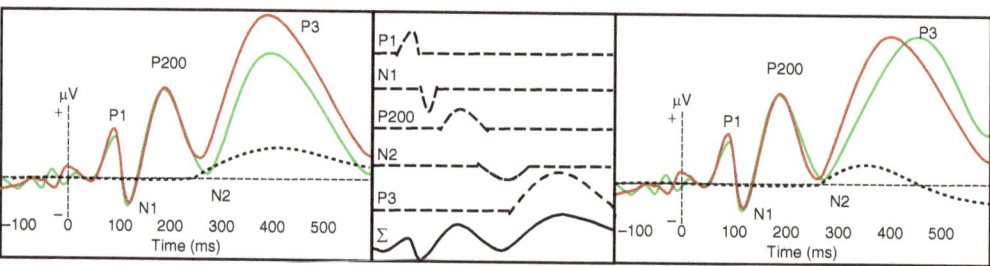

Figure 10.16 Two hypothetical ERPs with red and green conditions and the difference wave in a dotted line. Left panel: is green due to increased negativity at the N2 followed by a smaller P3? Right panel: is red giving enhanced P3 activity followed by an extra 'N500'? Middle panel: a schematic representation of how latent ERP components sum up to a measured ERP.

Difference waves sometimes help, but not always. Consider the right panel of Figure 10.16. Again, a simple calculation of N2 as the mean amplitude (p. 246) between 250 and 350 would suggest the green condition has a lower average amplitude than the red one. Here, the difference wave doesn't help, implicating an even more complex interpretation: an early, amplified P3 is followed by an additional, negative component. Maybe I can coin the special red-induced N500? Again, it makes more sense to apply Occam's Razor (choosing the more parsimonious explanation) and interpret the difference in ERPs as the green condition having a *delayed* peak, rather than inferring two effects.

Comparing ERPs with different numbers of epochs

As illustrated in the right panel of Figure 10.16, ERP components can show differences in time as well as in amplitude. There are two basic causes for ERP latency effects: neural (or psychological) and physical (or artificial). In terms of the former, it is a well-established find-

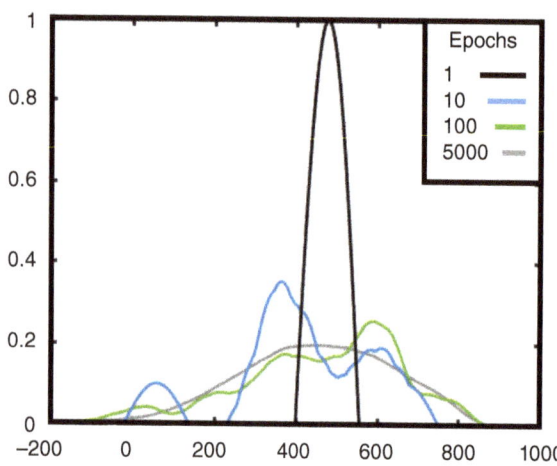

Figure 10.17

ing that certain (usually later) components are affected by task differences, 'cognitive load', and response preparation (Verleger, 1997). In terms of the latter, timing problems and error may cause even the earliest ERPs to occur earlier or later than expected. For example, if a stimulus is shown on the screen at 0 ms but the trigger only arrives at 10 ms, then the P1 may occur 10 ms earlier than expected.

Both causes can result in problems interpreting the averaged ERPs. In Figure 10.17, I show a simulation of what happens if an idealised 1 μV peak occurs at 500 ms, randomised with a standard deviation of 200 ms. Making ERPs over 1, 10, 100, or 5000 'epochs' like this causes the ERP to gradually become flatter, its highest point (peak magnitude, p. 246) going from its latent score of 1 to about 0.4 with 10 epochs, 0.3 with 100, and barely over 0.2 with 5000 epochs. Note that this is purely due to the latency shifts of the peak. If it were always exactly occurring as the ideal '1 epoch' peak shows, it would not matter if you had 5 epochs or a million.

This means that, especially for peaks that do not have a consistent latency, you should avoid comparing ERPs that are the average of varying numbers of epochs. As you can see above, not only does the ERP become less positive, it also smooths out, *smearing* in latency.

Or to put it as an easy-to-remember rule: the more epochs within an ERP, the fatter and flatter it will be.

Consequently, be careful when it comes to interpreting differences between conditions that have different 'cell-counts' (number of epochs within a condition). For us, for example, there is one relevant epoch for every five irrelevant ones, so the cell-counts will be different. What do you do in this case? One way to deal with this is randomly removing as many irrelevant epochs as necessary until the cell counts are equalised. Another is to use single trial measures, as these should not suffer from the smearing problem. They are, however, much more noisy than ERPs, so some use clever tricks to distinguish instead between latency variations that are stimulus-locked, response-locked, or neither (cf. Ouyang et al., 2016). Finally, some measures, like mean amplitude (p. 246), are much less affected by latency variations than peak magnitude. While the 5000 epoch ERP is indeed flatter and fatter than the 10 epoch ERP, its *volume* is not all that different. We will turn to different ways of measuring peaks and areas in the next chapter.

ESSENTIAL EXERCISES

Essential exercise 1: Testing the main hypotheses in EEGLAB

In the last chapter, you used single subject data to detect ERP components. Now, use the same approach and the STUDY design in EEGLAB to find out what the main, visual evoked potentials are based on multiple subjects.

Is there a clear P1, N1, and so on? These ERP components should be visible even without contrasting two conditions. Define all visual evoked potentials you can distinguish both in time and space:

- In terms of time, look up from the STUDY plots their range as well as their peak latencies.

- In space, note where the components look strongest and if there are any neighbouring electrodes showing similar peaks.

- Also note any electrodes where they show up with an inverse polarity (if the peaks show positivity in some electrodes, it will likely show negativity in another or some others).

In Chapter 2, well before I even coded the experiment, I set up a set of hypotheses. The full explanation can be found on page 33, but the gist of it was this:

I.   Relevance – the target or 'to be kept in mind' vs non-target 'other faces' dimension – should *not* affect early potentials or anything before c. 150 ms.

II. Relevance should affect the P3 component of the ERP, as has been found since the 1960s. The P3 starts normally around 300 ms, and commonly has a topography with a maximum over the centro-parietal electrodes (CPz, Pz).

Using the approach from the previous chapter but with the STUDY design, find out how the contrast between relevant and irrelevant conditions affects the ERP.

•   Are there any clear differences in the VEPs you defined above?

•   Does relevance affect the ERPs occurring before 150 ms?

•   Does relevance affect the P3 component?

Essential exercise 2: Testing oddball pets

In Chapter 2, I made a number of additional, somewhat more exploratory hypotheses related to some of the more novel aspects of the experiment. Some of these are harder than others to analyse, though, so I would start with the ones that refer to the 'pet' condition. While it is possible to compare the pets with relevant *and* irrelevant conditions, this can make interpretation challenging, as EEGLAB will now conduct 1-way 3-level repeated measures ANOVAs. This means that any observed significant difference could be between pets and relevant, pets and irrelevant, and relevant and irrelevant. To make it slightly easier, contrast the pets with, first, the irrelevant condition, and then the relevant condition. With each contrast, try to figure out what is unique to the visual differences between pets and human faces, the probabilistic differences (pets occur less often), before explaining anything at a 'higher' level, involving cognition or affect.

•   Does the pet ERP show different early potentials? Is there, perhaps, a pet-related early ERP?

•   Do the pet ERPs look more like the irrelevant ERP or more like the relevant ERP? Both seem a priori reasonable: pets are never kept in mind (thus irrelevant), but they are strikingly odd (most images are humans, so the pets occur less), which evokes oddball P3s.

•   If the pet ERPs show marked differences in the ERP, conduct a time/frequency analysis to investigate whether there are special brain frequencies evoked by seeing pets.

•   Finally, reflect on your observations here by referring to the final part of this chapter's 'Interpreting differences in the ERP'. Are there any alternative explanations for your ERP-related observations?

Additional exercise 1: Effect of relevance on seeing yourself

In Chapter 9, you sorted the epochs into datasets in which people saw themselves or saw others, and that were relevant and irrelevant. This makes for a 2 × 2 type of design that is much more complex than the 1 × 2 (essential exercise 1) and 1 × 3 (essential exercise 2). It is also more practically challenging, as both seeing yourself and seeing a relevant image occurs infrequently, so the occurrence of both happening is very rare indeed. Consequently, you should reflect on how best to analyse this with the tools currently at your disposal:

- Do you enter all conditions as a single Condition marker in the Study?

- Do you enter 'self' and 'other' as groups, or why should that be avoided?

- Do you follow a design as with essential exercise 2, making selective contrasts?

- Do you keep all subjects even if they have very few epochs in some design cell?

We will return to the question of what to do with the 2 × 2 design in the next chapter, but for now it may be useful to explore this analysis, keeping the drawbacks in mind.

Additional exercise 2: Plotting grand averages in MATLAB

In Chapter 9, I discussed for-loops and if-conditionals in the context of creating datasets from code, batch-processing the sorting of epochs into datasets with only certain types of trials. Let's take this knowledge further by doing much of the rest of the analysis discussed in this chapter by MATLAB code as well. As you may have noticed, EEGLAB can create some spectacularly ugly figures if you let it, so I will also show that with a bit of effort MATLAB can make substantially prettier plots.

Let's say we want to create figures with 2 channels, Cz and Pz, showing the ERPs of 2 conditions, relevant (rel) and irrelevant ('irr'). After the work in the last chapter, I have all the data from 7 subjects within Chapter 10's 'Data' directory, with names like: 'umeeg101_11_irr' (the 11 referring to the operation step, as explained in Chapter 7). Have a look at the following code:

```
eeglab
indir = 'Data';
participants = {'101', '102', '103', '104', '105', '106', '107'};
conditions = {'irr', 'rel'};
EEG = pop_loadset('filename', ['umeeg' participants{3} '_11_' conditions{1} '.set'], 'filepath', indir);
```

Is it clear to you that this should load file 'Data\umeeg103_11_irr.set'? If not, please look back to the additional exercises in Chapter 9.

Otherwise, run the code and have a look in MATLAB's variable view by double-clicking on EEG in the workspace. Note the data are a three-dimensional variable, for me a $33 \times 512 \times 1346$ single. These numbers may seem random but if you look around a bit more in the EEG struct, you will see them repeating over and over. For example, 33 also comes after 'nbchan', while 'chanlocs' seems to be a 1×33 struct. Double-clicking on chanlocs gives a table with all the channels – for example, for me channel 24 seems to be Pz, 512 repeats as the number of 'pnts', and times is a 1×512 double array. Double-clicking on times shows a grid starting at -199.2 and ending at 798.82 – indeed, the time-points of our epoch. Finally, 1346 is the number of epochs (and trials) within this dataset.

So, if I want to calculate the mean Pz voltage across the 1346 trials – in other words, the ERP for Pz – you might think the following code works:

```
Pzmean = mean(EEG.data(24,:,:));
```

…but you'd be mistaken. Try it and notice MATLAB shows in the workspace that Pzmean is a three-dimensional array of $1 \times 1 \times 1346$ in size. In other words, it calculated the mean across the second dimension, rather than over the number of trials, which was the third dimension. So, we need to specify which dimension is averaged:

```
Pzmean = mean(EEG.data(24,:,:),3);
```

This returns the mean of the EEG data for channel 24, averaged across the third dimension. Which can be plotted simply by:

```
plot(Pzmean);
```

…or, better, which can be plotted with the actual epoch times on the X axis:

```
plot(EEG.times, Pzmean);
```

If you grasp the above on dimensionality, you should be prepared for the following statement: To calculate the grand average ERP with 2 conditions and 2 channels, we need to load our data of 7 subjects to a four-dimensional matrix of size $7 \times 2 \times 2 \times 512$ (the order is relatively unimportant at this stage). So we create such a variable, filling it with zeros, then load all the data in and average as we go:

```
channels2plot = {'Cz', 'Pz'}; %channels we aim to plot
ERP = zeros(7,2,2,512); %files, conds, chans, times
```

```
for cfile = 1:size(ERP,1)
    for ccond = 1:size(ERP,2)
        EEG = pop_loadset([datain 'umeeg' participants{cfile} '_11_' conditions{ccond} '.set']);
        for cchan = 1:size(ERP,3),
            ERP(cfile,ccond,cchan,:) = mean(EEG.data(24,:,:),3);
        end %for cchan
    end %for ccond
end %for cfile
```

Note the comments: While declaring ERP, I specify what I aim the dimensions to be. For-loops use the size of each dimension, for example cfile is run from 1 to 7, as the 'size' of the first dimension of ERP is 7. Variable names in for-loops I start with a 'c' to denote 'current' – e.g. current File – and are ended with a comment to clarify what part of the for-loop is ended. Note that this means that the line starting 'ERP(cfile, ccond, cchan,:)' is run 7 × 2 × 2 = 28 times.

Once you get this running without errors, you will notice it is nevertheless not correct. It takes channel 24 for each subject, rather than Cz *and* Pz. The problem is that due to two different channel configurations (some subjects recorded with 32 channels, others with 64), and removed channels in earlier preprocessing, we cannot simply use a static value (24 here). To solve that we could either:

- Look up the Cz and Pz for each subject and use a variable to save these
- Interpolate the channels with EEGLAB, which conveniently causes the variables to be sorted in the same way, or
- Look it up with another for-loop.

The first option I am far too lazy for, and the second I am not fond of as it obscures which channels are interpolated and which ones are just resorted, so I suggest the third path. As a hint to get you started, see the following:

```
for cchancmp = 1:length(EEG.chanlocs)
    if strcmp(EEG.chanlocs(cchancmp).labels, channels2plot{cchan})
        ERP(cfile,ccond,cchan,:) = mean(EEG.data(cchancmp,:,:),3);
    end %if strcmp
end %for cchancmp
```

Note the strcmp, which is the MATLAB function that compares strings. In this case, it compares each label, as defined in EEG.chanlocs, with the current channel within channels2plot. So, for example, if I write in the MATLAB command:

```
EEG.chanlocs(14).labels
```

MATLAB will tell me 'Cz'. So, if each channel in chanlocs is compared first with channels2plot{1} (which is Cz), then with 2 (Pz), it will sooner or later detect that EEG.chanlocs(14).labels is indeed channels2plot{1}. Only at that point will the part underneath the IF-conditional be run: ERP(cfile,ccond,1,:) is then filled by the answer of mean(EEG.data(14,:,:)).

Once you get that working, you have all data ready for plotting. You just need to average over the number of subjects, which as you recall is the first dimension of the ERP matrix (files × conds × chans × times). For example, the ERP of Cz with both conditions would be:

```
Czmean = mean(ERP(:,:,1,:),1);
```

If you try this, note that MATLAB – somewhat confusingly – returns a four-dimensional matrix, of size $1 \times 2 \times 1 \times 512$. So even if it averaged over the subjects, its dimension is still there. It's a bit like having 7 paper Post-Its with stick figure drawings: after we calculate the 'average stick figure' of the 7, drawing it on another, we can keep on referring to this average as being on 1 paper, but at some point, you stop mentioning that. In MATLAB, we solve that (very MATLAB) problem with a function called 'squeeze', which just removes all singular dimensions:

```
Czmean = squeeze(mean(ERP(:,:,1,:),1));
```

…will therefore turn the $1 \times 2 \times 1 \times 512$ matrix into one of 2×512, and store the result into Czmean, which then has a size of number of conditions by number of time-points, which makes much more sense.

If you plot this (and try that), you will notice that MATLAB will show an incredible mess of lines (Figure 10.18). This is because MATLAB treats the first dimension by long-standing mathematics convention as the X axis. So instead of plotting a 2×512 figure, you will need to plot a 512×2 figure:

```
plot(Czmean')
```

…only adds the single accent, and Bob's your uncle, matrix rotated!

Finally, take a look at how to create a double figure with both the Cz *and* Pz plotted in different panels. For this we can use the function *subplot*, which makes a figure-of-figures as a raster with a number of rows and columns, then designating a specific currently active plot with a single counter.

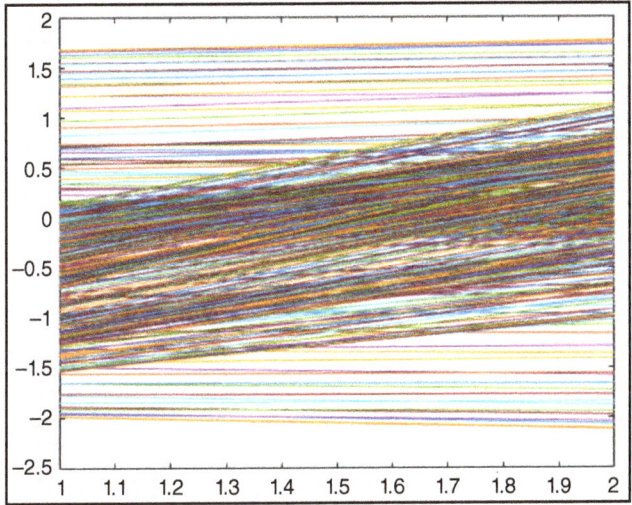

Figure 10.18

So, if we want to plot Cz above Pz, we can define a subplot as having 2 rows and 1 column, then say where we want to plot:

```
figure; %note: start by creating a new figure 'sheet'
subplot(2,1,1); %now plot to a 2 row, 1 col fig, panel 1.
plot(Czmean'); %plots to assigned position.
subplot(2,1,2); %now plot to same 2 row/1 col fig, panel 2
plot(Pzmean'); %given that you have calculated this
```

Now, create a two-panel grand average for Cz and Pz, with the following additional features:

- Include the odd condition as a third line.
- Include EEG.times on the X axis.
- Add, using the eponymous MATLAB functions, a title for each channel's subplot, and a legend only to the second.
- Add lines through the X axis and Y axis at their 0 points using functions line, xlim, and ylim.

Use, whenever necessary, MATLAB's extensive help documentation (help [function_name]).

EXPLORE: ANALYSES BEYOND POWER SPECTRA

You might have noticed EEGLAB references other measures in its STUDY design and PLOT menu item. What more could you explore with EEGLAB and beyond?

With the same interface allowing you to compute event-related spectral perturbation, or time/frequency power analysis in other words, you can get EEGLAB to compute inter-trial coherence. Intriguingly, clicking ERSPs pops up a warning that feels more like an ad: 'Checking both "ERSP" and "ITC" does not require further computing time. However, it requires disk space'. Well, we all have enough disk space but not time, so basically you get ITC for free. Being Dutch, that is simply irresistible for me. But is it useful? Coherence, put simply, is like correlation for time series data, such as EEG signals or time/frequency transformed EEG signals. As more technically described in the original EEGLAB-announcing paper (Delorme & Makeig, 2004), inter-trial coherence describes the degree of synchronisation of activity to the external events to which epochs are time-locked. Remember, as described in Chapter 8 (pp. 167–168), an ERP can only be detected if it occurs at a predictable time from an event: if a P3 is unrelated to the event it will necessarily flatten out. In other words, the activity as such is not nearly as important as the fact that it is synchronised across trials. Therefore, the inter-trial coherence is calculating the degree of similarity between epochs in time and frequency band. Thus, this measure, which indicates the strength of phase-locking as a measure between 0 and 1, may give a more precise causal reference point for observed ERPs than spectral power. A good introduction of the theory and its applications may be found in Makeig et al. (2004).

If you grasp the idea of inter-trial coherence, you may be interested in continuing the exploration down the rabbit-hole of time/frequency transformation. For example, one of the grand theories of consciousness, by Crick (yes, *that* Crick of the DNA) and Koch (1990) is that the higher, gamma frequency spectrum is related to the attentional mechanism by which different features of a single object are kept together as a coherent whole, such as we consciously experience. Such theories seem to have kick-started a frantic search for measures that can indicate **functional connectivity** between different brain areas, as opposed to simple indices of activity or structural connectivity (the literal connection between neurons or brain regions). In EEG, one of the ideas is that if activity is not merely phase-locked to an external event, but similarly *between* electrodes (surface connectivity) or modelled brain regions (neuronal connectivity), then these two areas might be 'talking to one another about the same event'. There are many different measures of connectivity (a useful taxonomy can be found in Bastos & Schoffelen, 2016), but EEGLAB provides a simple measure of inter-trial cross-coherence of channels or ICA components via EEGLAB > plot > time/frequency transforms). But, be warned, if you thought that having 64 channels and 512 time-points already gives a dazzling amount of data, imagine the number of potential number of type I errors (discussed in the next chapter) one can make if there are also, say, five frequency bands and each channel can be compared with every other channel: times (512) × frequencies (5) × channel-comparisons (0.5 × 64 × 63) = 5,160,960. And that is before we even start with the various different indices and that recent findings suggest neural areas may also engage in cross-frequency coupling (Jensen & Colgin, 2007). Some people find that super-exciting rather than headache-inducing, but I have the suspicion that those people are more often biophysicists rather than psychologists.

And if you think 2016 channel comparisons is not enough, my absolute hero, Roberto Pascual-Marqui, is here to the rescue. As discussed in Chapter 1 (p. 8), the inverse problem makes it impossible to measure just what brain source activity gives rise to the EEG activity as measured from the scalp. Impossible, or just really hard? By making the reasonable assumption that neighbouring neurons fire synchronously, low resolution electromagnetic tomography (LORETA) calculates the smoothest source distribution possible as the one that is most likely to have given rise to the scalp distribution. This results in theoretically exact localisation (Pascual-Marqui, 2007), albeit at the rather low spatial resolution of 6,239 *voxels* (3D 'pixels'). Various commercial software packages (e.g. BrainVision Analyzer) have implemented LORETA and associated (sLORETA, eLORETA) algorithms, but a little-known fact is that you can freely download a version of Pascual-Marqui's idiosyncratic yet functional EEG/MEG software from his website: https://www.uzh.ch/keyinst/loreta.htm, including fully functional ERP analysis tools, time-frequency transforms, connectivity algorithms, and oodles of other measures I have never even heard of. It's very personal in a way you will by now realise I am rather partial to, being as it is programmed in PASCAL, nerdy (we learn his Windows username is 'Gandalf the Grey'), and oddly jingoistic (Zero-Error Forever!). It also comes with an easy tutorial that makes the basics of getting into source activity analysis low threshold and fun.

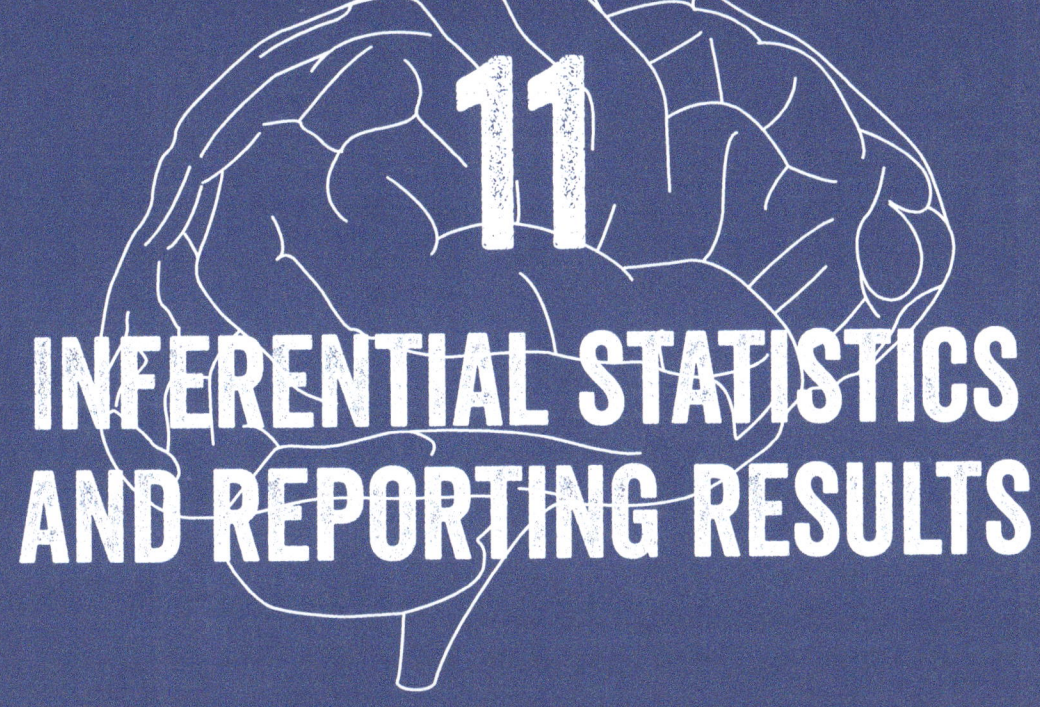

11

INFERENTIAL STATISTICS AND REPORTING RESULTS

IN THIS CHAPTER, YOU WILL LEARN:

- How to calculate area averages and peak amplitudes
- How to export data from EEGLAB/MATLAB
- How to report your results
- How to avoid questionable research practices in cognitive neuroscience research.

Inferring is what my cats do around 6 pm. They normally get their wet food around 7 pm, you see, and they use their internal clock, or the emptiness of their bellies as data to infer that it's dinner time. They do not, I now know, use the outside light, given that in Finland it gets dark stupidly early at some point, while in the winter, there's almost no night at all, so the light does not provide diagnostically useful data. I reckon they do feel we are stupid enough to use the daylight or other bad data, though, as the cats have been jumping on top of my keyboard to help me in my most important job, for how would I otherwise know it is time?

I move the next cat gently off my desk. They get their food around 7 pm anyway, because if not, they will next start to harass me around 5 pm. If they have an internal clock, I suppose it runs at a circadian rhythm of 23 hours.

Scientists are like cats in using data to infer what they hope are facts about the world. Psychological scientists use data, generally behavioural data, self-reports, and nowadays neuroimaging data, to infer facts about people. There are two kinds of scientists in this regard. The first group include clinical psychologists and neuropsychologists, who generally aim to make inferences about individuals: given that this person shows these symptoms or test scores, we might validly make a diagnosis that said person has some mental disorder or is 'neurotypical'. The second include experimental psychologists, cognitive neuroscientists, and the rest of us, who stick to the vastly safer goal of inferring about populations: given that this sample shows behaviour X or test scores Y, what does that mean for people in general? We sometimes call this approach the nomothetic one, which tries to establish general *laws*, as opposed to the idiosyncratic one, which is focused on the specific and even unique. Indeed, *psychonomy* was used to dissociate the former from the latter, rather condescendingly differentiating it from psychology 'as astronomy is from astrology'.[1]

Generally speaking, we do inference on 'what makes people tick' using quantitative methods. Put simply, if 15 out of 20 people show some experimental manipulation is effective – be it a therapeutic intervention reducing symptoms, or a person who is kept in mind evoking an amplified P3 – there is a good chance of it constituting a general rule.[2] We do this with inferential statistics, the statistical methods that allow us not only to describe samples, but to make inferences at the population level. Having taught statistics for most of my adult life, I am aware that these stats are also the stuff of nightmares for many undergraduates and not very few of my colleagues besides. Even as statistics may therefore be as vital to EEG research as to any other empirical science, I present them fairly gently here, sandwiched between practical tutorials and a little bit of critical discussion.

In that final discussion, I will explain how you might think stats allow us to do valid, objective inference, but unfortunately, scientists are a bit like cats and use what they *hope* will be a good conclusion (it is dinner time) as they analyse their results. This is true even in cognitive neuroscience, fancy pictures of coloured *brainblobs* notwithstanding, and despite ostensible motivation lacking. Or is dinner at stake for cognitive neuroscientists as well? I will explain why neuroimaging tools provide easy temptation for committing Questionable Research Practices, and why cognitive neuroscience may well be the next frontier in the Wild West of what is now called the replication crisis.

Before you begin: Make sure you have data from the Chapter 9. See Chapter 10 introduction for a summary of the prerequisites and how they appear in the online material (https://study.sagepub.com/Spape).

---

[1] Cf. Theo Mulder in an interview with *De Psychonoom*. Although the magazine no longer exists, Psychonomic Societies still do, with many wonderful journals and the great Psychonomics Annual Meeting.

[2] Of course, depending on the validity of the design, the representativeness of the sample, and so on.

EXPORTING DATA FROM MATLAB/EEGLAB

In the last chapter, I used EEGLAB to do an analysis over seven participants (four of whom were me), in which the conditions were compared against one another for every time point. We did this not, however, to immediately answer our research questions, but to present a region of interest (ROI): a particular area of our electrophysiological data defined in time (latency) and space (channels) from which we compute the final measurements. These, in turn, can be fed into spreadsheets and statistics software, giving some of those nice p values we use to make inferences: given the null-hypothesis, what is the probability that our data should look as it does? We will turn to that question soon enough, even bringing out the dreaded many-ways ANOVA interaction effects. But let's start gently by first considering how to make ROIs before I describe how to export the data, how they can be inspected using simple spreadsheet software, and then finally we'll get on with the dirty business of statistics.

The idea of the ROI is to operationalise something abstract like 'a P3' into a metric – a concrete number. There are two ways of making ROIs: literature-based and data-driven. If you are doing an exact replication of a previous study, it would make sense to look up the EEG pre-processing and analysis section of manuscripts and use the same techniques. The problem is that large differences in equipment, references, and many specifics in preprocessing usually exist, so in practice this is rarely ideal. Indeed, I often wonder whether people who report this actually blindly followed the venerable study as much as they said, or rather conducted a data-driven analysis before matching it with a selective literature search. Then again, what did the previous study base their operationalisations on? Often enough, following this line of questioning results in a likely 'this is what our lab has always done'. It is not inconceivable that even a respected lab has always done something wrongly.

A data-driven analysis is therefore not necessarily worse, even though it opens up the question whether you are not inadvertently engaging in **questionable research practices** (QRPs). Indeed, we might take our analysis from Chapter 10 and say 'the P3 will be measured as the average amplitude between 371 and 763 over Pz'. Then, computing the amplitude for every subject, and doing a paired-t test between relevant and irrelevant conditions, we find the result will be *insanely* significant ($p < .0000001$). This is because we just **double-dipped** our analysis, defining it in such a way as to ensure a positive response. I will continue to cover QRPs like double-dipping at the end of this chapter, explaining both why they are so tempting and how to avoid them. For now, we can define the measure based on our analysis, with the ultimate aim of extending the analysis to what is actually interesting, such as how much the P3 is affected by seeing oneself.

A good ROI, then, should involve an answer to the questions *where*, *when*, and *what*. Where and when we have already answered, defining two ROIs. ROI 1 involved frontocentral electrodes F2, FCz, FC2, and Cz between 270 and 350 ms while ROI 2 was between 370 ms and 800 ms, in centroparietal electrodes CP1, CPz, P1, and Pz. I will, however, redefine the specific electrodes later while talking about **conforming datasets**, with an eye on the fact that not all my recordings have all these channels available.

Then the remaining question is *what*. If you see an ERP, it feels like it should be easy to calculate: It has never occurred to me that the height of Mount Everest is up to debate. I am no mountaineer, but I would stick a flag on top and measure the vertical distance from sea level with a *lot* of string. Where exactly *is* the top, though? This question is much easier to answer with EEG as we have knowledge of every single timepoint's voltage, which is analogous to having vertical measurements of the entire mountain range. And yes, it is also much more hospitable inside the office if we disregard that the romantic geographer of my dreams wouldn't have to deal with the horribly distracting social media climate.

Back to the point, which is that there are many ways of calculating what exactly the size of the ERP is. First, you could always just calculate the **global peak**, or maximal amplitude (minimal amplitude, if we're discussing negativity and an intrepid spelunker), but it turns out this is rarely a good idea. After you calculate grand average ERPs, one usually gets the idea that the similarity across participants is far larger than it really is, and, as illustrated in the middle panel of Figure 11.1, this often results in maxima being registered at the borders of your **peak detection interval** (the region of interest in which you search for the presence of peaks). To account for that, a **local peak** is calculated as occurring only if the surrounding data are 'lower' (or higher, if it is a lower valley, though we continue to call this a local peak calculation). Both the time at which a peak occurs (**peak latency**) and its amplitude (**peak magnitude**) are commonly calculated.

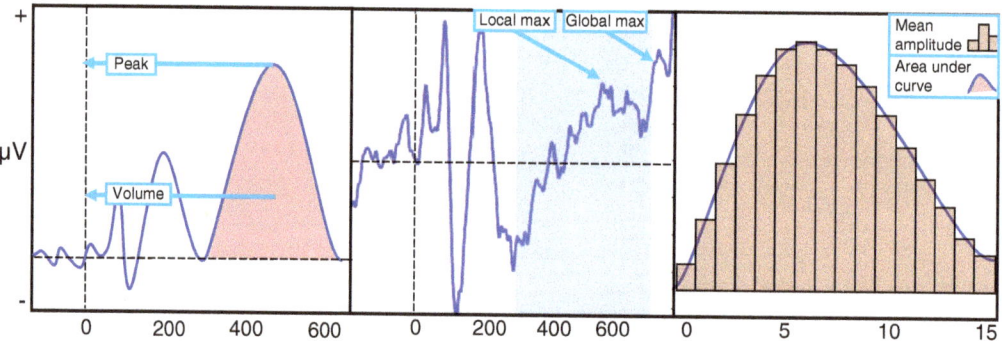

Figure 11.1 ERP peak and volume measurements. Left panel: A peak can be described by its highest point (peak magnitude measurements) or its volume (area measurements). Middle panel: Peaks can be computed as the maximum or minimum voltage within a region or the highest point within the surrounding area. Here, only the local maximum is a real peak. Right panel: Mean amplitude is the mean calculated across time-points, i.e. the average over the 15 bars. The smaller the width of each bar, the more precise the measurement of the actual area of the component will be. This is the central idea behind area under curve measurements.

Peak measurements are, however, easily affected by high frequency noise. A random spike in the data easily gives very strong values that distract from the real shape of the ERP. To avoid this, a low-pass filter can help, but it is then vital to determine the degree to which the filter affects the component's latency. **Volume** or **area measurements** such as **mean amplitude** and **area under curve** (AOC) avoid this problem as a spike does not nearly affect an entire range as much as it does a single point. The idea is that rather than defining the mountain by its top, area measures measure the volume of the mountain. Taking the average across all time points is a very simple way of approximately doing this, while area under curve involves doing a *definite integral* over the data. If you haven't taken your A-levels in mathematics or physics, you could see that like this: if the sum across the 16 values from 0 to 15 ms in Figure 11.1 gives a good estimate of the volume of the curve, imagine chopping up each bar to 1,000 microseconds. Thus, the smaller the step size (the width of the bar), the more the sum of bars approach the true volume. Thus, the area under the curve computes the volume of the ERP, which is technically more precise than simply the sum of the limited number of bars. Then again, practically, millisecond steps are already pretty small, so your measures are not going to be vastly different whether you choose the one or the other.

Mean and sum

Note that I am conflating the volume as the sum of the bars with the mean amplitude. If you use the same region of interest across participants, as is the default, this does not affect the results, since the mean is the same as the sum divided by the number of timepoints (16 in the example above) and a linear transformation does not have any impact on significance. Be clear, however, in describing your results, as mean amplitudes will be much smaller than 'sum amplitudes'.

Now let's get started in extracting the data out of EEGLAB and into something familiar!

Conforming datasets

To begin with, however, there is the issue of **dataset conformity**. Essentially, to compute averages or do any other analysis across datasets, they have to be commeasurable – i.e. comparable across their measures. Here, for example, we are running into the issue that some datasets were recorded with 32 and others with 64 channels. It is unlikely you will have such an extreme situation, but nevertheless, you will usually have the issue that some participants will have missing channels. We can fix that to some extent. As described before, EEG data are imprecise measures with regards to location, such that two channels that are near to one another will pick up very similar activity. Finally, this bug becomes a feature: it means that we can repair the 'hole in the EEG canvas' that is the EEG by stitching together the surrounding electrodes. **Interpolating channels** works relatively well, but note that (1) you do not regain any lost data, and (2) it is easier to interpolate central channels than peripheral ones. To see what I mean, imagine trying to reconstruct the temperature in London from weather stations in surrounding locations (e.g. Cambridge, Southend, Brighton, and Reading). This will be much easier than guessing the temperature in Brighton, given that you can't place weather stations inside the sea.[3] However, if a spike in the weather is unique to London, say, a microclimate surface temperature fluctuation due to a convening of internet celebrities, it will not be so easily reconstructed.

---

[3] I am no meteorologist, but I *reckon* you can and they do. It just sounded slightly less naïve than saying 'given that you can't place weather stations beyond the edge of a map' like some flat-Earther.

To interpolate, in EEGLAB, go to Tools > interpolate electrodes. In the menu, you get the options to interpolate channels within your current dataset or from another dataset. The former option only allows you to interpolate pre-existing channels, which means you overwrite their data by interpolating from the surrounding channels. The latter means you select one of the preloaded datasets (in EEGLAB>Datasets by their number), and either specify which ones from that other dataset should be added (by interpolation) to the currently active dataset *or* to interpolate every channel available in the other dataset that is currently missing in the current. If you have your own datasets, and they are all recorded with the same amplifier, I would suggest you use this last option. Load all files you need for your final analysis (also of all conditions), then additionally load a single dataset that has all channels – for example, a raw file. Then activate dataset 1, and interpolate based on the last, continue with 2, and so on. I also wish there were faster ways, and there are, as described in the additional exercises of the previous chapter, but these require a bit of MATLAB coding.

STUDY interpolation

Note that EEGLAB helpfully points out that it will interpolate with the STUDY design. Which it does, as shown in Chapter 10, but they are then *only* available within the STUDY design, and not as separate EEGLAB datasets, which makes exporting harder.

For the rest of the chapter, if you have your own files, the best way would be to do the above. Then, whether you use your own files or the ones that come with this book, go through 1 file of each subject (e.g. the 'rel' condition) and find out which channel numbers are in which ROI. If you are following the essential track, use as ROI 1 Cz (270–350), and as ROI 2 Pz (370–600). If you are following the additional track, use as ROI 1 FC1, Cz, and FC2, and as ROI 2, CP1, Pz, and CP2. Of course, if you used your own data recordings, you should use adjusted values. Note, I slightly changed the latency of the second ROI so that it does not overlap the successive stimulus too much. I suggest you fill out a table such as the following to keep track:

Subject	ROI 1 (FC1, Cz, FC2) channel numbers	ROI 2 (CP1, Pz, CP2)
1		
2		

…and so on. If you are following the additional track, make sure you always keep the same orders of channels within ROIs. You can find out which channel is which number in EEGLAB by going to Edit > Channel Locations and finding the one (essential) or three (additional) channels. Another way is using Plot > Channel data (scroll), and simply counting channels from the top until you find the one you're looking for.

Using MATLAB to export data to a spreadsheet

I will be showing you how to use MATLAB to export EEGLAB data with an in-between spreadsheet step, before the ultimate 'real statistics software' kicks in. This is not necessary, strictly speaking, as you could easily compute area averages and peak amplitudes using MATLAB, and export them to SPSS or a similar format. However, like certain dictators, I firmly believe in the power of looking at things,[4] or at least I believe it helps you to get a good grasp of where your data are coming from. Furthermore, any EEG/ERP paper worth its salt will present an ERP figure, and these are a lot easier to produce in spreadsheets software such as Excel than in MATLAB.

But before you begin with this, it is a good exercise to plan out just what you want your exported data to look like. Remember, spreadsheets look a bit like tables comprising small, rectangular cells that are wider than high. Rows are numbers and columns are letters. This perhaps does not come as a surprise, but it bears keeping in mind because EEG data have high temporal accuracy, which in our case specifically means that we have 512 time points from -200 to 800. Scrolling around from left to right is always a bit harder, but it is even harder since cells are wider than high, and unintuitive since column 512 translates to 'SR'. Of course, you could change all that, but for now let us just take as a given that we want to see the time dimension arranged vertically. It isn't critical, most shapes will work, but I find it is often useful to sketch out what I *want* something to look like, before making it happen. Remember, we have 512 timepoints, 2 or 6 channels of interest, a number of conditions (I use 3 for the example), and a number of subjects. So I drew Figure 11.2.

Now we have a clear goal, although there are many ways of achieving it. Much like with interpolating and sorting out your electrodes, there are primarily easy-but-tedious ways and harder-but-faster ways. I will explain one of each, with the idea that if you have been following the essential track in this book, you may want to use the first, whereas if you worked through the additional exercises, you may prefer the second.

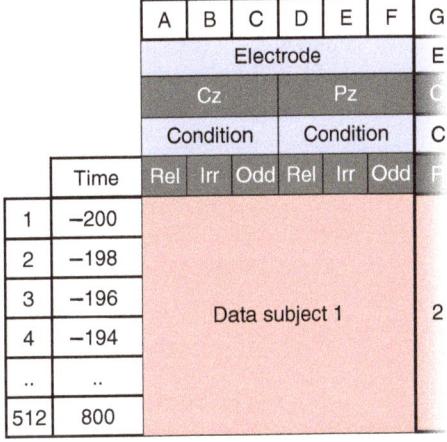

Figure 11.2 Sketch of spreadsheet data to be exported. Time is arranged vertically, while within subject factors (in blue, the *levels* within the factors are grey) are arranged left to right, and all subjects are in big blocks that stack horizontally.

Essential exporting

Start by loading into EEGLAB all the data you want to do analysis over from one subject. For example, I would like to do an analysis over the relevant, irrelevant, and odd conditions, so I load up umeeg101_11_rel.set, umeeg101_11_irr.set and umeeg101_11_odd.set. Make sure you adhere in the ordering of loading to the sketch you made (or if you didn't bother, *I* made).

Now, let's focus on using a tiny bit of MATLAB code to get the first column (A) exported. Remember, as discussed in Chapter 7 (Starting MATLAB), datasets can be accessed with MATLAB code as ALLEEG(dataset number). Given that my EEGLAB was clean to begin with,

---

[4]kimjongunlookingatthings.*tumblr.com*

ALLEEG(1) thus refers to the first EEG dataset in memory, which is umeeg101_11_rel.set. Double-click in the MATLAB workspace on ALLEEG to show a summary (Figure 11.3).

	filename						nbchar	trials	pnts		data		srate		xmin
'u...	'umeeg101_11_rel.set'	'H:...	'u...	[] 'rel'	[] Po...		34	293	512	34x512x293 single			512		-0.1992
'u...	'umeeg101_11_irr.set'	'H:...	'u...	[] 'irr'	[] Po...		34	1516	512	34x512x1516 single			512		-0.1992
'u...	'umeeg101_11_odd.set'	'H:...	'u...	[] 'odd'	[] Po...		34	59	512	34x512x59 single			512		-0.1992

Figure 11.3

Source: Reprinted with permission of The MathWorks, Inc.

Each row is a different ALLEEG, for example, row 1 is ALLEEG(1). Each column is a different *field* within ALLEEG, and these are accessed by using a period. For example, typing ALLEEG(2).trials into MATLAB's command window returned 1516 for me.

Notice ALLEEG(1).data equals 34 × 512 × 293 single: this means a *three-dimensional array*. It sounds complicated, but you'll get used to it – much like you got used to living in 3D space as well. Notice the values: 34 is the same as the 'nbchan' field of every dataset, 512 as the number of 'pnts', 293 as the 'trials'. It is not a coincidence: ALLEEG(1).data contains all data as a channels × times × trials variable. Again, that is pretty hard to visualise, but let's say we enter into MATLAB's command window:

```
ALLEEG(1).data(1,2,3)
```

...then MATLAB will return the EEG potential of the first dataset for channel 1 (FP1) at time 2 (−198 ms), of trial 3.

If that makes sense, remember that we want data for channel Cz for the first column, which I just looked up and Cz was channel 14 for my dataset. So the line becomes the following:

```
ALLEEG(1).data(14,2,3)
```

And we want all times, not just times 2, which is written in MATLAB code like so:

```
ALLEEG(1).data(14,:,3)
```

Then, finally, we want the data not for trial 3, but *averaged across trials*. All trials is again :, and average is done in this way:

```
mean(ALLEEG(1).data(14,:,:))
```

However, now MATLAB does not know *what* to average: it could be the channels, the times, or the trials. Instead, we need to specify the dimension to be averaged, which is the third:

```
mean(ALLEEG(1).data(14,:,:),3)
```

And that's 1 line of data. In fact, just changing it into:

```
plot(mean(ALLEEG(1).data(14,:,:),3))
```

…will give a plot of the participant's Cz ERP.

Now, you can extend this line of reasoning to the next two datasets, by **concatenating** datasets, or stacking them together.

```
[mean(ALLEEG(1).data(14,:,:),3); mean(ALLEEG(2).data(14,:,:),3)]
```

…concatenates datasets 1 and 2, the semicolon meaning they are stacked vertically, the second right below the first.

Follow this logic by concatenating 1, 2, and 3, then in the MATLAB workspace double-click on 'ans'. You should see the table in Figure 11.4.

1	2	3	4	5	6	7	8
-0.1554	-0.1971	-0.1734	-0.1072	-0.0469	-0.0270	-0.0637	-0.1281
0.2146	0.1628	0.1082	0.0612	0.0267	0.0062	-0.0066	-0.0294
-0.7895	-0.7355	-0.6724	-0.5992	-0.5318	-0.4549	-0.3536	-0.2556

Figure 11.4

Source: Reprinted with permission of The MathWorks, Inc.

That already looks quite spreadsheet-like, right? The only issue is that it's the wrong direction: there are three datasets vertically, but time is going horizontally. To switch it, simply add a parenthesis to your last command:

```
[mean(ALLEEG(1).data(14,:,:),3); mean(ALLEEG(2).data(14,:,:),3)]'
```

…which transposes the data, as you will see by inspecting 'ans'. Select all 'ans' data, copy them into memory (e.g. ctrl+c), open your favourite spreadsheet software, paste the data, and save the data.

That's that for Cz of subject 1. Do the same for the other electrode (Pz), and you have the first block of data in my sketch complete.

Note: if you got stuck anywhere along the way, please see the online material (https://study.sagepub.com/Spape).

Additional exporting

In the last chapter (pp. 235–238), I already discussed how to:

1. Load datasets specific to conditions within a nested for loop, then to

2. Loop through the channels, selecting a specific one to

3. Calculate a mean ERP value.

So, we have very few new operations to learn:

a. To capture all timepoints

b. To concatenate these data to match the sketch

c. To export the data

The easiest way to go about this is to use the basic building blocks we made in last chapter's additional exercise 2 and fix them to work for this scenario. Copy the file you made there (or download it from the website, https://study.sagepub.com/Spape) and look up where I previously showed the line of code that selected the correct channel before computing the mean:

```
ERP(cfile,ccond,cchan,:) = mean(EEG.data(cchancmp,:,:),3);
```

Spreadsheets are not customarily four-dimensional but two-dimensional, so the strategy I will be showing here is to create single columns for each subject–channel–condition combination. Then, we only need to stack the columns horizontally to follow the basic design.

```
outcol = mean(EEG.data(cchancmp,:,:),3);
```

I created this new variable to hold a column (remember, variables in MATLAB do not need to be declared in advance, they will be created whenever [var name] = uses a new name). But contrary to the name, this gives a row, rather than a column, so we need to transpose it. By adding a simple ' to the end, its dimensions are flipped, changing its columns to rows and rows to columns:

```
outcol = mean(EEG.data(cchancmp,:,:),3)';
```

Now we just need to stack everything. First, I will create an empty variable to replace the previous line saying ERP = zeros(7,3,2,512):

```
outdat = [];
```

And then we add a line underneath outcol =, which stacks files together:

```
outcol = mean(EEG.data(cchancmp,:,:),3)';
outdat = [outdat outcol]; %concatenate column with all data.
```

The square brackets in MATLAB cause matrices to be stacked horizontally, so by stacking outcol to the right of outdat, it continues to expand outdat with each subject–condition–channel combination. However, if you run this code now, it will work but not recreate the sketch, because its for-loop structure goes from subject to condition to channel, while the sketch goes from subject to channel to condition. This requires a bit of reordering, giving the following code:

```matlab
indir = 'filepath/'; %change to where your data are

pps = {'101', '102', '103', '104', '105', '106', '107'}; %subjects

cnds = {'rel', 'irr', 'odd'}; %conditions to export

chns2exp = {'FC1', 'Cz', 'FC2'}; %channels to export

outdat = []; %empty variable

for cfile = 1:length(pps)

    for cchan = 1:length(chns2exp)

        for ccond = 1:length(cnds)

            EEG = pop_loadset('filename', ['umeeg' pps{cfile} '_11_' cnds{ccond} '.set'], 'filepath', indir);

            %get correct chan

            for cchancmp = 1:length(EEG.chanlocs)

                if strcmp(EEG.chanlocs(cchancmp).labels, chns2exp {cchan})

                    %get mean activity of each timepoint

                    outcol(:,1) = mean(EEG.data(cchancmp,:,:),3)';

                    %concatenate column right of the outdat

                    outdat = [outdat outcol];

                end %if strcmp

            end %for cchancmp

        end %for cchan

    end %for ccond

end %for cfile
```

Note: This is not the most elegant bit of code, for example, it requires the loading of every subject as many times as you have electrodes to export. In the additional exercises, I will demonstrate an easy way to first loading all data *before* exporting them, which will run much faster.

And then we just need to export outdat. There are many ways of exporting data from MATLAB and the one explained in essential exercise 1 (copy–paste from workplace) is probably the worst. That is, if there is a lot to export, then MATLAB will take forever to copy the data into memory, so it is usually much easier to export to disk. To write files to text format, you usually use:

```
save ('outdat.txt', '-ascii', 'outdat')
```

…with the first parameter being the filename, the second saying you want to export to text (ASCII) file format, and the third being the actual variable name. MATLAB is a bit finicky when it comes to text files: it will only easily export up to two-dimensional variables that are either only text or only numbers, and the numbers need to be of double format (higher precision) rather than EEGLAB's default single format (which is easier on memory). So you might need to first convert outdat to double:

```
outdat = double(outdat);
```

Another way that I find usually easier and often faster is to directly convert to spreadsheet '.xls' or '.xlsx' format. It does require having Microsoft Excel installed on your computer, though, as it accesses Excel routines:

```
xlswrite('outdat.xlsx', outdat);
```

SPREADSHEETS

Spreadsheets come in handy when you want to keep track between the data as they are and the various measures you ultimately use in your statistics software. It is, of course, possible to use MATLAB to perform the necessary statistics (see additional exercise 2), removing the need for dedicated statistics software altogether. Likewise, you could, if you are so inclined, import the data you just exported in your statistics software, and do your analysis with some fancy multi-level modelling. However, that requires either an excellent grasp of either MATLAB or the more advanced statistics, which makes things unnecessarily difficult.

In the following, then, I will explain how to get your data into such a shape that a relatively simple repeated measures ANOVA in any statistics software will enable you to answer the research questions. For this, I will use the ubiquitous spreadsheet software Microsoft Excel, but I do not believe any of its competitors will cause obstacles in your learning. OpenOffice and its cousin LibreOffice give good options for working with spreadsheets, while Google Sheets allows very similar work and makes collaboration with your peers usually smoother than Microsoft's online versions.

To begin, open the file you just exported in your spreadsheet software, select all data, then cut and paste them in such a way that there are now 6 rows down and 2 columns to the right. Then, add headers directly above the data. Note that we made sure in the previous tutorial that they were designed to go from top to bottom: Subject, Electrode, Condition (meaning $7 \times 2 \times 3$ columns if you use my seven datasets and the essential two electrodes). I follow the classic SPSS design in which the further down means the faster it varies, so subject 1, electrode 1, condition 1 will be the first column, subject 1, electrode 1, condition 2, the second, and so on (why this is convenient will become clear further on). Now, edit cells so you can see which cell means what, adding column headers first for each condition directly above the data, followed by electrode, followed by subject, more or less like Figure 11.5.

	A	B	C	D	E	F	G	H	I	J
1			101						102	
2			Electrode						Electrode	
3			Cz			Pz			Cz	
4			Condition						Condition	
5			Rel	Irr	Odd	Rel	Irr	Odd	Rel	Irr
6	Time	-200	-0.1554	0.2146	-0.7895	-0.2450	0.1023	-0.7044	-0.9337	-0.5918
7		-198	-0.1971	0.1628	-0.7355	-0.3043	0.0906	-0.6585	-0.8805	-0.5538
8		-196.1	-0.1734	0.1082	-0.6724	-0.3154	0.0828	-0.5658	-0.8369	-0.5070
9		-194.1	-0.1072	0.0612	-0.5992	-0.3050	0.0795	-0.4730	-0.7831	-0.4625
10		-192.2	-0.0469	0.0267	-0.5318	-0.3038	0.0765	-0.4140	-0.7047	-0.4297

Figure 11.5

Notice I coloured some cells and added borders between subjects to match my sketch and keep things organised.

Note: To create a 'time' variable, write down -200 (or the first time point of your epoch) in the cell left of the first datapoint. Go one step down and insert the formula: '=B7+(1000/*samplerate*)', with samplerate being replaced with your sample rate (which is, for me, 512). Press ENTER, then hover with your mouse above the bottom-right corner of the new cell, such that the normal cursor becomes a filled, narrow cross, then double-click. The last formula will be extrapolated downwards, so all time values will be there. Alternatively, you could just copy–paste the formula of B8 a couple of hundreds of times, but that is slightly exhausting.

Now we have all the data we need to work with, and it's looking reasonably organised, so let's start with creating area averages.

Computing area averages

Since we have the data arranged vertically by time, we can create area averages (as mean amplitudes) by simply computing means within the ROI across all the rows. To do this, create a little table below your current data with headers for each electrode and condition combination (example shown in Figure 11.6), only for the first subject. Then add two rows for each latency window corresponding to the ROI.

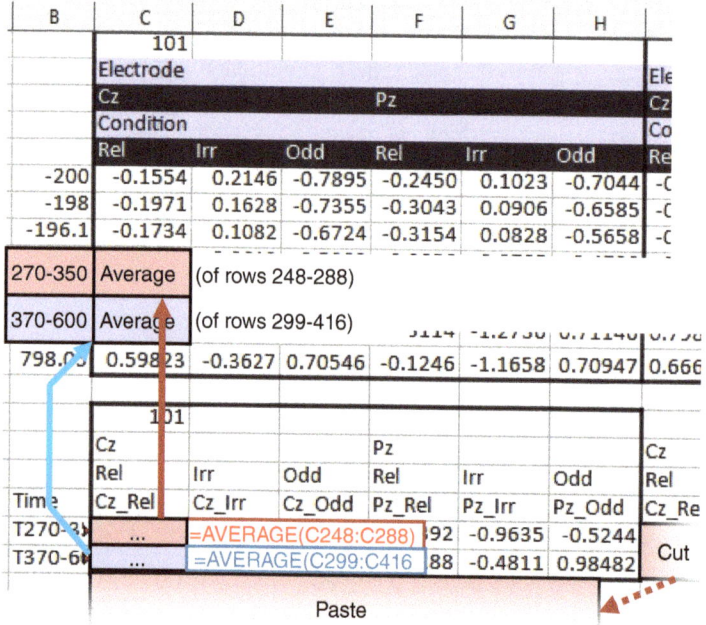

Figure 11.6

As shown in Figure 11.6, look up the time windows manually in your data that reflect the latencies of the ROI: for example, for me all times between 270 ms and 350 ms correspond to rows 248–288. To compute the mean amplitude for the 270–350 window of the first column (C), I enter the formula =AVERAGE(C248:C288). Follow this logic for the second ROI, then extrapolate the cells to continue until the last column of the last subject (or use copy-pasting). Once all area averages are computed, *cut* the cells from every subject to the right of subject 1, then paste the data below this subject, then repeat the trick for every new subject, culminating in having all subjects one below the other, with each taking two rows. Finally, name the cell to the left of Time with 'Participant', and add the number of each particular subject below (the first two rows being 101 for me, the second two 102, and so on). What we have now is a very typical set of data for statistics software with a within subject 2 (electrodes) × 3 (conditions) design, and two different measures (by Time).

Note: This works because for no obvious reason, copy–paste in spreadsheets means translating (reinterpreting) an existing function to the new location, while cut–paste means moving it (not changing). You will see what I mean if you try copy–pasting a subject's averages instead of cut–pasting.

Peak magnitudes and latencies

As explained (p. 246), peak magnitudes can be computed as maximum amplitudes within a range, although this often leads to poor results. Let's first look at what would happen here if we try to do so anyway. You can use the same workflow explained with computing area averages, but now calculating the =MAX(range), with range corresponding to the area. Here, it would be preferred to keep a longer range given that there can be quite some interindividual variation, so I'm using the entire range between 270 and 600 (rows 248–416 above). So subject 1's Cz maximum voltage is calculated in cell C534 as $=MAX(C248:C416)$ for me. To get the matching peak latency, we can then add a lookup function directly below the =MAX, which searches in the same range to find the matching value, which we just calculated, and to return the corresponding latency from the times column. For example, in Excel, this can be achieved by entering '=XLOOKUP(voltage_to_look_for, range_of_voltages, range of latencies)'.

Note: if you run into any problems with the spreadsheet, please have a look at the online material (https://study.sagepub.com/Spape), where you'll find completed versions of this tutorial.

However, if you were now to extrapolate to the right both the maximum and lookup function, you would run into problems. That is, every next subject has different peaks, and voltages, but the range of latencies is always the same -200 to 800. To avoid that, add dollar signs ($) to the range of latencies. For example, my function in C535 now becomes:

```
=XLOOKUP(C534,C238:C416,$B238:$B416)
```

As soon as I then copy–paste this cell one to the right, it will say:

```
=XLOOKUP(D534,D238:D416,$B238:$B416)
```

Notice how the Cs all changed to Ds, while the B remained the same. Now you can copy–paste (or extrapolate) the cells to the right across all participants, and then cut–paste them one below another. As with the area averages, make sure you give each row a clear name such as peak magnitude and peak latency.

We now have the peak amplitudes and latencies, but notice how often there are strange values. For example, we have five latencies at the latency detection boundaries (270–600), which I explained is because a maximum does not always coincide with a local peak (see Figure 11.7). While fancier algorithms are available (google MATLAB Central's peakfinder), you can also apply some manual labour and obtain peak maxima and latencies by visually inspecting each participant's ERP. To do this, select both the time, and the first data column, and insert a scatter plot (Google Docs: Insert Chart, Excel: Insert: Scatter XY). After that, move the selected data around to find the subject/condition specific ERP and the reason why the peak detection failed. For example, creating a figure of one peak that was detected at 598.83 ms, it's easy to see that the reason is 'localised' at the edge of the window since the ERP keeps on drifting upwards from 300 ms to c. 700 ms. The closest to a local peak seems to be either some 50 ms before 600 ms, or just after. It thus makes sense to use, for this subject (across conditions), either a longer or a shorter peak detection interval. I am afraid I cannot give a clear answer to what is best in these ambiguous scenarios, but try to be as consistent as possible, and never let it depend on the final results (see the end of this chapter).

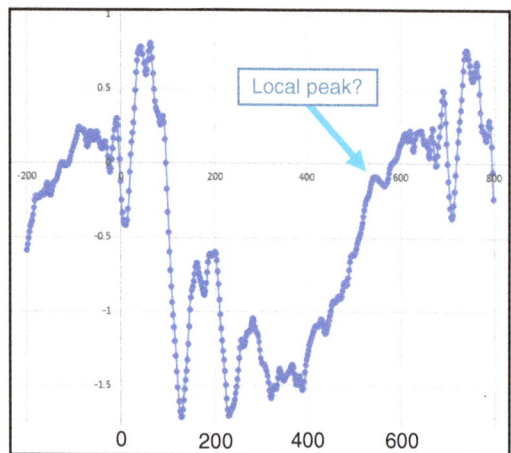

Figure 11.7

Crafting ERP figures

As you can see above, it is not hard to create some-what reasonable ERP figures with spreadsheets. You can also make a nice grand average ERP figure with spreadsheets with far less effort than by using MAT-LAB or SPSS. Here is how I do this.

Much like I calculated area averages below each column, we can also compute ERPs across partici-pants by averaging horizontally (Figure 11.8):

Note how I added new columns to the right of the data that each compute the average for an elec-trode × condition combination across all subjects. For example, if I had only two participants, the above scheme shows how to compute the grand average by calculating the mean over the two values of Cz-Rel. Following that, you can extrapolate the single cell downwards (across all time points) and to the right (across the other conditions and electrodes). As long as you keep the same ordering – do not suddenly have the 'odd' condition next to 'rel' for the grand averages – this makes computing grand averages a breeze. And simply selecting the time column and the three condition columns to the right of it, then clicking insert graph (make a scatter), will generally give you a nice grand average figure (see Figure 11.9).

A	B	C	D	E	F	G	H	I	J	K	L	M	N
	Subject 1						Subject 2						Grand average
	Cz			Pz			Cz			Pz			Cz
Time	Rel	Irr	Odd	Rel	Irr	Odd	Rel	Irr	Odd	Rel	Irr	Odd	Time	Rel
4	-200	...					...							=AVERAGE(A4, G7)

Figure 11.8

Nice? Bafflingly, there are many students at every level and even researchers out there who do the above, are pleased with the result, and go and get pizza or something. Or at least, I suppose that is what happens, because the work I mark and review often has figures in there that do not do justice to the research that has gone into it. It does not take that much extra time to make what should be the pinnacle of your research effort into something that does not look hopelessly amateurish! OK, maybe there is some pointing and clicking involved, but it does not require much cognitive resources to apply some basic rules of graphic design. To help, I'll give a few tips.

Tip 1: Keep the printed end-result in mind

ERP figures are small but contain a lot of data. Expect them to be at most 1 column on a sheet of A4 paper – so about 8.5 cm in width. To keep track of this, I find it helpful to arrange my figures in PowerPoint on a slide that is the size of a sheet of A4 paper. Everybody hates PowerPoint, but I know even some computer scientists swear by it for quickly creating and arranging complex figures. While designing, keep in mind that text should be readable and never be smaller than 7 pt.

Note: To export high resolution pictures from PowerPoint, save the presentation as .PDF, then import the pdf to Photoshop (or free software GIMP), at which point you can blow up the resolution to as many megapixels as you like. Then export to .TIFF, which is what publishers usually like to see.

Tip 2: Never waste paper space

Compare, for example, the charts in Figure 11.9. Chart titles are unnecessary (they go in descriptions), intervals with no data (before -200 ms, after 800 ms) are useless, and the legend can usually fit somewhere without blocking interesting data.

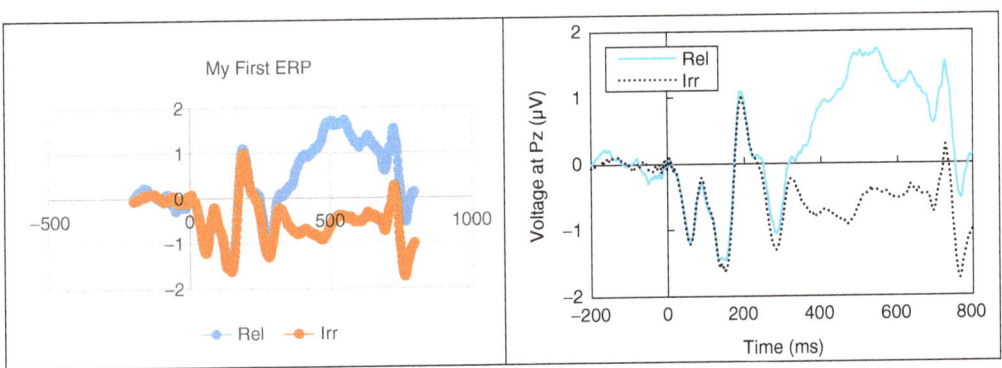

Figure 11.9 Before and after applying some basic rules

Tipe 3: Do add useful elements

For example, in Figure 11.9 (above, right) I added axis names (units), and small tick marks. Usually, however, I find it a struggle to arrange these exactly as planned within a spreadsheet, so I add them manually (again, in PowerPoint).

Tip 4: Use no more than three ERP lines (try not to, anyway)

Sometimes there is a very clear double pair of conditions, or a clear relationship between conditions (a 100/80/60/40 dB sound), but even then, question whether clarity might not be enhanced with using multiple panels or difference waves.

Tip 5: Arrange ERPs from different channels sensibly

Follow the 20/20 system as much as possible (C3 goes to the left of Cz, Fz above Cz, etc.). Such complex figures quickly take up more space, so in the literature it is not uncommon that individual channel graphs are c. 4 cm in width. Rule 2 then starts to matter even more because things quickly look cramped. To avoid that, remove *all* redundant elements. For example, if you have three figures of ERPs stacked below one another, you will only need the one labelled X axis since they all use the same one. You will also likely only need one legend as long as the same conditions are plotted and rule 6 is followed.

Tip 6: Use consistent formatting

This is important between figures as much as within them. Keep in mind that if you do not specify the size of your figures, you will likely only adjust them bigger and smaller at the final step, which means that some font-sizes that looked legible before may now be unreadable. I have a few basic rules: ERP lines should have the same width (2 pt makes them appear smoother), I like some fonts more than others, and so on. The details are less important than being consistent about them, although avoid defaults for they tend to be awful (I can recognise MS Office blue as a distinct colour!). Whatever they are, follow your own rules and the end-product will most certainly look better.

It may be obvious, then, that I find ERP figures important. However, some people insist that there's only one number that adequately summarises the worth of research, and that number is *p*. Let's talk *significance*.

ANALYSING THE DATA

Following the in-between step of extracting four different measures (mean amplitude 270–350 ms, mean amplitude 370–600 ms, peak magnitude, and peak latency), we can now do the rest of the analysis in statistics software. For most psychologists, that will mean SPSS (after all, it is an abbreviation for the Statistical Package for the Social Sciences) as no matter how much negative reaction it provokes from beginning students and expert statisticians alike, its educational licence is pretty cheap. Or perhaps it is just because, like some sort of obscene hazing ritual, we feel it is necessary to undergo the same hardships we did in our academic childhoods. Indeed, I have been using it since version 8 and I am amazed how simple it is to pick up and continue using it like I was when I was still dancing at Woodstock (1999, not the *real* one, where I wasn't either), with confusing data entry, baffling 'wizard' dialogues, and ugly graphics. So, no matter how much of a concept SPSS is, I try to avoid talking about it and will assume you have a passing knowledge as to how to enter data and conduct basic statistical analysis in whatever *statistical package* you might be working with.

In the *package that shall not be named*, then, the norm is to outline single subjects as rows, placing levels of within-subject factors ('within-participant variables') as columns. We have already done most of the work in the spreadsheet, so we can now simply copy all data over to the stats software, and rename the columns there. I then have the following columns (I changed 'time' to Measure), with each participant taking up four rows:

Measure	Participant	Cz_Rel	Cz_irr	Cz_Odd	Pz_Rel	Pz_Irr	Pz_Odd

Again, in *that package*, you can split file, organise output (or compare output) by Measure, in order to do the same analysis separately for each measure. After that, there are two easy analyses possible: one-way repeated measures ANOVAs investigating the effect of the condition on either the Cz (blue) or Pz (red) electrode, or two-way+ repeated measures ANOVAs that also estimate the degree to which the conditions affect the electrodes differently. If such interaction effects leave you confused beyond repair, I would suggest sticking with the one-way designs, because it is going to get a lot more complicated than just two-way interactions.

One-way designs

One-way repeated measures ANOVAs estimate the variance accounted for by a single within-subjects factor. Factors, in this context, are simply categorical variables, but it seems we don't call them that so as not to confuse variables as they are understood in experimental design with variables as they appear in statistics software, which is usually one variable = one column. Thus, with repeated measures ANOVAs, the three columns Cz_Rel, Cz_Irr, and Cz_Odd need to be specified as corresponding to three levels of a single within-subject experimental design variable, i.e. factor. In *that package*, then, as well as with great alternatives such as JASP (see Explore at the end of the chapter), we perform a General Linear Model, Repeated Measures analysis, specifying a factor with three levels. It has never been possible in *that package* to name the actual levels, nor for it to find it out from their very obvious names, and so I often name the levels in the name of the factor as well. Thus, I name the factor 'Condition_rel_irr_odd', such that 'condition' is the name of the factor, and level 1 is 'rel', 2 is 'irr', and 3 is 'odd'. Get the analysis to also give you a bunch of figures, tables with means, post hoc comparisons (in SPSS confusingly presented under Options > Display Means > Compare main effects, select Bonferroni), and estimates of effect sizes. My output looks like Table 11.1 for the mean amplitude between 270 ms and 350 ms.

Table 11.1 Tests of within subject effects. Note: Mauchly's W = .35, p = .07.

	Sphericity	SS	df	MS	F	P	η_p^2
Condition	Sphericity Assumed	22.83	2	11.41	63.70	4.07E-07	0.91
	Greenhouse-Geisser	22.83	1.21	18.89	63.70	5.56E-05	0.91
Error	Sphericity Assumed	2.15	12	0.18			
	Greenhouse-Geisser	2.15	7.25	0.30			

Remember that ANOVAs assume that the variance is similar between conditions or, strictly speaking, that the variance of the differences between all the levels of the factors (e.g. similar variance between 1–2, 1–3, and 2–3) is equal at a population level, which is called the sphericity assumption. *Mauchly's test* is therefore by default calculated to test whether they are not, with significance on the test indicating that the sphericity assumption is violated. Classically, if Mauchly's test is not significant, we assume sphericity, and use only those tests from Table 11.1 that start with 'sphericity assumed',[5] while if it is not, we turn towards the Greenhouse-Geisser adjustment, ignoring the first and third line of Table 11.1. Now, given that Mauchly's test was not significant (p = .07), we can draft the following line for the results section of a paper:

‘*In repeated measures ANOVA with relevance (relevant vs irrelevant vs odd) as factor and Cz mean amplitude as dependent, relevance had a significant effect, F (2, 12) = 63.70, MSE = 0.18, p < .0001, η_p^2 = 0.91.*’

…while if Mauchly had been significant, it would be F (1.21, 7.25) = 63.70, MSE = 0.30, p < .0001, η_p^2 = 0.91. Either way, it is a beautiful line of words and numbers that makes me think of a magic formula. It also makes me yearn for a future in which that package will just use the correct sphericity assumption or even completely parse the magic itself instead of making us piece together the puzzle.

So now we know there is a difference between the conditions, but it remains unclear whether they all differ from one another, or whether there's just the odd one out. For that we go into the post-hoc tests (Figure 11.2).

Table 11.2   Post hoc differences between conditions. Note: 1 = relevant, 2 = irrelevant, and 3 = odd.

Condition (i)	Condition (j)	Mean difference (i-j)	p
1	2	.712*	0.001138
	3	2.480*	2.36E-04
2	1	−.712*	0.001138
	3	1.768*	0.001977

This is where the named variable (Condition_rel_irr_odd) can come in useful, because otherwise it would be hard to see what 1, 2, and 3 are. In this case, however, we don't even need to think since all levels differed from one another. We therefore add another couple of words to the results:

‘*…Bonferroni-adjusted post-hoc comparisons suggested relevant stimuli had higher mean amplitwudes than both irrelevant (D = 0.71, p = .001), and odd (D = 2.48, p < .001) ones. Irrelevant stimuli also had a higher mean amplitude than odd (D = 1.77, p = .001) ones.*’

---

[5]Which does not entirely make sense, as non-significance does not equal proof for the null hypothesis.

But depending on the obstinacy of your tutor or reviewer #2,[6] I prefer to reduce the complexity of such sentences as much as possible, so something like this also works:

> *...Bonferroni-adjusted post-hoc comparisons showed all levels differed significantly, ps < .002, with relevant stimuli having the highest amplitudes and odd stimuli the lowest.*

Of course, that makes your Results section very short, and I know students often worry about that, because since tutors enjoy boring people with statistics, they must evidently love nothing better than reading the number-heavy section of a thesis, right? Perhaps so, but if you do EEG, your results section easily explodes due to the number of tests. For example, the above lines should be written out for every measure you used, so at least four times in our example. You should tweak them a bit so as not to repeat them exactly, of course, but it doesn't require much literary talent to write a Results section.

However, where possible, it is more elegant to analyse the data within a single statistical design and combine the various measures, enabling us to see how one effect (such as the condition) plays out differently across another effect (such as the electrodes). Let's look at this in detail.

Multifactorial designs

As a rule, unifactorial designs are for statistics newbies. I suppose it is possible to have a solid experimental design in which a single factor neatly describes control vs experimental conditions, but in reality it rarely happens. Here, for example, we might want to test something like the hypothesis hinted at in the additional exercise of Chapter 2:

> *If, for example, seeing yourself is really automatic, it might show a reduced target effect (after all, irrelevant targets will also be slightly relevant).*

That leads to the following prediction: *The effect of relevance should be smaller for seeing oneself than for seeing someone else.* A two-way interaction effect, such as defined here, requires you to name both factors – or the levels within them – and their relationship as somehow depending on one another. This is different from what we call **independent effects** (sometimes called *additive* effects), which is when both effects are present, but they do not influence one another. For example, two independent effects are described if seeing someone you had to keep in mind amplifies the P3 *and* seeing yourself also amplifies the P3. An **interaction effect**, on the other hand, is if seeing someone you had to keep in mind amplifies the P3 *but this effect is smaller* if you see yourself. Typically, this distinction is explained visually with graphs, and graphs showing lines that either go parallel (additive effects) or to some extent cross. Personally, however, I would prefer it if you would understand the distinction rather than merely be able to spot it, but Figure 11.10 is what the above conclusions would look like in visual form.

---

[6]I recommend the Facebook group *Reviewer 2 must be stopped* for early-stage postgraduates – the rest of you are probably already on there.

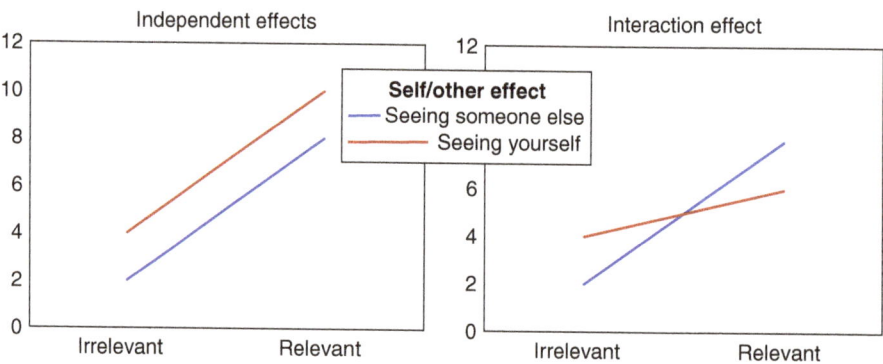

Figure 11.10

Of course, such interaction effects are normal in most experimental designs, but in EEG research they are ubiquitous, and higher-order (three way and up) interactions are not uncommon. The reason is that EEG adds *space* and *time* to most designs, so, if you were only interested in relevance, you might add the following questions:

1. Does relevance affect the entire scalp area (very unlikely), or does it depend on the specific area?

2. Does relevance affect all potentials (very unlikely) or does it affect a specifically time-related potential?

Combine the questions of space and time with relevance, and you get a three-way interaction. These are almost always complicated to explain, as they need to include a relationship between all three factors. For example: is the difference in scalp distribution caused by relevance particular to a time-window, or does it affect all measured potentials equally? Is the scalp distribution of the effect of relevance different in the first measured interval from the second? Of course, if you have relevant and irrelevant conditions, one frontal and one parietal electrode, one early and one late interval, there just *might* be a single electrode that shows a relevance effect only in one time interval. Then, if all seven other combinations are equal, the three-way interaction would be significant but nothing else is. But that, sadly, rarely happens and three-way interactions are usually hard to interpret.

Let's have a look at our own. Following the tutorial type of approach of the one-way designs, we had the factor of relevance arranged as the narrowest column definition and the electrode at the wider, and had four measures (mean amplitude 270–350, mean amplitude 370–600, peak latency, peak amplitude) for Cz_Rel, Cz_Irr, Cz_Odd, Pz_Rel, Pz_Irr, and Pz_Odd. Now, either change the table or create a new one so that the mean amplitudes of the first interval (T1) are followed by the amplitudes of the second (T2), such as illustrated in Figure 11.11. Make sure you give sensible names to each column and have one subject per row.

As before, copy the data (everything below the headers) into your stats software. Hint: If you don't feel like naming all the columns in your software again, you can also copy them in the spreadsheet, paste-special, select 'values' and 'transpose' as special options, and paste. The neat column of variable names can typically be copied straight into the stats package.

	T1						T2					
	Cz			Pz			Cz			Pz		
	Rel	Irr	Odd	Rel	Irr	Odd	Rel	Irr	Odd	Rel	Irr	Odd
Subject Time	Cz_Rel_T1	Cz_Irr_T1	Cz_Odd_T	Pz_Rel_T1	Pz_Irr_T1	Pz_Odd_T	Cz_Rel_T2	Cz_Irr_T2	Cz_Odd_T	Pz_Rel_T2	Pz_Irr_T2	Pz_Odd_T2
101 AverageAmp	-0.088	-1.060	-3.289	-0.339	-0.963	-0.524	0.645	-0.383	-0.645	1.329	-0.481	0.985
102 AverageAmp	-1.094	-1.334	-3.787	-1.475	-1.005	0.444	-0.598	-0.687	-1.737	1.395	-0.467	2.171

Figure 11.11

Again, do a repeated measures ANOVA, but now specify three factors: Time_T1vsT2, Electrode_CzvsPz, and Relevance_RelvsIrr. I am leaving out the *odd* level from the design here, so as not to spoil essential exercise 1. The results are shown in Table 11.3.

Table 11.3 ANOVA of Time (T1 vs T2) × Elec (Electrode: Cz vs Pz) × Rel (Relevance: Relevant vs Irrelevant) with average amplitude as dependent. ns = not significant.

Source	SS	df	MS	F	p	η^2_p
Time	9.82	1	9.82	40.22	.001	0.87
Error(Time)	1.47	6	0.24			
Elec	1.29	1	1.29	2.54	ns	0.297
Error(Elec)	3.05	6	0.51			
Rel	13.86	1	13.86	70.05	.0002	0.921
Error(Rel)	1.19	6	0.20			
Time * Elec	0.00	1	0.00	0.001	ns	0
Error(Time*Elec)	2.01	6	0.33			
Time * Rel	3.10	1	3.10	101.32	<.0001	0.944
Error(Time*Rel)	0.18	6	0.03			
Elec * Rel	0.24	1	0.24	6.17	.048	0.507
Error(Elec*Rel)	0.23	6	0.04			
Time * Elec * Rel	1.43	1	1.43	21.85	.003	0.785
Error(Time * Elec * Rel)	0.39	6	0.07			

Now, your average stats book will tell you to report (1) all significant main effects, and (2) every significant interaction, and (3) every non-significant interaction you predicted. In EEG research, that quickly leads to a lot of numbers absolutely no one is interested in. For example, electrode is not significant, so what? If it were significant, would it tell us anything? I typically leave non-significant, non-interesting main effects out if I can avoid it, particularly if I have multiple analyses like this, so as not to repeat boring information *ad nauseam*. But it's still quite a lot of information, especially if your reviewers or tutors insist on getting effect sizes and mean squares, as is customary. To spare my editor, I will keep it relatively short.

In repeated measures ANOVA with relevance (relevant vs irrelevant vs odd), time (T1 vs T2), and electrode (Cz vs Pz) as factor, and average amplitude as dependent, relevance and *time had significant main effects, ps < .002, while electrode did not. More interestingly, time and relevance also interacted, F (1, 6) = 101.32, p < .0001, with the later effect of relevance being far larger (T2 rel = 0.90 μV, T2 irr = -0.57 μV) than the early effect (-0.41 vs -0.93). Relevance also interacted with electrode, with the effect of relevance being more visible on the Pz (Pz: Rel 0.46 μV vs irr -0.66 μV) than Cz (0.03 μV vs -0.84 μV). Finally, the three-way interaction was also significant, F (1, 6) = 21.88, p = .003. This showed that the early relevance effect was more pronounced in Cz (rel–irr = 0.71 μV) than Pz (rel-irr = 0.34 μV), while the later relevance effect was reversed, and weaker over Cz (1.02 μV) than Pz (rel-irr = 1.92 μV).* 〟

Note: Because there are only two level factors, univariate post-hoc tests are pointless (they are equivalent to univariate F-tests), and I'm leaving out multivariate post-hoc tests as they tend to involve cherry-picking (see next section).

Of course, that last sentence before the note is the more interesting of the bunch, as it suggests there are two relevance effects: a central, early one, and a parietal, late one. How did I come up with that one 'this showed' sentence, though? In general, it becomes easier to understand interaction effects as *differences of differences*, and arithmetically, we can literally do that. So, I calculated from the estimated marginal means provided by the statistics package relevance *effects* instead of simple means, by subtracting irrelevant from relevant conditions, getting these two pairs: 0.71 and 0.34 (in T1) and 1.02 and 1.92 (in T2), which is much easier to read than 8 values. A three-way interaction, then, is not so difficult as such, as it is simply the difference in T1 (0.71 – 0.34 = 0.37) compared to the difference in T2 (1.02 – 1.92 = -0.90), which is equivalent to a simple paired T-test.

But if the underlying logic of stats is not your thing, try to move slowly from the parts you still understand. A relevance effect is clear, right? A relevance effect that happens more in the front than in the back should also be clear, as is a relevance effect that is bigger in the late EEG than in the early one. The three-way interaction is just the early relevance effect being more central, and the late relevance effect being more parietal.

Does this whole data exercise add anything, though? Arguably, you had already seen what happened in the frontal electrodes and the parietal and noticed that the frontal relevance effect occurred a bit earlier and the parietal later. In fact, didn't we choose the ROIs specifically with that in mind? How obvious is that an effect we specify based on what the data look like then turns out to be a 'real effect'? The following section will explain why the approach presented over the previous sections is a fantastic example of double-dipping at its worst. I will explain the problem and how to avoid it in the future, before I let you start to get on with doing a better job yourself.

EXPLORATION AND CONFIRMATION: METHODOLOGICAL ISSUES IN EEG RESEARCH

EEGLAB makes it very easy to create and test simple statistical designs, but this makes it also dangerously easy to come up with nonsensical findings. Yes, I am still upset about the last chapter's suggestion that I am not special. Friends, *frenemies*, and – I reckon – total strangers alike agree I am, though they may differ in their definition of 'special'. So, it must be true that something went wrong in the analysis in terms of statistics, and I will go back until the truth as I know it comes out!

You will realise that such reasoning is a tempting sort of confirmation bias: I already knew I was special so anything that invalidates my truth gets a special treatment of tinkering at my data until some electrode at some point agrees with me. EEG presents countless possibilities for that sort of thing, but even if you do reaction time experiments or survey studies, there are many ways of tinkering at your command: Do you log-transform your data to *normalise* them or perhaps remove outliers? Do you compute missing values? Do you drop some items to increase the inter-item reliability? We all know the tools of the trade that are supposed to improve the validity of the study, but the freedom they provide is tempting to misuse when the reality doesn't conform to our preconceptions.

So, what is the difference between what is now called a questionable research practice (or QRP) and a solid, statistical choice? Clearly, if I do all my data collection and apply reasonable if not very advanced statistics and find a final *significance* level of $p = .052$, I will utter some profanities and sit back. Now what? Publishing non-significant results is hard, and so the temptation of going back to my data and giving an extra hard look is immense. 'Aha', I might think. 'If these two people were found to be outliers (and now that I look back at them, they *were* weird!) and were removed from the analysis, my significance level would drop *just* below the critical threshold!' This constitutes the problem of **cherry-picking**: we try out a whole bunch of special manoeuvres, or select only the best items in our survey, or keep only the preferred participants, and then maintain that the selection is somehow more representative for the population than the full set.

This is tempting because it can mean the difference between months of work being tossed aside and a nice crown of achievement in the form of a publication. Essentially, however, we know that this *backtracking* is the path of Evil Statistics and although there is large statistical power to be got from the Dark Side[7] (in affecting p), it comes at a tremendous peril (to get retracted). The path of righteousness is essentially straightforward, running in the following, idealised form:

> 'From theory, we derive a hypothesis, which we can test using data. Once the data are in, we obtain the test measurement, which verifies or falsifies the hypothesis, increasing or reducing support for the theory.'

Within that framework, cherry-picking means generating lots of different measurements and selecting the best one – probably the one that has the lowest p value. As I pointed out before, cognitive neuroscience in general suffers from the problem that, as a field, it is still rather young and, as such, experimental (in the engineering sense: with rapid advancement and new

---

[7]I promise this is the first and last *Star Wars* reference, which I put in there for a colleague. But for those of you who are into that sort of thing, I recommend *Star Wars Psychology: Dark Side of the Mind* (Langley & Goldman, 2015), which I got as a Secret Santa gift.

statistical technologies). That means cherry-picking is often easy: we can easily select among the many electrodes a favourite, such as Pz in the present study. And we do not need to stick with raw amplitudes, but can calculate peak magnitudes, perhaps log-transform, or suddenly switch to an entirely different analysis, such as based on spectral analysis. Because it is so experimental, however, it is also sometimes necessary to do *some* cherry-picking, as there is not always a gold standard on where a potential is or how the analysis should be done. How do you avoid the danger of becoming a star in the halls of scientific infamy? I will outline typical problems to watch out for and finish this chapter with strategies to avoid pitfalls.

HARKing

Instead of selecting the one useful measurement, it is also possible that none of the measurements gives the result you were looking for, or you had only one to begin with, but instead of being honest and admitting this, you simply rewrite your predictions so the problematic result fits right in. Such hypothesising after the results are known (or HARKing) is naturally very tempting given that you normally write the final manuscript after obtaining the data and doing the analysis, especially if the theory was pretty vague to begin with. In our study, for example, we expect a P3 response to seeing a relevant person, which should be an enhanced amplitude from c. 300 ms particularly over the parietal area, but do you have a clear prediction of what happens with regards to seeing yourself? To me, the following hypotheses are all plausible:

1. Seeing yourself, even if irrelevant, will produce a P3 response vs seeing an irrelevant other.

2. Seeing yourself relevant increases the P3 response relative to seeing a relevant other.

3. Seeing yourself irrelevant produces the same P3 response as seeing a relevant other.

Then combining 1 and 3, we could predict:

4. Seeing yourself relevant minus seeing yourself irrelevant is a smaller difference than seeing another relevant minus another irrelevant.

HARKing, then, would be the QRP of testing all hypotheses and only reporting those that worked out. Even worse would be to then read up on all sorts of literature that will tell you why these were the only reasonable hypotheses to begin with, causing a methodological revisionism of starting with the results and ending with predictions and theory.

Double-dipping or 'The Texas Sharpshooter' fallacy

Of course, it is very well possible that the basic effect of relevance is located somewhere else than the effect of seeing yourself. Maybe seeing yourself causes a more frontal P3 than seeing just any relevant picture? This is similar to HARKing, but the scattershot approach to methods is commonly known as double-dipping, or committing The Texas Sharpshooter fallacy.

The Texas Sharpshooter

While the outline of the cautionary tale of the Texas Sharpshooter is known by most scientists, only EEG researchers are privy to the full story. At the risk of excommunication from the Order of the Mind Electric, let me divulge our secret. Grey Walter, EEG researcher *extraordinaire*, dapper gentleman and swashbuckling gunslinger, once spectacularly failed to locate the Electroencephalographic Society's Annual Meeting and found himself in the sleepy town of Cut and Shoot, slightly north of Houston. Naturally, he decided to ask for the way back to Bristol over at the saloon. 'Howdy, partner', said he in fluent Wild West Talk to a random sample of individuals. But before he could further explain the circumstances of his present confusion, a gruff man answered: 'We don't take kindly to electrophysiologists in this town, sir! It's going to be one of these shootouts at dawn we do in these parts, or your turtle is gonna get it', pointing at Grey Walter's robot turtle, Elmer. 'Humbug, Sir Cowboy,' said Walt. 'Academics simply do not do dawns. But what if I shoot five times a bullet within an inch of another while being blindfolded? Surely you would not risk such a straight shot.' Thus agreed, the scientist was swiftly blindfolded, unceremoniously swirled around like a fidget spinner, and told to show his trick. Slowly, as the embarrassed laughter of inconsequential extras died to a slow murmur, he took out his M134 minigun and shot straight ahead at what might well have been a wall. Taking his blindfold off, he strode to the wall, and peered at it for some time. Finally, finding from across the hundreds of fired shells a nice cluster of bullets, he took out his whiteboard marker and drew a perfect circle around them. 'Well, gentlemen', said he, 'Will you now show me the way to Bristol?' None answered. 'How rude', he shrugged, got on his trusty mechanical turtle, and rode off into the sunset.

And we can do that not only at the level of electrodes. Our Texas Sharpshooter, for example, might test all electrodes, finding Pz the best to test, and fail to report the full search. But even in Pz, we do essentially 512 tests. Looking at the ERP plots, only one is necessary, so if one would look at last chapter's ERP figure (p. 226), it would be most strategic to do the test right where the difference looks biggest, say in Pz at about 500 ms. Indeed, if I take that exact point and then look at the actual tests, I got a p of .00047 – more than 100 times smaller than the conventional critical $p < .05$.

But no, this is what statisticians call 'double-dipping'. You first do one test to figure out where to conveniently place a second one, and thus maximise significance. Of course, you might do a first test without real statistics, but as I argued elsewhere, looking at a graph is a kind of informal, 'visual' test, which means we are basically doing the same as cowboy Walt above. Cheating, in other words.

Type I and type II errors

Instead, if we had been fair, we would recognise the fact that there were 61 channels × 512 time-points of tests being run, increasing the chance of committing a *type I* error by about 31,232. As you will remember from your statistics class, a type I error is wrongly rejecting the null hypothesis (there is no difference between conditions). The probability of observing a difference of at least 1.2 µV given the data, and given that there is no difference, may be .00047 for testing a single time point, but if we try that same test 31,232, it is obviously much more likely to happen at least once. Indeed, Bonferroni correction would mean reducing the chance of committing a type I error by decreasing the critical *alpha* level by the number of tests, so that only a *p* value below .0000016 (.05 / 31,232) will count as significant.

But that is not entirely fair. Given volume conductance over the skin (p. 8), the voltages in one channel closely resemble the neighbouring ones, and due to the low-pass filtering, any observed voltage will be very similar to the surrounding few milliseconds. As such, you might test only, say, 16 electrodes and 50 timepoints, and end up with a vastly smaller number of tests. And here's a thought experiment: let's say you'd know where to look (not unreasonable) and had simply attached only that electrode, and never bothered to record any data but the single datapoint at 500 ms (with baseline subtracted). Does that mean that the chances you observed a statistical significance were now 31,232 smaller?

I will leave the philosophy of statistics to those enjoying this discussion[8]. The point is that while making a type I error is anathema to any scientist, being so conservative with your adjustments that you make it completely impossible to ever find something is hard-headed. That is to say, the chance of committing a type II error – of wrongly accepting the null-hypothesis – can be a danger in its own right that increases by statistically controlling for type I errors. Here, that would mean ignoring the evidence clearly indicating that we do see a healthy sort of P3 response in our data, replicating hundreds, if not thousands, of other EEG studies. And, obviously, if we only had taken the time to make a forward prediction – before obtaining the data – we would have avoided the entire debacle. But how does the reader know that you've been honest in such a case, or simply HARKened back to the olden days while writing your paper?

Strategies to avoid cherry-picking, HARKing, and related QRPs

Indeed, as a reviewer, almost all papers make the following claim: 'We found an effect here, just like Pre-eminent Scientist (1981), who did a similar study, thus seeing apples produces the same type of effect as seeing pears.' Such reasoning is already problematic, as very different brain processes can well give rise to the same observed measurements, but it also clearly cherry-picks from the evidence. Given that a whole lot of research has already been done, you are pretty much guaranteed someone, somewhere, somewhen found related effects. This can then easily suggest consensus that is unlikely to reflect the literature. Moreover, it may allow you to suppose a much more forgiving statistical threshold than would be required on the basis of looking at the totality of your data. You and I naturally are honest people, so we know that we first looked up PS (1981),

---

[8]See, for example, the Psychological Methods Discussion Group on Facebook.

then committed to data analysis, and then tested a specific region of interest (ROI) in the data, but how can a reviewer know this? These non-exclusive strategies will help you look better in the eyes of reviewers and tutors:

- Preregister your research. Replication studies used to be very unpopular with journals, which is remarkable given that science was supposed to be all about creating trustworthy, replicable results. However, once multiple labs signed up to test some of the big findings in the field, it became clear that many effects failed to replicate: unconscious priming with elderly stereotypes did not slow movement (Doyen et al., 2012); power-posing did not improve confidence (Davis et al., 2017); and controlling oneself once did not result in a subsequent loss of self control (Lurquin et al., 2016). Such studies were conducted using preregistration reports, which means one publishes a sort of plan of action for your study: What are your predictions? How will you test them? And so on. By uploading these to a publicly accessible repository, such as https://osf.io, everyone can see at what stage exactly you made your predictions. osf. io can also hold all your data (certainly a huge quantity of EEG datasets), further enhancing the transparency of your research by allowing others to see just how your results match up with the data.

- Distinguish between confirmatory and exploratory analyses in your paper. Even if you preregister your work, chances are that you did not fully consider how complex real data are. For example, our P300 effect might well not only be localised parietally, but have an additional frontal element. Even if your preregistration then only mentions the parietal test, and even if that part fails the *a priori* tests you specified – in other words, in your *confirmatory* analysis – you can easily add a section with *exploratory* analysis. While this may retract from your image as the all-knowing scientist who set up the predictions *just so* beforehand, there is something to be said for being honest in admitting your ignorance. And, after all, if you already knew everything to start with, what was really the point of doing so much research?

- Distinguish confirmatory from exploratory data in your analysis. The classic way of doing research and coming up with a solid set of predictions is the noble *pilot study*. Get some data in, analyse them to obtain a region of interest, estimate an effect size (e.g. https://stats.idre. ucla.edu/other/gpower/), then obtain an additional group of participants to acquire confirmatory data. It is a great idea, but few people actually do this, because it takes twice as long as just guessing everything (perhaps based on literature or rules of thumb that are prevalent in many labs), and it makes for a difficult write-up in the final paper. Another tactic I have seen used is simply to recruit a few more participants and set these aside. So, let's say we have 30 subjects, let 10 subjects be the 'pilot' group and use these to do as much cherry-picking, HARKing, and crazy weird spectral analysis as you feel necessary, just to get the very best sort of effect you can. Following that, use the other group to confirm the analysis, hoping that the result should indeed replicate. This essentially removes the whole problem of double-picking from your analysis, as the two groups are independent.

- Follow an exploratory search for a region of interest by a confirmatory analysis. A somewhat similar approach to the exploratory/confirmatory distinction is followed in this book. Many studies take it as a given that previously found effects can be readily replicated: the P3 wave by contrasting rare vs frequent stimulus, the MMN (p. 13) with repeating vs mismatching tone, the N400 (p. 14) with a semantically fitting vs nonsensical word in a sentence.

There being already much literature and even some consensus on the interpretation of such ERP components, relatively few of us are excited about again replicating these effects. Instead, we are more often concerned with how a secondary factor interacts with the primary effect. One strategy to avoid double-dipping is to maximise the primary effect by optimising (i.e. cherry-picking) channels and locations, keeping an eye on the literature to make sure the final choice is also defensible. This measure then becomes effectively like a region of interest (ROI) as is common in fMRI analysis (although in fMRI a region of interest can be defined even as a contrast with a resting state baseline task). After doing the exploratory search for the ROI, the analysis of the secondary effect can then be conducted in purely confirmatory mode. The main problem of this strategy is, again, in the possibility that it can overlook any effect that occurs outside the region of interest, for example if seeing yourself affects not the P3, but some earlier or non-parietal electrode. The temptation is then strong to make this strategy into some sort of back-and-forth optimisation technique, which is exactly what we were trying to avoid. For example, maybe one first finds an ROI, then notes that some frontal electrode in the confirmatory analysis is *almost* significant, then going back to the ROI to somehow include this electrode, just so it makes the write-up easier. In that case, a more honest approach would be to involve an extra, exploratory analysis in the paper, as explained above.

- Report those null effects. Sometimes, however, one may search for an effect in the brain activity as one may for the monster in Loch Ness. Maybe it's just not there. And yes, this always makes writing a paper harder, and reviewers will have an easy time remarking that some tiny detail in your experiment caused the problem. To make it easier for you against your critics, I recommend learning Bayesian statistics (see Explore section) so you can point out just how much your data improves the likelihood of the null-hypothesis. And, against the feeling that somehow you need to confirm previous studies to maximise impact, I suggest the alternative image of yourself as the rebel scientist of the Revolutionary Null Party, doggedly pursuing scientific truth in the midst of spurious correlations, marginal significance, and non-replicable results.

ESSENTIAL EXERCISES

Essential exercise 1: Odd pets

Now that you know how to analyse the data, you can do these analyses yourself:

1. Define a region of interest based on your data.

2. Export the relevant data to a spreadsheet and organise it so they can be analysed.

3. Include the odd condition in your analysis and write a results section.

4. For the cherry on top, create a good-looking figure that shows the effects of relevance (including oddball stimuli) on the ERP of all electrodes you used as your region of interest.

Essential exercise 2: Two- or three-way interactions

Once you are comfortable doing the above, do the same but keeping a two- or three-way interaction in mind. Use multiple electrodes (e.g. Fz, Cz, and Pz) and create an ERP figure showing two difference scores: between relevant and irrelevant, and between odd and irrelevant, showing the time-course of both effects. In the accompanying written Results section, also perform the analyses required to determine how latency and peak magnitudes are affected by the conditions and electrodes (note: time is not conveniently included as a factor when it comes to peak magnitudes and latencies as they use a wider region of interest).

Additional exercise 1: More elegant exporting

Previously, I showed how to export the data in a way that works but is not exactly elegant. First, all data were loaded many times, rather than just the once, which is slow, especially if you extend the method to more conditions, electrodes, and so on. Second, you were then required to edit the data afterwards in some spreadsheet, which is also tiring. Why not copy all the data to a spreadsheet in one go?

We have previously discussed the cell structure as a convenient way to organise multiple string values: conditions = {'condition1', 'condition2'}; that sort of thing. But cells are far more versatile than that: numbers, single letters, strings, can all be inserted within cells, as can structs (like the EEG variable we use to contain EEG datasets) and even cells themselves. In the following two exercises, I'll show how this works.

First, edit the part of additional exporting by splitting it into two parts. I left out the initial variables, but otherwise the code below shows how to load all data:

```
EEGs = {}; %declare an empty cell

indir = 'filepath'; %change to location of data

pps = {'101', '102', '103', '104', '105', '106', '107'}; %subjects

cnds = {'rel', 'irr', 'odd'}; %conditions to export

chns2exp = {'FC1', 'Cz', 'FC2'}; %channels to export

outdat = []; %empty variable

for cfile = 1:length(pps)

    for ccond = 1:length(cnds)

        %Load data into cell:

        EEGs{cfile,ccond}=pop_loadset('filename',[datain 'umeeg' pps{cfile} '_11_' cnds{c}
        '.set' 'filepath',indir]);

    end %for ccond

end %for cfile
```

Now, if you double-click EEGs in MATLAB's workspace, you will note that it is a cell of seven rows (or however many subjects you have) and three columns (if you are using three conditions). Double-clicking on any of the cells will retrieve a complete EEG struct. So, EEGs{3,2} contains the dataset of subject 3, condition 2. If your PC has enough RAM – and they are starting to do so by default – this is a very fast, if not very memory-efficient, way of loading the data. After this, you can refer to each EEGs, which will still be in fast access memory, so exporting will go a lot faster. You could simply copy–paste the entire previous export script and replace EEG with EEGs{cfile,ccond} and it will use the specific subject/condition combination for its export.

But while you're at it, you might as well also use cells for exporting the data. Our previous export variable 'outdat' was only able to use numbers, but a cell can be easily exported to spreadsheet format, as long as it only contains numbers and letters. What I normally do is to use two counters, a row counter and a column counter, and work my way from top left to bottom right through what I want my data to look like. So, from top to bottom is defined by time (512 values), so I count my row counter up by 1 every time I loop through each subject, each channel, each condition – in that order.

```matlab
crow = 1; %create a row counter

for ctime = 1:length(EEGs{1,1}.times) %do a loop through all timepoints

    ccol = 1; %create and reset a column counter

    crow = crow + 1; %add 1 to the row counter – every row is 1 timepoint

    for cfile = 1:length(pps)

        for cchan = 1:length(chns2exp)

            for ccond = 1:length(cnds)

                for cchancmp = 1:length(EEGs{cfile,ccond}.chanlocs)

                    if strcmp(EEGs{cfile,ccond}.chanlocs(cchancmp).labels,chns2exp{cchan})

                        ccol = ccol + 1; %add 1 to the column counter

                        %get mean activity for the single timepoint

                        outdat{crow,ccol} = mean(EEGs{cfile,ccond}.data(cchancmp,ctime,:));

                    end %if strcmp

                end %for cchancmp

            end %for cchan

        end %for ccond

    end %for cfile

end %for ctime
```

There is very little additional code, but note the placement of column and row counter. Because ccol is at the deepest level, it updates with every new channel, condition, and file. Only after all that is done, and ccol has updated 63 times (7 (subjects) × 3 (channels for the essential exercise still!) × 3 conditions), do we move on to the next row, also resetting the column counter. Now, to top this and make it a bit easier to read after exporting, add the following line just below ccol = 1 so it updates the first column with times:

```
outdat{crow,1} = EEGs{1,1}.times(ctime);
```

And add the following line just below ccol = ccol + 1, so you also get column headers:

```
outdat{1,ccol} = [pps{cfile} '_' chans{cchan} '_' cnds{ccond}];
```

You might want to add a single conditional, like if crow == 2, ... to only write the headers once, but I will leave that to you. xlswrite (see p. 254) will create a spreadsheet from outdat much as it did while it was still purely numeric.

So, where's the exercise? I would suggest you first try to get this kind of export procedure running, and then apply the idea as you work through additional exercise 2.

Additional exercise 2: Four-way and five-way interactions

As with essential exercise 1, perform either a four-way or five-way ANOVA and write a complete Results section based on an analysis that combines some or all of the following factors:

1. Electrode (left vs centre vs right)

2. Electrode (frontal vs middle vs parietal)

3. Time (N2, early P3, late P3)

4. Relevance

5. Selfishness (self vs other)

Note: depending on the number of datasets you have available, you might need to calm down with the number of levels in each factor.

EXPLORE: JASP AND MATLAB ANALYSIS

If you don't like SPSS, and few people do, I can wholeheartedly recommend JASP (https://jasp-stats.org/) by some of the same people who brought us OpenSesame. It is free, runs on Windows, Mac, and Linux, and is designed with people in mind who would like to kick the SPSS habit. That means it can do pretty much all obvious inferential statistics I can think of and quite a few more. It is also designed with the idea in mind that the *frequentist* approach, such as I have followed throughout this chapter, is fundamentally limited or even flawed. That is to say, proof of a null-hypothesis being false does not mean the alternative hypothesis is correct. Bayesian inference techniques estimate the likelihood that one or other model of the truth, be it the null-hypothesis or alternative hypothesis, is true given the data. The idea is then that rather than something is false as in null-hypothesis significance testing, the evidence collected in an experiment increases the odds that our inference is correct. This kind of theory and the underlying statistics have long bedazzled psychologists, many of them thinking it is probably a better way of doing stats but they can't be bothered with doing the maths. JASP makes it very easy, giving you the choice of doing stats in a classic way or using alternative, Bayesian methods, requiring little more than a press of the button.

MATLAB and indeed EEGLAB itself also provide many useful tools for doing inferential stats. In particular, I have found it useful to present ERP figures such as I showed in the last chapter, made from within MATLAB, with parts of the figure shaded with the output from repeated measures ANOVAs. I think I have shown enough about creating figures and exporting means to give you a basic idea of how to implement this. Let's say we have 1 electrode and want to test the entire range: (1) select bins of data (e.g. from 0–20, 20–40, 40–60, etc.) and calculate means per subject for the first bin (0–20), per condition, (2) organise these in one cell featuring 3 arrays: {all means condition 1, all means condition 2, all means condition 3}, (3) run anova1rm_cell – which is included as EEGLAB function, so writing 'help anova1rm_cell' in MATLAB command gives good documentation. If the repeated measures ANOVA is significant, shade part of the graph. I normally just put a line underneath the timepoint. (4) Go to the next bin and repeat. There are also functions like anova2rm_cell (two-way repeated measures ANOVAs), ttest (for t tests), and many, *many* more can be downloaded from the mathworks.com forum.

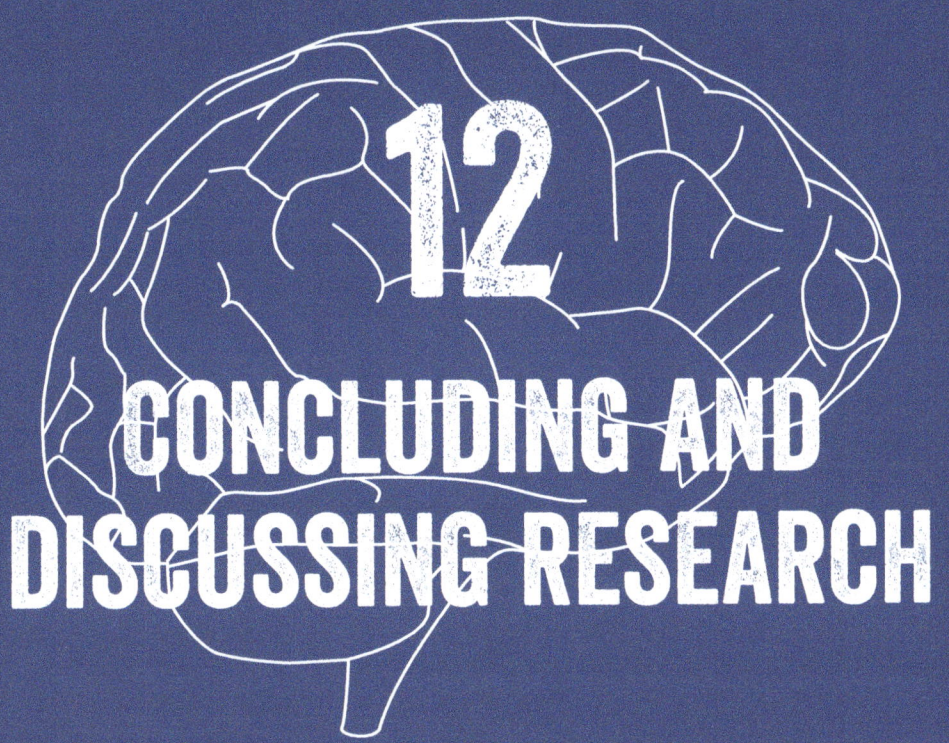

12

CONCLUDING AND DISCUSSING RESEARCH

IN THIS CHAPTER, YOU WILL LEARN:

- How to write a satisfying conclusion
- How to discuss your findings in three so-whats
- That there are always ways your research can be applied
- What reasonable concerns you could discuss in a limitation section

I hate writing conclusions. When you first start doing research, there are all these ideas, theories, predictions, and putting them together as a coherent story is great. Then you start designing your experiment and collecting results, and when you do the analysis, it's either great, or it isn't. Or perhaps it's something in between, suggesting, in the words of Ben Goldacre, 'I think you'll find it's a bit more complicated than that'.

Research, I hope, you do to find answers, but as you write a conclusion, you should already know the answer, or know that you don't have any. All that is left is convincing others that this is the case, and that somehow the research was still worthwhile. You already know that, so why bother?

Clearly, I am not a PR kind of person. Indeed, as I was dating my then-future spouse, I knew she was quite impressed with a certain band, having various albums and even knowing their cellist personally, and I knew I had some sort of role also. It took me several weeks to inform her that the slight role was having composed all the music she was listening to. I suppose I have somehow taken in the entire Disney-princess adage: people should appreciate you for who you 'really' are, while informing someone you're royalty or have achieved something worthwhile stops them from seeing the intrinsic beauty at best, or amounts to bragging at worst. Terrible dating advice, she told me.

And I suppose she's right. You can do the greatest research project and hope your tutor, PhD supervisor, or line manager will recognise its merits, but telling them about it helps. Try to look at the longer-term goal of finding the answer: it's about more than your finding it, but for the further enlightenment of the world.

That sounds a little too sincere for my liking, to be honest, so if that is not your style, then think of it as an essential part of the whole *stand back I am doing SCIENCE* deal. It's simply something we must do.

Well, normally we, but now just *you* must do. Seeing as this is my book, I feel completely in my right to avoid writing a conclusion or a discussion, and just discuss how a discussion should be concluded. In other words, in this final chapter, I give you a strategy for writing conclusions and discussions so as not to have to deal with them myself. As a published author, I hope my insights are good for something. Perhaps they are not, and are just terrible advice, but then you may get some comfort out of the fact that even someone who's completely out of his depth manages to get things published.

CONCLUDING A PAPER

Whenever I am procrastinating, I tell myself that writing a paper is really pretty simple. Far easier than cleaning the house, alphabetising my bookshelves, writing an incisive critical comment on social media – so let's do those *hard tasks* first. It *should* be simple, though, as usually you have already done the work and know what you are going to write – certainly if, like me, you left this to the end – and now it's just a question of explaining what you did. Remember, it's not like you get literary awards for this, so the aim is to have the necessary content and to make it readable. Having a structure helps with both.

Think of a paper as a more detailed abstract. I like writing abstracts for the same reasons I hate writing conclusions, so let's start with that. There are many guides to writing abstracts, but you cannot get much higher impact than *Nature*, so if it's good enough for *Nature*, it should be OK for your tutor, supervisor, or even Reviewer #2. Here, I adapted *Nature*'s guide for How to construct a Nature summary paragraph[1] as a colour-coded guide to writing a typical EEG-style abstract.

1. Basic introduction to the field.

2. More detailed background.

3. The general problem or 'gap' in the literature.

4. The approach or method used to fix the problem.

5. Main results.

6. What the main result reveals in direct comparison with the gap.

7. Putting the results in a more general context and providing a broader perspective.

Detecting important stimuli is a critical cognitive function. Psychophysiological studies show a common brain response known as the P3 both to stimuli that match a predefined target and to stimuli that stand out from others. However, it remains unclear whether the P3 to relevant stimuli differs from the P3 to odd stimuli. Using a rapid serial visual presentation, we asked participants to keep in mind one face while viewing irrelevant faces, relevant faces, and odd (pet) stimuli. ERP analysis showed both relevant faces and odd stimuli evoked strong positive activity over parietal areas, suggesting a common P3 response. However, important differences between relevance and oddness were found in topography and latency. We therefore inferred that the P3 to relevant stimuli can be dissociated from the P3 to odd stimuli. The findings were further discussed as suggesting a common mechanism of 'tagging' stimuli as either important-because-relevant or important-because-unexpected. In this framework, either tag is similarly important, but for different reasons.

Figure 12.1

Notice how the abstract is conceptually structured like an hourglass, going from generic background to specific question, present study, and then back again to answering the question, and explaining what it means in general. This is a type of reductionism, exploring a general idea or theory by testing specific hypotheses, much as we use a sample of people and a limited set of stimuli to operationalise our general population and focus our research question. In the conclusion and discussion, it is then up to us to go back again and interpret the specific finding within the general context.

Writing a conclusion

It is not uncommon to start your conclusion by summarising the study. Maybe that is because some of us cannot handle reading a whole paper in one go, so the reader could use a recap from time to time to remind them of what the paper was about. Or maybe it is because the conclusion comes just after the Results, which can be rather tiring to read, unless you love stats and significance (and why not? Give *p*s a chance, to paraphrase John Lennon). Write a summary of the study, if you feel it serves a purpose, but keep it short and concise, as the obvious place for any person who needs the information quickly is in the Abstract at the beginning of the paper.

---

[1]nature.com/nature/authors/gta Note: *Nature* uses a summary rather than an abstract, which is longer (300 rather than 200 words) and includes 'here we show' as a statement. If an article in another journal uses those exact words, it has probably previously been rejected by *Nature*.

Alternatively, write a summary merely as a writing practice: now that you finished the study and the results, ask yourself again, what was the basic question, what was the study design, and what was the main result. Now that you have the first half of your abstract, it's as good a time as any to (re)-write that!

The main part of a conclusion, however, should be to answer the research question. This should follow clearly from the results, but it should not *be* the result. Remember, the results comprise statistical analyses in our reductionist scheme, while the conclusion is what we infer from the statistics at a population level.

Let's take a silly example, of asking whether men are taller than women. The results may say that with a random sample of 100 men and 100 women, you showed the former were significantly taller, $p < .05$, although there was some overlap between the distributions, pooled SD = 2'8" (7.1 cm). Therefore, men are generally taller than women, which should be the essence of your conclusion. Avoid statistic terms, like ps and 'significant'. Also avoid stating the obvious and redundant, like that not every man is taller than every woman, as the term 'generally' already adequately captures that. Finally, some of the challenges to the conclusion, such that the study probably involved self-reports rather than, say, hormonal tests, and that sex and gender do not necessarily overlap, are best left to the discussion. Traditionally, the conclusion sticks to giving the more immediate answer to the research question.

Concluding EEG studies

What should you conclude in EEG studies, such as the present study? First, answer the main research question, as defined in your hypotheses. I defined two (see p. 33). The first was a kind of sanity check: relevant faces should not give any effect before 150 ms. In a classic null-hypothesis testing scheme, this would be hard to test, but even if it would (Bayesian inference, p. 276), based on looking at the ERPs, I'd bet there wouldn't be any differences.[2] The second was the more critical one, predicting a P3 in response to relevance, even specifying it to start at c. 300 ms, and having a topography over centro-parietal electrodes. As we first noticed in Chapter 10 and then further analysed in Chapter 11, this seemed indeed to be the case, which makes writing this part of the conclusion a breeze:

> *Perceiving relevant faces indeed evoked a late positive potential. Contrasting relevant with irrelevant faces, the former was found to evoke relative positivity on central sites between 270 and 350 ms and on parietal sites between 370 and 600 ms.*

Within the same paragraph, you may want to argue how that relates to previous studies ('this replicates…') and whether this could indicate a common P3 response or perhaps dissociable P3a and P3b responses (Polich, 2007). Either way, it is pretty clear that something along the lines of a late positive potential or P3 was found, and a final line of the paragraph may be a good place to justify this conclusion in the light of the arguments you just presented. I like to avoid speculations here as much as possible, as the discussion gives ample space for that.

---

[2]You may have tested this and found I was wrong – let me know!

Once an answer is provided to the main research question, you have a clear starting point from which to explore the other conclusions. In Chapter 2, I suggested more hypotheses, and in the abstract above I give an indication of what a conclusion to a further research question could be. Maybe the oddballs (pets) and the relevant stimuli (faces to be kept in mind) can be dissociated in terms of their temporal and topological characteristics. Maybe seeing yourself gives a P3 even if the face is irrelevant? In other words, the rest of the conclusion can be devoted to answering all the other research questions, gradually putting them in the context of the wider literature, while still avoiding speculation.

WRITING A DISCUSSION

If your conclusion presents your main message, but in a narrow way, your discussion presents the context. A good discussion can be written by asking '*so what?*' as many times as necessary to impress a tutor or gag even the most obstinate critic who believes the research was done just to satisfy your curiosity. In my mind, that'd be a great reason, but you do not need to convince me. Instead, your task is to trace your way back from the narrow practical answer you gave to the research question, all the way to the basic question of what makes humans tick. The most satisfying discussion ends like a fine novel, bringing you back to the starting point of the introduction, and giving you the wholesome feeling that the conclusion was inevitable, yet still enlightening.

Not that I normally manage that.

The first so what?: How does the conclusion fit in the literature?

The first *so what?* generally puts the conclusion in the context of the literature. In cognitive neuroscience in general, it is common to explain what the current findings tell us about the brain measurement that is being used. Thus, in our EEG research, you could explain what the current findings tell us about the P3, putting the research perhaps in the context of some of the classic P3 theories. As I explained in the beginning of this book (p. 14), the context-updating hypothesis (Donchin, 1981), suggests that the P3 is related to task-relevance and a continuously updating model of how probable various stimuli are.

That can explain the basic effects, but then it is time to discuss your other conclusions, which are often harder to align with existing theory. For instance, images of yourself in the present design were as probable as any other and, depending on the task, as task-relevant, yet I bet that they show some effect. Do other theories give a better account, or perhaps are the theories compatible? Ideally, there are two opposing theories in bitter dispute, trenches dug below – hopefully not too high – piles of evidence, waiting for your decisive bit of evidence. This is rarely the case, of course, so we have three options:

- One or more theories may *assimilate* the findings. Someone was already quite right, and now we have shown their theory also holds true in this other domain. This means a minor adjustment or contextualisation of the theory may be needed, which can still be necessary and useful.

- One or more theories needs to be *accommodated* as the findings simply don't fit. Of course, accommodated could also mean the entire hotel needs dramatic restructuring or even demolition, so to speak, or scientifically terminated. For example, if you show a dead fish has brain activity in an fMRI, then the methodological theory that allows you to do this analysis is dead in the water.

Accommodation

I took this from Piaget of course, although I could have called it a paradigm shift in research programming, or just scientific trolling. My PhD supervisor used to call it 'sock-pissing', with the following remarkable story: 'Imagine some guy has a favourite pair of socks and wears them every day. Waking up, he finds out you urinated in them. Now, even if he launders them hundreds of times, he will always remember and the socks will never feel quite right again.' The socks are supposed to be the favourite theory, and the act is the extremely off-putting bit of evidence you produced, which follows the theory's originator around like a bad smell. But it remains a bizarre analogy.

- Or you have something entirely new. This is very tempting as it makes you into some sort of originator of a theory, which gets many citations. In EEG research, it is also very easy to do because your potential is likely manifested just a little earlier or later than another one. The only reason it doesn't happen more often than it already does is because reviewers have read the literature the 'originator' obviously did not.

The second so what?: What is the wider relevance of the finding?

If the first 'so what?' explains what the meaning of the finding is for the specific field, the second 'so what?' should discuss the merits in the wider context of a more general science. Here, that means we move onwards from psychophysiology as such towards psychological science. Basically, what do the findings say about the human condition? This can be difficult, as once you are so much into the experiment (faces, oddballs) and the cognitive neuroscience (P3, orbitofrontal activations), you may find it hard to escape the comfortable trappings of the field. We all know some colleagues who no longer manage at all and can only relax and talk at science conferences.

Try to ask yourself, then, what do our present findings say about human beings, their behaviour, and cognition? For the present study, remember that a P3 is not just there to amuse us, but signifies the occurrence of something that is uncommon and relevant. In other words:

an important yet scarce resource. That is one technique, to put an evolutionary spin to your findings, although that is a slippery slope that can easily lead to frankly absurd theories about gendered colour preferences. So, let's say our findings suggest seeing your own face evokes a P3 potential even if it is otherwise irrelevant. Well, that is weird, isn't it? It's not like we are evolutionarily adapted to spotting selfies in the wild. Some theories out there suggest that anything owned by the self is recalled better (Turk et al., 2011), recruiting additional neural structures, which perhaps leads to seeing your own face also leading to additional 'mirror' network activity. At least, it fits if you don't think about it for too long, because why indeed would your brain especially need to note its frontal façade? Logically, there are only two reasons: seeing your own face is a sign of the matrix glitching; OR, we are evolutionarily endowed to spot our time-travelling selves so as to avoid having the universe collapse upon itself.

In other words, it is not always obvious what your findings *really* mean. If you feel the explanatory gap is simply too large, it is not criminal to skip this part and rather extend the former 'so what?', or leap over it onto the next one. That is, if someone else already published a large theory of how the psychophysiological theory relates to cognition and behaviour, a summary of the theory may suffice, and your only job is to say how well everything fits together. Alternatively, if you don't have much to say about the human condition, you could still make a solid point that there is clear utility in your research, as in some application or even innovation just waiting for neuroscience inspiration.

The third *so what?*: Are there any practical implications?

Does it seem like there are any practical applications for your research? Perhaps not. Most people I know zigzag through motivational phases of hope and cynicism throughout their career. A cynical view is particularly disheartening, because not only do you feel that your work in research is pointless, but it also places a huge burden on others: what about their tax-moneys? Of course, maybe this won't be the case for you, for example if you are just learning the tools of psychophysiology to give you an edge on the job market. Indeed, you might be the next hotshot in marketing, bringing in pure corporate gold by detecting which brand logo pops out most using the science of brain waves!

That is not precisely what I mean by usefulness, though. The truth is that we do a lot of research out of simple curiosity, a need to know, and a larger quest to increase knowledge. In exactly the way the financial fortunes of the insanely rich do not trickle down to the poor, so scientific knowledge does cross-pollinate in culture and affect society. However, that does not mean it is obvious how the knowledge will affect day-to-day life in the far future, even if you are the one accumulating said knowledge. Ask people at CERN when exactly we can have our pet Higgs Boson.

In other words, don't worry too much when you think about applications, as it's not like the innovators of tomorrow are reading your paper and go KA-TCHING!

Think of it instead as science fiction: What could the world be like? My human–computer interaction colleagues over at Computer Science are great at this. I contribute psychophysiology expertise to their work and they can paint a lovely picture of a not-too-far future where everybody walks around with an EEG all the time. Does that sound likely, convenient, or desirable? Sure, it'd be great for researchers, but I envision walking around the mall while a rapidly blinking ad-wall keeps spamming me with face images until it evokes a P3, whereupon it will give me shopping suggestions. I thought this was unlikely with current scalp-based EEG for many of the reasons explained in Chapters 1 and 7, but then along comes Elon Musk's Neuralink and some commercialised electrocorticography, and I'm starting to worry. Not, as some futurists and the popular media do, about their being able to 'decode the mind', but the present study indicates P3-driven advertisement is feasible. Welcome to Black Mirror! On the plus side, it means you can easily discuss the ethical implications of your research.

However, it is not only various new and exciting forms of dystopia that computer science applications can come up with. In fact, EEG research has enabled a very benevolent application of their research in brain–computer interfacing: the idea is that by discriminating mental states, we can enable communicating with a computer and others. The initial idea, by Farwell and Donchin (1988) was to show a grid of letters, with rows and columns flashing while the subject concentrates on specific letters. By detecting which row and column evoked a P3, they were able to use the EEG to enable people to spell single letters, potentially allowing someone to communicate who is otherwise unable to move a single muscle in their body (e.g. locked-in syndrome). In our lab, we have recently replaced the grid of letters with a current state-of-the-art artificial neural network model, showing people faces in a design much like we used in this book, allowing them to generate novel faces that match what they were looking for (Kangassalo et al., 2020). We have even been able to do this with a task in which participants looked for attractive faces, using Hans Berger's Brain Mirror (p. 2) as a sort of Snow White-style device for visualising someone's perfect ideal of attraction (Spapé et al., 2021).

Limitations and future research

The final part in a discussion section is often the limitations and/or future research section. This is where students commonly write 'we had a small sample' and 'we did not test anyone from Greenland'. As I mention in the exercises, this should generally be avoided as it is always true (unless you are at the University of Nuuk) and misses what such a section should be about: reasonable concerns.

What are *reasonable concerns*? The most serious type of concern is the confound (p. 25), typically a variable that correlates with your independent variable and can account for the observed differences. In other words, an alternative explanation of your results. These can be purely methodological: for example, if you contrast relevant with irrelevant stimuli without accounting for having many more epochs in the latter condition (generally resulting in this ERP having stronger peaks, see p. 233). Or they can be more theoretical, for example, if we have required participants to respond to relevant images, we would be unable to decide whether observed differences were to 'relevance' or to the motor activity associated with the response. In either case, do as much as possible to address the limitation, explaining how you still think the research is valuable, or how an alternative theory may compete with your preferred one.

Sometimes reviewers also come up with *unreasonable concerns*. For example, they themselves might be interested in sex differences and would like to know whether there are any inter-individual differences in the ERPs, especially if a male or female face is shown. And indeed, that could well be interesting, but EEG is generally not great for individual differences research (unless you are extraordinarily careful), for which reason a particular design was chosen for the particular research question, which had nothing to do with sex differences. It is also an unreasonable concern, because sex differences are not correlated with the independent variable (male and female images could both be relevant and irrelevant), and male and female participants did the same experiment. There is therefore no confound. The reviewer might be interested, but they are welcome to do their own study. In the meantime, the danger of adding a variable to the analysis without clearly pre-specifying its place in the model will strongly increase the complexity of the design and plenty of chance for type I errors to occur.

There are three ways of dealing with such a reviewer, other than just giving into the request. One, such a comment from a reviewer could perhaps be moved to a *future studies* part of this section, with a few references and suggestions as to how the question could be tackled. Two, particularly for this example, you may counter (in the rebuttal, not the paper) that you disagree that this is a reasonable question; after all, you only have access to data regarding gender identity, not sex. Three, you can strategically appease the reviewer, adding the proposed analysis as supplementary material, knowing that most people don't read these anyway.

The third type of concern is about the generalisability of the data. In our reductionist scheme, for example, we started with relevant stimuli and we ended up with faces. What about smiling faces, moustachioed faces, baby faces, non-faces, phases, houses, gardens, and familiar scents? What about other populations, clinical populations, Greenland populations, unconscious populations? This is often a very hard question to answer, but you may be able to argue that the selection is representative for the whole. Alternatively, if others have used other stimuli or populations, then you can point those out. If not, it seems reasonable that you *had to start somewhere*. That is hardly a satisfying answer during review, but as a last resort, you can always add the proposed alternative to your *future research*.

While you go over all the various ways your experiment fails to be perfect, try to pre-empt criticisms, but do not shoot your own foot. Remember that the reader knows far less than you about the actual study, having just read about it. It is just like cleaning a house before having visitors: it is much easier to spot the remaining spots if your face has been in them the last few hours than if you just arrived. If such a visitor is a friend, then it makes sense to say you are not entirely satisfied with the cleanliness. If it's a landlord doing a house inspection, that would be counterproductive. The key is to be honest without oversharing.

Our future research part of this section may already be overly long if you made a point of addressing all reasonable and unreasonable concerns and offering ideas to improve the generalisability. It is, however, possible to explain what the logical next step for this line of research is. Again, don't only think of what you personally would want to do, but base this on the rest of the discussion. For example, if your findings equally fit two theories very well, can you think of a design that could prove the superiority of one theory over the other?

Ideally, if you could make that last part look a bit grand, it would almost sound like a conclusion. With the proposed design, we would be able to find out why so much of EEG concerns either our detecting something that is rare and relevant (the P3), or different and problematic (the N2). How is this related to goals and motivations, of problem-solving and coping? Such questions are of interest to any psychologist, and as they ponder along with you, they might feel satisfied in the knowledge that there is something very human about the electric study of the mind.

The end.

EXERCISES

There are very few good exercises other than the obvious: Write that paper or report, or edit and finish the blog or vlog, or other way of sharing the research. The amount of work depends on whether you have been following the essential or additional track in the exercises, with the latter generally involving more exploration, additional hypotheses, and hence more to be reported. Everything here depends on the tutor, supervisor, or journal you're going for, but to give you something of a cheatsheet, I collected a couple of do's and don'ts for discussions and beyond.

- Avoid the following clichés in your conclusion and discussion. *We only had a small sample size*: given 7 billion people on Earth, this is generally the case. Or a more recent version: *we only tested WEIRD populations* (Western, Educated, Industrialised, Rich, Democratic). I have nothing against wokeness at all, and it is good to have a diverse sample (and stimulus material), but basic functions (such as relevance detection) tend to work very similarly, and it is rather condescending to imply that other populations can't keep a face in mind.

- Make your work readable. Most of my colleagues avoid silly metaphors, flowery language or any other sign that a writer enjoys their work. A proper scientist, apparently, must combine clinical precision in research with sterile presentation in writing. I clearly overcompensate in this regard, but if you lose your audience, you will lose in your grade or in your citation score, because, all evidence to the contrary, even teachers and scientists don't like being bored. If I read your thesis and you sound like Jane Austen, I promise I will give you an A.[3]

- However, keep in mind your audience, because they give the final judgement on the matter. So even if you, I, and Mr Darcy think you're great, that is not enough. Some of my colleagues definitely feel that serious work needs a serious tone and a sufficiently formal writing style. Since you need to keep them happy as well, it may be helpful to adhere at least sometimes to their demands, and be aware of their rules. Some of these can be boldly broken, such as split infinitives, but it is a good exercise to make your writing clear and descriptive. A good, modern style guide is Steven Pinker's (Pinker, 2014).

- Avoid both oversimplifying and overcomplicating. Simple is not necessarily best, but always beats being wrong or unreadable. This goes for style as much as theory. There seems to be a certain type of student who takes a thesaurus and goes from being short and wrong to sesquipedalian and fallacious, which does not help. In theory and evidence, oversimplifying can mean cherry-picking the evidence to only support one theory, while overcomplicating is failing to properly synthesise different theories into a coherent whole. For example, if you have a structure in which x says y, but a says b, while p says q, it is time to group the various theories into clusters and explain how these are or are not related.

- Avoid 'more research must be carried out'. We all know this is true – we would otherwise be out of jobs – so it doesn't need saying. Instead, suggest some good ways for others to carry out studies with specific recommendations that follow from the present study.

---

[3]Seriously, don't try this at home. Instead, do read some William James to get a sense that high-impact science writing used to be much more dramatic.

EPILOGUE: SO YOU WANT TO BE A COGNITIVE NEUROSCIENTIST?

You got the brains, but have you got the books? Right, you just read this one, so I guess that's in order. There are certainly more prominent voices in EEG/ERP research and analysis, but let's imagine you already went through the more Serious Works I mentioned (on p. 22, and Cacioppo et al., 2017): now what? It's not like you can just walk up and down EEG avenue, look at the signs, and knock on the door where they're looking for a guild apprentice. I know, because I have tried, as well as many other unlikely places (see p. 166) on- and offline.

To give you a good start of where to find a funded position as a PhD student, postdoctoral researcher, or anything higher, consider the following tips:

- Not as good as fabled EEG avenue with its quaintly wiggly street and evocative shopping windows, but certainly more useful is **your local lab**. I do not know exactly how it works, but somehow, if you spend enough time doing experiments, helping others, perhaps doing the occasional research assistant contract, you will likely learn about opportunities. It is how I heard about my future PhD position (though not advertised under that name) and I know many like me. Of course, there may be none available, but in such a case you could ask about possibilities of helping someone write a grant application. There is very little risk involved if you do most of the writing, so don't feel shy about asking.

- The more obvious place to look for advertisements is, of course, ye olde **internet**. Here, I will suggest jobs.ac.uk for the UK, higheredjobs.com for the USA, both conveniently giving options for searching *psychology*. Most sites present themselves as global even if they concentrate on certain areas: timeshighereducation.com/unijobs/ in the UK, for example, but also East Asia, and academictransfer.com in the Netherlands, but also somewhat around Europe. If you would only ever consider a few universities, of course, you may also look at whatever their websites are mentioning directly. There are also various online boards that concentrate on neuroscience specifically: the Federation of European Neurosciences has one, for example (fens.org/careers/job-market), and the aforementioned EEGLAB list (eeglab.org/others/EEGLAB_mailing_lists.html) has regular advertisements posted by email as well.

- Finally, there's **social media**. Here, I should mention that LinkedIn.com is not just boring news of people you vaguely remember celebrating anniversaries for jobs where you never quite know whether they mean condolences instead. In fact, it has an incredibly large database of jobs going for it, with many PhD, postdoc, and faculty positions around the world. I imagine there are also other social media places for jobs, but these do tend to be rather in flux, so you might as well just ask there. Or me: I'm on Instagram (@msovspape) if you want me to congratulate you on finishing this book.

REFERENCES

Bargh, J. A., Chen, M., & Burrows, L. (1996). Automaticity of social behavior: Direct effects of trait construct and stereotype activation on action. *Journal of Personality and Social Psychology*, 71(2), 230.

Barnhoorn, J. S., Haasnoot, E., Bocanegra, B. R., & van Steenbergen, H. (2015). QRTEngine: An easy solution for running online reaction time experiments using Qualtrics. *Behavior Research Methods*, 47(4), 918–929.

Bastos, A. M., & Schoffelen, J.-M. (2016). A tutorial review of functional connectivity analysis methods and their interpretational pitfalls. *Frontiers in Systems Neuroscience*, 9, 175.

Bennett, C. M., Wolford, G. L., & Miller, M. B. (2009). The principled control of false positives in neuroimaging. *Social Cognitive and Affective Neuroscience*, 4(4), 417–422.

Bentin, S., Allison, T., Puce, A., Perez, E., & McCarthy, G. (1996). Electrophysiological studies of face perception in humans. *Journal of Cognitive Neuroscience*, 8(6), 551–565.

Berger, H. (1904). *Über die körperlichen Äusserungen psychischer Zustände: Weitere experimentelle Beiträge zur Lehre von der Blutzirkulation in der Schädelhöhle des Menschen* (Vol. 1). G. Fisher.

Berger, H. (1910). *Untersuchungen über die Temperatur des Gehirns*. G. Fischer.

Berger, H. (1929). Über das Elektrenkephalogramm des Menschen. *European Archives of Psychiatry and Clinical Neuroscience*, 87(1), 527–570.

Bestmann, S., & Krakauer, J. W. (2015). The uses and interpretations of the motor-evoked potential for understanding behaviour. *Experimental Brain Research*, 233(3), 679–689.

Bladin, P. F. (2006). W. Grey Walter, pioneer in the electroencephalogram, robotics, cybernetics, artificial intelligence. *Journal of Clinical Neuroscience*, 13(2), 170–177. https://doi.org/10.1016/j.jocn.2005.04.010

Blau, V. C., Maurer, U., Tottenham, N., & McCandliss, B. D. (2007). The face-specific N170 component is modulated by emotional facial expression. *Behavioral and Brain Functions*, 3(1), 1–13.

Blum, Deborah. (2006). *Ghost hunters: William James and the search for scientific proof of life after death*. Penguin Press.

Boksem, M. A., & De Cremer, D. (2010). Fairness concerns predict medial frontal negativity amplitude in ultimatum bargaining. *Social Neuroscience*, 5(1), 118–128.

Bramon, E., McDonald, C., Croft, R. J., Landau, S., Filbey, F., Gruzelier, J. H., Sham, P. C., Frangou, S., & Murray, R. M. (2005). Is the P300 wave an endophenotype for schizophrenia? A meta-analysis and a family study. *NeuroImage*, 27(4), 960–968. https://doi.org/10.1016/j.neuroimage.2005.05.022

Brazier, M. A. B. (1959). The EEG in epilepsy: A historical note. *Epilepsia*, 1(1–5), 328–336.

Cacioppo, J. T., Tassinary, L. G., & Berntson, G.G. (Eds.). (2017). *Handbook of Psychophysiology*, 4th edition. Cambridge: Cambridge University Press.

Carey, D. P., & Johnstone, L. T. (2014). Quantifying cerebral asymmetries for language in dextrals and adextrals with random-effects meta-analysis. *Frontiers in Psychology*, 5, 1128.

Chapman, R. M., & Bragdon, H. R. (1964). Evoked responses to numerical and non-numerical visual stimuli while problem solving. *Nature*, 203, 1155–1157. https://doi.org/10.1038/2031155a0

Coan, J. A., & Allen, J. J. (2004). Frontal EEG asymmetry as a moderator and mediator of emotion. *Biological Psychology*, 67(1), 7–50.

Cohen, M. X., Elger, C. E., & Ranganath, C. (2007). Reward expectation modulates feedback-related negativity and EEG spectra. *NeuroImage*, 35(2), 968–978. https://doi.org/10.1016/j.neuroimage.2006.11.056

Coles, M. G. (1989). Modern mind-brain reading: Psychophysiology, physiology, and cognition. *Psychophysiology*, 26(3), 251–269.

Colzato, L. S., Slagter, H. A., Spapé, M. M., & Hommel, B. (2008). Blinks of the eye predict blinks of the mind. *Neuropsychologia*, 46(13), 3179–3183.

Conway, A. R., Cowan, N., & Bunting, M. F. (2001). The cocktail party phenomenon revisited: The importance of working memory capacity. *Psychonomic Bulletin & Review*, 8(2), 331–335.

Courchesne, E., Hillyard, S. A., & Galambos, R. (1975). Stimulus novelty, task relevance and the visual evoked potential in man. *Electroencephalography and Clinical Neurophysiology*, 39(2), 131–143.

Crick, F., & Koch, C. (1990). Towards a neurobiological theory of consciousness. *Seminars in the Neurosciences*, 2, 203. http://papers.klab.caltech.edu/22/1/148.pdf

Davis, M. L., Papini, S., Rosenfield, D., Roelofs, K., Kolb, S., Powers, M. B., & Smits, J. A. (2017). A randomized controlled study of power posing before public speaking exposure for social anxiety disorder: No evidence for augmentative effects. *Journal of Anxiety Disorders*, 52, 1–7.

Debener, S., Minow, F., Emkes, R., Gandras, K., & Vos, M. (2012). How about taking a low-cost, small, and wireless EEG for a walk? *Psychophysiology*, 49(11), 1617–1621.

Delorme, A., & Makeig, S. (2004). EEGLAB: An open source toolbox for analysis of single-trial EEG dynamics including independent component analysis. *Journal of Neuroscience Methods*, 134(1), 9–21. https://doi.org/10.1016/j.jneumeth.2003.10.009

Di Pellegrino, G., Fadiga, L., Fogassi, L., Gallese, V., & Rizzolatti, G. (1992). Understanding motor events: A neurophysiological study. *Experimental Brain Research*, 91(1), 176–180.

Donchin, E. (1981). Surprise!… surprise? *Psychophysiology*, 18(5), 493–513.

Doyen, S., Klein, O., Pichon, C.-L., & Cleeremans, A. (2012). Behavioral priming: It's all in the mind, but whose mind? *PloS One*, 7(1), e29081.

Duncan-Johnson, C. C., & Donchin, E. (1979). The time constant in P300 recording. *Psychophysiology*, 16(1), 53–55.

Eimer, M., Kiss, M., & Nicholas, S. (2010). Response profile of the face-sensitive N170 component: A rapid adaptation study. *Cerebral Cortex*, 20(10), 2442–2452.

European Council (2016, April). *General Date Protection Regulation* (EU 2016/679). (GDPR). European Council. Implemented 25 May 2018.

Falkenstein, M., Hohnsbein, J., Hoormann, J., & Blanke, L. (1991). Effects of crossmodal divided attention on late ERP components. II. Error processing in choice reaction tasks. *Electroencephalography and Clinical Neurophysiology*, 78(6), 447–455.

Farwell, L. A. (2012). Brain fingerprinting: A comprehensive tutorial review of detection of concealed information with event-related brain potentials. *Cognitive Neurodynamics*, 6(2), 115–154.

Farwell, L. A., & Donchin, E. (1988). Talking off the top of your head: Toward a mental prosthesis utilizing event-related brain potentials. *Electroencephalography and Clinical Neurophysiology*, 70(6), 510–523.

Feuillet, L., Dufour, H., & Pelletier, J. (2007). Brain of a white-collar worker. *The Lancet*, 370(9583), 262.

Folstein, J. R., & Van Petten, C. (2008). Influence of cognitive control and mismatch on the N2 component of the ERP: A review. *Psychophysiology*, 45(1), 152–170. https://doi.org/10.1111/j.1469-8986.2007.00602.x

Fridlund, A. J., & Cacioppo, J. T. (1986). Guidelines for human electromyographic research. *Psychophysiology*, 23(5), 567–589.

Gehring, W. J., Goss, B., Coles, M. G., Meyer, D. E., & Donchin, E. (1993). A neural system for error detection and compensation. *Psychological Science*, 4(6), 385–390. https://doi.org/10.1126/science.1066893

Gehring, W. J., & Willoughby, A. R. (2004). Are all medial frontal negativities created equal? Toward a richer empirical basis for theories of action monitoring. Errors, Conflicts, and the Brain. *Current Opinions on Performance Monitoring*, 14–20.

Ghuman, A. S., Brunet, N. M., Li, Y., Konecky, R. O., Pyles, J. A., Walls, S. A., Destefino, V.,

Wang, W., & Richardson, R. M. (2014). Dynamic encoding of face information in the human fusiform gyrus. *Nature Communications*, 5, 5672.

Graham, F. K., & Clifton, R. K. (1966). Heart-rate change as a component of the orienting response. *Psychological Bulletin*, 65(5), 305.

Gratton, G., Coles, M. G., & Donchin, E. (1983). A new method for off-line removal of ocular artifact. *Electroencephalography and Clinical Neurophysiology*, 55(4), 468–484.

Greicius, M. D., Krasnow, B., Reiss, A. L., & Menon, V. (2003). Functional connectivity in the resting brain: A network analysis of the default mode hypothesis. *Proceedings of the National Academy of Sciences*, 100(1), 253–258.

Hagoort, P., Brown, C., & Groothusen, J. (1993). The syntactic positive shift (SPS) as an ERP measure of syntactic processing. *Language and Cognitive Processes*, 8(4), 439–483.

Hajcak, G., Moser, J. S., Holroyd, C. B., & Simons, R. F. (2006). The feedback-related negativity reflects the binary evaluation of good versus bad outcomes. *Biological Psychology*, 71(2), 148–154. https://doi.org/10.1016/j.biopsycho.2005.04.001

Hayward, R. (2001). The tortoise and the love-machine: Grey Walter and the politics of electroencephalography. *Science in Context*, 14(4), 615–641.

Holroyd, C. B., & Coles, M. G. (2002). The neural basis of human error processing: Reinforcement learning, dopamine, and the error-related negativity. *Psychological Review*, 109(4), 679. https://doi.org/10.1037/0033-295X.109.4.679

Hyvärinen, A., & Oja, E. (2000). Independent component analysis: Algorithms and applications. *Neural Networks*, 13(4–5), 411–430. https://doi.org/10.1016/S0893-6080(00)00026-5

Jeffreys, D. A. (1989). A face-responsive potential recorded from the human scalp. *Experimental Brain Research*, 78(1), 193–202.

Jeffreys, D. A., & Axford, J. G. (1972). Source locations of pattern-specific components of human visual evoked potentials. I. Component of striate cortical origin. *Experimental Brain Research*, 16(1), 1–21.

Jenkins, R., Dowsett, A. J., & Burton, A. M. (2018). How many faces do people know? *Proceedings of the Royal Society B*, 285. DOI: 10.1098/rspb.2018.1319

Jensen, O., & Colgin, L. L. (2007). Cross-frequency coupling between neuronal oscillations. *Trends in Cognitive Sciences*, 11(7), 267–269.

Joyce, C., & Rossion, B. (2005). The face-sensitive N170 and VPP components manifest the same brain processes: The effect of reference electrode site. *Clinical Neurophysiology*, 116(11), 2613–2631. https://doi.org/10.1016/j.clinph.2005.07.005

Jung, T.-P., Makeig, S., Humphries, C., Lee, T.-W., Mckeown, M. J., Iragui, V., & Sejnowski, T. J. (2000). Removing electroencephalographic artifacts by blind source separation. *Psychophysiology*, 37(2), 163–178.

Kaan, E., & Swaab, T. Y. (2003). Repair, revision, and complexity in syntactic analysis: An electrophysiological differentiation. *Journal of Cognitive Neuroscience*, 15(1), 98–110.

Kangassalo, L., Spapé, M., & Ruotsalo, T. (2020). Neuroadaptive modelling for generating images matching perceptual categories. *Scientific Reports*, 10(1), 1–10.

Kayser, J., & Tenke, C. E. (2010). In search of the Rosetta Stone for scalp EEG: Converging on reference-free techniques. *Clinical Neurophysiology: Official Journal of the International Federation of Clinical Neurophysiology*, 121(12), 1973.

Knispel, J., & Wright, G. (1989). Method and apparatus for translating the EEG into music to induce and control various psychological and physiological states and to control a musical instrument. Google Patents.

Kornhuber, H. H., & Deecke, L. (1965). Hirnpotentialänderungen bei Willkürbewegungen und passiven Bewegungen des Menschen: Bereitschaftspotential und reafferente Potentiale. *Pflüger's Archiv Für Die Gesamte Physiologie Des Menschen Und Der Tiere*, 284(1), 1–17.

Kotchoubey, B. (2006). Event-related potentials, cognition, and behavior: A biological approach. *Neuroscience & Biobehavioral Reviews*, 30(1), 42–65.

Krause, F., & Lindemann, O. (2014). Expyriment: A Python library for cognitive and neuroscientific experiments. *Behavior Research Methods*, 46(2), 416–428.

Kutas, M., & Federmeier, K. D. (2011). Thirty years and counting: Finding meaning in the N400 component of the event related brain potential (ERP). *Annual Review of Psychology*, 62, 621.

Kutas, M., & Hillyard, S. A. (1980). Reading sense-less sentences: Brain potentials reflect semantic incongruity. *Science*, 207(4427), 203–205.

Lang, P. J., Bradley, M. M., & Cuthbert, B. N. (1999). *International affective picture system (IAPS): Instruction manual and affective ratings*. The Center for Research in Psychophysiology, University of Florida.

Langley, T., & Goldman, C. (2015). *Star Wars psychology: Dark side of the mind*. Sterling.

Larson, S. J., Sances, A., & Christenson, P. C. (1966). Evoked somatosensory potentials in man. *Archives of Neurology*, 15(1), 88–93.

Lega, B. C., Jacobs, J., & Kahana, M. (2012). Human hippocampal theta oscillations and the formation of episodic memories. *Hippocampus*, 22(4), 748–761.

Lehmann, A. (1912). *Grundzüge der Psychophysiologie. Eine Darstellung der normalen, generellen und individuellen Psychologie*. Reisland.

Lepage, K. Q., Kramer, M. A., & Chu, C. J. (2014). A statistically robust EEG re-referencing procedure to mitigate reference effect. *Journal of Neuroscience Methods*, 235, 101–116.

Leslie, G., & Mullen, T. R. (2011). MoodMixer: EEG-based Collaborative Sonification. *NIME*, 296–299.

Libet, B., Gleason, C. A., Wright, E. W., & Pearl, D. K. (1983). Time of conscious intention to act in relation to onset of cerebral activity (readiness-potential) the unconscious initiation of a freely voluntary act. *Brain*, 106(3), 623–642.

Lindell, A. K. (2006). In your right mind: Right hemisphere contributions to language processing and production. *Neuropsychology Review*, 16(3), 131–148.

Linden, D. E. J. (2005). The P300: Where in the brain is it produced and what does it tell us? *The Neuroscientist*, 11, 563–576. https://journals.sagepub.com/doi/10.1177/1073858405280524

Lopez-Calderon, J., & Luck, S. J. (2014). ERPLAB: An open-source toolbox for the analysis of event-related potentials. *Frontiers in Human Neuroscience*, 8, 213.

Luck, S. J. (2014). *An introduction to the event-related potential technique* (2nd ed.). MIT Press. http://mitpress.mit.edu/books/introduction-event-related-potential-technique-1/

Luck, S. J., & Kappenman, E. S. (2020). Resources to assist EEG/ERP researchers during the COVID-19 pandemic. *Psychophysiology*, 57(9), e13659. https://doi.org/10.1111/psyp.13659

Lurquin, J. H., Michaelson, L. E., Barker, J. E., Gustavson, D. E., Von Bastian, C. C., Carruth, N. P., & Miyake, A. (2016). No evidence of the ego-depletion effect across task characteristics and individual differences: A pre-registered study. *PloS One*, 11(2), e0147770

Makeig, S., Debener, S., Onton, J., & Delorme, A. (2004). Mining event-related brain dynamics. *Trends in Cognitive Sciences*, 8(5), 204–210.

Mangun, G. R. (1995). Neural mechanisms of visual selective attention. *Psychophysiology*, 32, 4–18.

Manola, L., Roelofsen, B. H., Holsheimer, J., Marani, E., & Geelen, J. (2005). Modelling motor cortex stimulation for chronic pain control: Electrical potential field, activating functions and responses of simple nerve fibre models. *Medical and Biological Engineering and Computing*, 43(3), 335–343.

Mathôt, S., Schreij, D., & Theeuwes, J. (2012). OpenSesame: An open-source, graphical experiment builder for the social sciences. *Behavior Research Methods*, 44(2), 314–324.

Matsumoto, C. S., Shinoda, K., Matsumoto, H., Seki, K., Nagasaka, E., Iwata, T., & Mizota, A. (2014). What monitor can replace the cathode-ray tube for visual stimulation to elicit multifocal electroretinograms? *Journal of Vision*, 14(9), 2–2.

Mehrabian, A. (1971). *Silent messages* (Vol. 8). Belmont, CA: Wadsworth.

Meijer, E. H., Ben-Shakhar, G., Verschuere, B., & Donchin, E. (2013). A comment on Farwell (2012): Brain fingerprinting: A comprehensive tutorial review of detection of concealed information with event-related brain potentials. *Cognitive Neurodynamics*, 7(2), 155–158.

MettingVanRijn, A. C., Kuiper, A. P., Dankers, T. E., & Grimbergen, C. A. (1996). *Low-cost active electrode improves the resolution in biopotential recordings*. Proceedings of 18th Annual International Conference of the IEEE Engineering in Medicine and Biology Society, 1, 101–102.

Millett, D. (2001). Hans Berger: From psychic energy to the EEG. *Perspectives in Biology and Medicine*, 44(4), 522–542.

Miranda, E. R. (2006). Brain-computer music interface for composition and performance. *International Journal on Disability and Human Development*, 5(2), 119.

Mischel, W. (1990). Personality dispositions revisited and revised: A view after three decades. In L. A. Pervin (Ed.), *Handbook of personality: Theory and research*. The Guilford Press..

Morris, J. S., Frith, C. D., Perrett, D. I., Rowland, D., Young, A. W., Calder, A. J., & Dolan, R. J. (1996). A differential neural response in the human amygdala to fearful and happy facial expressions. *Nature*, 383(6603), 812–815.

Näätänen, R., Gaillard, A. W., & Mäntysalo, S. (1978). Early selective-attention effect on evoked potential reinterpreted. *Acta Psychologica*, 42(4), 313–329.

Näätänen, R., & Picton, T. (1987). The N1 wave of the human electric and magnetic response to sound: A review and an analysis of the component structure. *Psychophysiology*, 24(4), 375–425.

Neininger, B., & Pulvermüller, F. (2003). Word-category specific deficits after lesions in the right hemisphere. *Neuropsychologia*, 41(1), 53–70.

Nieuwenhuis, S., Ridderinkhof, K. R., Blom, J., Band, G. P., & Kok, A. (2001). Error-related brain potentials are differentially related to awareness of response errors: Evidence from an antisaccade task. *Psychophysiology*, 38(5), 752–760.

Ninomiya, H., Onitsuka, T., Chen, C., Sato, E., & Tashiro, N. (1998). P300 in response to the subject's own face. *Psychiatry and Clinical Neurosciences*, 52(5), 519–522.

Nitschke, J. B., Miller, G. A., & Cook, E. W. (1998). Digital filtering in EEG/ERP analysis: Some technical and empirical comparisons. *Behavior Research Methods, Instruments, & Computers*, 30(1), 54–67.

Oldfield, R. C. (1971). The assessment and analysis of handedness: The Edinburgh inventory. *Neuropsychologia*, 9(1), 97–113.

O'Regan, J. K., & Noë, A. (2001). A sensorimotor account of vision and visual consciousness. *Behavioral and Brain Sciences*, 24(5), 939–973.

Ouyang, G., Sommer, W., & Zhou, C. (2016). Reconstructing ERP amplitude effects after compensating for trial-to-trial latency jitter: A solution based on a novel application of residue iteration decomposition. *International Journal of Psychophysiology*, 109, 9–20. https://doi.org/10.1016/j.ijpsycho.2016.09.015

Palahniuk, C. (1996). *Fight club*. W. W. Norton.

Pascual-Marqui, R. D. (2007). Discrete, 3D distributed, linear imaging methods of electric neuronal activity. Part 1: Exact, zero error localization. *ArXiv Preprint* ArXiv:0710.3341. http://arxiv.org/abs/0710.3341

Peirce, J. W. (2007). PsychoPy—Psychophysics software in Python. *Journal of Neuroscience Methods*, 162(1–2), 8–13.

Peirce, J. W., Gray, J. R., & MacAskill, M. (under contract). *Programming experiments in Python*. Sage.

Peirce, J. W., & MacAskill, M. (2018). *PsychoPy: Building experiments in PsychoPy*. Sage.

Petrov, Y., Nador, J., Hughes, C., Tran, S., Yavuzcetin, O., & Sridhar, S. (2014). Ultra-dense EEG sampling results in two-fold increase of functional brain information. *NeuroImage*, 90, 140–145. https://doi.org/10.1016/j.neuroimage.2013.12.041

Pfurtscheller, G., & Da Silva, F. L. (1999). Event-related EEG/MEG synchronization and desynchronization: Basic principles. *Clinical Neurophysiology*, 110(11), 1842–1857.

Pfurtscheller, G., & Neuper, C. (1994). Event-related synchronization of mu rhythm in the EEG over the cortical hand area in man. *Neuroscience Letters*, 174(1), 93–96.

Picton, T. W., Hillyard, S. A., Krausz, H. I., & Galambos, R. (1974). Human auditory evoked potentials. I: Evaluation of components. *Electroencephalography and Clinical Neurophysiology*, 36, 179–190. https://doi.org/10.1016/0013-4694(74)90155-2

Pinker, S. (2014). *The sense of style: The thinking person's guide to writing in the 21st century*. Penguin Books.

Polich, J. (2007). Updating P300: An integrative theory of P3a and P3b. *Clinical Neurophysiology*, 118(10), 2128–2148. https://doi.org/10.1016/j.clinph.2007.04.019

Pravdich-Neminsky, W. (1912). Ein versuch der registrierung der elektrischen gehirnerscheinungen. *Zentralbl Physiol*, 27, 951–960.

Ramachandran, V. S., & Oberman, L. M. (2006). Broken mirrors: A theory of autism. *Scientific American*, 295(5), 62–69.

Ramautar, J. R., Kok, A., & Ridderinkhof, K. R. (2006). Effects of stop-signal modality on the N2/P3 complex elicited in the stop-signal paradigm. *Biological Psychology*, 72(1), 96–109.

Ravaja, N., Harjunen, V., Ahmed, I., Jacucci, G., & Spapé, M. M. (2017). Feeling touched: Emotional modulation of somatosensory potentials to interpersonal touch. *Scientific Reports*, 7, 40504.

Rossion, B. (2014). Understanding face perception by means of human electrophysiology. *Trends in Cognitive Sciences*, 18(6), 310–318.

Schupp, H. T., Junghöfer, M., Weike, A. I., & Hamm, A. O. (2003). Attention and emotion: An ERP analysis of facilitated emotional stimulus processing. *Neuroreport*, 14(8), 1107–1110.

Serrien, D. J., & Sovijärvi-Spapé, M. M. (2013). Cognitive control of response inhibition and switching: Hemispheric lateralization and hand preference. *Brain and Cognition*, 82(3), 283–290.

Shaffer, F., Combatalade, D., Peper, E., & Meehan, Z. M. (2016). A guide to cleaner electrodermal activity measurements. *Biofeedback*, 44(2), 90–100.

Shalgi, S., Barkan, I., & Deouell, L. Y. (2009). On the positive side of error processing: Error-awareness positivity revisited. *European Journal of Neuroscience*, 29(7), 1522–1532.

Spanos, N. P. (1986). Hypnotic behavior: A social-psychological interpretation of amnesia, analgesia, and 'trance logic'. *Behavioral and Brain Sciences*, 9(3), 449–467.

Spapé, M., Davis, K., Kangassalo, L., Ravaja, N., Sovijarvi-Spape, Z., & Ruotsalo, T. (2021). Brain-computer interface for generating personally attractive images. *IEEE Transactions on Affective Computing*. In press.

Spapé, M. M., Harjunen, V., Ahmed, I., Jacucci, G., & Ravaja, N. (2019). The semiotics of the message and the messenger: How nonverbal communication affects fairness perception. *Cognitive, Affective, & Behavioral Neuroscience*, 19(5), 1259–1272.

Spapé, M. M., Harjunen, V., & Ravaja, N. (2017). Effects of touch on emotional face processing: A study of event-related potentials, facial EMG and cardiac activity. *Biological Psychology*, 124, 1–10.

Spapé, M. M., Verdonschot, R., & van Steenbergen, H. (2019). *The E-Primer: An introduction to creating psychological experiments in E-Prime. Second edition updated for E-Prime 3* (2nd ed.). Leiden University Press.

Spehlmann, R. (1965). The averaged electrical responses to diffuse and to patterned light in the human. *Electroencephalography and Clinical Neurophysiology*, 19(6), 560–569.

Squires, K. C., Donchin, E., Herning, R. I., & McCarthy, G. (1977). On the influence of task relevance and stimulus probability on event-related-potential components. *Electroencephalography and Clinical Neurophysiology*, 42(1), 1–14.

Squires, N. K., Squires, K. C., & Hillyard, S. A. (1975). Two varieties of long-latency positive waves evoked by unpredictable auditory stimuli in man. *Electroencephalography and Clinical Neurophysiology*, 38(4), 387–401. https://doi.org/10.1016/0013-4694(75)90263-1

Sutton, S., Braren, M., Zubin, J., & John, E. R. (1965). Evoked-potential correlates of stimulus uncertainty. *Science*, 150(3700), 1187–1188.

Sutton, S., Tueting, P., Zubin, J., & John, E. R. (1967). Information delivery and the sensory evoked potential. *Science*, 155(3768), 1436–1439.

Turk, D. J., Van Bussel, K., Waiter, G. D., & Macrae, C. N. (2011). Mine and me: Exploring the neural basis of object ownership. *Journal of Cognitive Neuroscience*, 23(11), 3657–3668.

Van Strien, J. W., Franken, I. H. A., & Huijding, J. (2014). Testing the snake-detection hypothesis: Larger early posterior negativity in humans to pictures of snakes than to pictures of other reptiles, spiders and slugs. *Frontiers in Human Neuroscience*, 8. https://doi.org/10.3389/fnhum.2014.00691

Van Voorhis, S., & Hillyard, S. A. (1977). Visual evoked potentials and selective attention to points in space. *Perception & Psychophysics*, 22(1), 54–62.

Verleger, R. (1997). On the utility of P3 latency as an index of mental chronometry.

Psychophysiology, 34(2), 131–156. https://doi.org/10.1111/j.1469-8986.1997.tb02125.x

Verleger, R., Jaśkowski, P., & Wascher, E. (2005). Evidence for an integrative role of P3b in linking reaction to perception. *Journal of Psychophysiology*, 19(3), 165–181.

Walter, W. G. (1936). The location of cerebral tumours by electro-encephalography. *The Lancet*, 228(5893), 305–308.

Walter, W. G. (1937). The electro-encephalogram in cases of cerebral tumour (Section of Neurology). *Proceedings of the Royal Society of Medicine*, 30(5), 579.

Walter, W. G. (1943). An automatic low frequency analyser. *Electronic Engineering*, 16, 9–13.

Walter, W. G. (1961). *The living brain*. W. W. Norton & Company, Inc.

Walter, W. G. (1968). The social organ. *Impact of Science on Society*, 18(3), 179–186.

Walter, W. G., Cooper, R., Aldridge, V. J., McCallum, W. C., & Winter, A. L. (1964). Contingent negative variation: An electric sign of sensori-motor association and expectancy in the human brain. *Nature*, 203(4943), 380–384. https://doi.org/10.1038/203380a0

Walter, W. G., & Dovey, V. J. (1944). Electro-encephalography in cases of sub-cortical tumour. *Journal of Neurology, Neurosurgery, and Psychiatry*, 7(3–4), 57.

Walter, W. G., & Shipton, H. W. (1951). A new toposcopic display system. *Clinical Neurophysiology*, 3(3), 281–292.

World Medical Association (2013). *Ethical principles for medical research involving human subjects*. WMA. Available at: https://www.wma.net/.

Yao, D. (2001). A method to standardize a reference of scalp EEG recordings to a point at infinity. *Physiological Measurement*, 22(4), 693.

Yuval-Greenberg, S., Tomer, O., Keren, A. S., Nelken, I., & Deouell, L. Y. (2008). Transient induced gamma-band response in EEG as a manifestation of miniature saccades. *Neuron*, 58(3), 429–441.

Zhou, D. M., McAdams, E. T., Lackermeier, A., & Jones, J. G. (1994). *AC impedance of Ag/AgCl reference electrodes for use in disposable biosensors*. Proceedings of 16th Annual International Conference of the IEEE Engineering in Medicine and Biology Society, 2, 832–833.

Zhu, X., & Luo, Y. (2012). Fearful faces evoke a larger C1 than happy faces in executive attention task: An event-related potential study. *Neuroscience Letters*, 526(2), 118–121.

INDEX

CPSIA information can be obtained
at www.ICGtesting.com
Printed in the USA
LVHW070723090723
750439LV00008B/34